Algorithms

Sanjoy Dasgupta
University of California, San Diego

Christos Papadimitriou
University of California at Berkeley

Umesh Vazirani
University of California at Berkeley

Boston Burr Ridge, IL Dubuque, IA New York San Francisco
St. Louis Bangkok Bogotá Caracas Kuala Lumpur Lisbon London
Lisbon London Madrid Mexico City Milan Montreal New Delhi
Santiago Seoul Singapore Sydney Taipei Toronto

ALGORITHMS

This book is printed on acid-free paper.

8 9 0 DOC/DOC 1 5 4 3 2 1

ISBN 978-0-07-352340-8
MHID 0-07-352340-2

Publisher: *Alan R. Apt*
Executive Marketing Manager: *Michael Weitz*
Project Manager: *Joyce Watters*
Lead Production Supervisor: *Sandy Ludovissy*
Associate Media Producer: *Christina Nelson*
Designer: *John Joran*
Compositor: *Techbooks*
Typeface: *10/12 Slimbach*
Printer: *R. R. Donnelley Crawfordsville, IN*

Library of Congress Cataloging-in-Publication Data

Dasgupta Sanjoy.
Algorithms / Sanjoy Dasgupta, Christos Papadimitriou, Umesh Vazirani.—1st ed.
p. cm.
Includes index.
ISBN 978-0-07-352340-8 — ISBN 0-07-352340-2
1. Algorithms—Textbooks. 2. Computer algorithms—Textbooks. I. Papadimitriou, Christos H. II. Vazirani, Umesh Virkumar. III. Title.

QA9.58.D37 2008
518′1—dc22

2006049014
CIP

www.mhhe.com

To our students and teachers,
and our parents.

Contents

Boxes

Preface

This book evolved over the past ten years from a set of lecture notes developed while teaching the undergraduate Algorithms course at Berkeley and U.C. San Diego. Our way of teaching this course evolved tremendously over these years in a number of directions, partly to address our students' background (undeveloped formal skills outside of programming), and partly to reflect the maturing of the field in general, as we have come to see it. The notes increasingly crystallized into a narrative, and we progressively structured the course to emphasize the "story line" implicit in the progression of the material. As a result, the topics were carefully selected and clustered. No attempt was made to be encyclopedic, and this freed us to include topics traditionally de-emphasized or omitted from most Algorithms books.

Playing on the strengths of our students (shared by most of today's undergraduates in Computer Science), instead of dwelling on formal proofs we distilled in each case the crisp mathematical idea that makes the algorithm work. In other words, we emphasized rigor over formalism. We found that our students were much more receptive to mathematical rigor of this form. It is this progression of crisp ideas that helps weave the story.

Once you think about Algorithms in this way, it makes sense to start at the historical beginning of it all, where, in addition, the characters are familiar and the contrasts dramatic: numbers, primality, and factoring. This is the subject of Part I of the book, which also includes the RSA cryptosystem, and divide-and-conquer algorithms for integer multiplication, sorting and median finding, as well as the fast Fourier transform. There are three other parts: Part II, the most traditional section of the book, concentrates on data structures and graphs; the contrast here is between the intricate structure of the underlying problems and the short and crisp pieces of pseudocode that solve them. Instructors wishing to teach a more traditional course can simply start with Part II, which is self-contained (following the prologue), and then cover Part I as required. In Parts I and II we introduced certain techniques (such as greedy and divide-and-conquer) which work for special kinds of problems; Part III deals with the "sledgehammers" of the trade, techniques that are powerful and general: dynamic programming (a novel approach helps clarify this traditional stumbling block for students) and linear programming (a clean and intuitive treatment of the simplex algorithm, duality, and reductions to the basic problem). The final Part IV is about ways of dealing with hard problems: NP-completeness, various heuristics, as well as quantum algorithms, perhaps the most advanced and modern topic. As it happens, we end the story exactly where we started it, with Shor's quantum algorithm for factoring.

The book includes three additional undercurrents, in the form of three series of separate "boxes," strengthening the narrative (and addressing variations in the needs and interests of the students) while keeping the flow intact, pieces that provide historical context; descriptions of how the explained algorithms are used in practice (with emphasis on internet applications); and excursions for the mathematically sophisticated.

Many of our colleagues have made crucial contributions to this book. We are grateful for feedback from Dimitris Achlioptas, Dorit Aharanov, Mike Clancy, Jim Demmel, Monika Henzinger, Mike Jordan, Milena Mihail, Gene Myers, Dana Randall, Satish Rao, Tim Roughgarden, Jonathan Shewchuk, Martha Sideri, Alistair Sinclair, and David Wagner, all of whom beta tested early drafts. Satish Rao, Leonard Schulman, and Vijay Vazirani shaped the exposition of several key sections. Gene Myers, Satish Rao, Luca Trevisan, Vijay Vazirani, and Lofti Zadeh provided exercises. And finally, there are the students of UC Berkeley and, later, UC San Diego, who inspired this project, and who have seen it through its many incarnations.

Chapter 0
Prologue

Look around you. Computers and networks are everywhere, enabling an intricate web of complex human activities: education, commerce, entertainment, research, manufacturing, health management, human communication, even war. Of the two main technological underpinnings of this amazing proliferation, one is obvious: the breathtaking pace with which advances in microelectronics and chip design have been bringing us faster and faster hardware.

This book tells the story of the other intellectual enterprise that is crucially fueling the computer revolution: *efficient algorithms*. It is a fascinating story.

Gather 'round and listen close.

0.1 Books and algorithms

Two ideas changed the world. In 1448 in the German city of Mainz a goldsmith named Johann Gutenberg discovered a way to print books by putting together movable metallic pieces. Literacy spread, the Dark Ages ended, the human intellect was liberated, science and technology triumphed, the Industrial Revolution happened. Many historians say we owe all this to typography. Imagine a world in which only an elite could read these lines! But others insist that the key development was not typography, but *algorithms*.

Johann Gutenberg
1398–1468

Today we are so used to writing numbers in decimal, that it is easy to forget that Gutenberg would write the number 1448 as MCDXLVIII. How do you add two Roman numerals? What is MCDXLVIII + DCCCXII? (And just try to think about multiplying them.) Even a clever man like Gutenberg probably only knew how to add and subtract small numbers using his fingers; for anything more complicated he had to consult an abacus specialist.

The decimal system, invented in India around AD 600, was a revolution in quantitative reasoning: using only 10 symbols, even very large numbers could be written down compactly, and arithmetic could be done efficiently on them by following elementary steps. Nonetheless these ideas took a long time to spread, hindered by traditional barriers of language, distance, and ignorance. The most influential medium of transmission turned out to be a textbook, written in Arabic in the ninth century by a man who lived in Baghdad. Al Khwarizmi laid out the basic methods for adding, multiplying, and dividing numbers—even extracting square roots and calculating digits of π. These procedures were precise, unambiguous, mechanical, efficient, correct—in short, they were *algorithms*, a term coined to honor the wise man after the decimal system was finally adopted in Europe, many centuries later.

Since then, this decimal positional system and its numerical algorithms have played an enormous role in Western civilization. They enabled science and technology; they accelerated industry and commerce. And when, much later, the computer was finally designed, it explicitly embodied the positional system in its bits and words and arithmetic unit. Scientists everywhere then got busy developing more and more complex algorithms for all kinds of problems and inventing novel applications—ultimately changing the world.

0.2 Enter Fibonacci

Al Khwarizmi's work could not have gained a foothold in the West were it not for the efforts of one man: the 13th century Italian mathematician Leonardo Fibonacci, who saw the potential of the positional system and worked hard to develop it further and propagandize it.

But today Fibonacci is most widely known for his famous sequence of numbers

$$0, 1, 1, 2, 3, 5, 8, 13, 21, 34, \ldots,$$

each the sum of its two immediate predecessors. More formally, the Fibonacci numbers F_n are generated by the simple rule

$$F_n = \begin{cases} F_{n-1} + F_{n-2} & \text{if } n > 1 \\ 1 & \text{if } n = 1 \\ 0 & \text{if } n = 0. \end{cases}$$

No other sequence of numbers has been studied as extensively, or applied to more fields: biology, demography, art, architecture, music, to name just a few. And, together with the powers of 2, it is computer science's favorite sequence.

In fact, the Fibonacci numbers grow *almost* as fast as the powers of 2: for example, F_{30} is over a million, and F_{100} is already 21 digits long! In general, $F_n \approx 2^{0.694n}$ (see Exercise 0.3).

But what is the precise value of F_{100}, or of F_{200}? Fibonacci himself would surely have wanted to know such things. To answer, we need an algorithm for computing the nth Fibonacci number.

Leonardo of Pisa (Fibonacci)
1170–1250

An exponential algorithm

One idea is to slavishly implement the recursive definition of F_n. Here is the resulting algorithm, in the "pseudocode" notation used throughout this book:

```
function fib1(n)
if n=0: return 0
if n=1: return 1
return fib1(n-1) + fib1(n-2)
```

Whenever we have an algorithm, there are three questions we always ask about it:

1. Is it correct?
2. How much time does it take, as a function of n?
3. And can we do better?

The first question is moot here, as this algorithm is precisely Fibonacci's definition of F_n. But the second demands an answer. Let $T(n)$ be the number of *computer steps* needed to compute `fib1`(n); what can we say about this function? For starters, if n is less than 2, the procedure halts almost immediately, after just a couple of steps. Therefore,

$$T(n) \leq 2 \text{ for } n \leq 1.$$

For larger values of n, there are two recursive invocations of fib1, taking time $T(n-1)$ and $T(n-2)$, respectively, plus three computer steps (checks on the value of n and a final addition). Therefore,

$$T(n) = T(n-1) + T(n-2) + 3 \text{ for } n > 1.$$

Compare this to the recurrence relation for F_n: we immediately see that $T(n) \geq F_n$.

This is very bad news: the running time of the algorithm grows as fast as the Fibonacci numbers! *$T(n)$ is exponential in n*, which implies that the algorithm is impractically slow except for very small values of n.

Let's be a little more concrete about just how bad exponential time is. To compute F_{200}, the fib1 algorithm executes $T(200) \geq F_{200} \geq 2^{138}$ elementary computer steps. How long this actually takes depends, of course, on the computer used. At this time, the fastest computer in the world is the NEC Earth Simulator, which clocks 40 trillion steps per second. Even on this machine, fib1(200) would take at least 2^{92} seconds. This means that, if we start the computation today, it would still be going long after the sun turns into a red giant star.

But technology is rapidly improving—computer speeds have been doubling roughly every 18 months, a phenomenon sometimes called *Moore's law*. With this extraordinary growth, perhaps fib1 will run a lot faster on next year's machines. Let's see—the running time of fib1(n) is proportional to $2^{0.694n} \approx (1.6)^n$, so it takes 1.6 times longer to compute F_{n+1} than F_n. And under Moore's law, computers get roughly 1.6 times faster each year. So if we can reasonably compute F_{100} with this year's technology, then next year we will manage F_{101}. And the year after, F_{102}. And so on: just one more Fibonacci number every year! Such is the curse of exponential time.

In short, our naive recursive algorithm is correct but hopelessly inefficient. *Can we do better?*

A polynomial algorithm

Let's try to understand why fib1 is so slow. Figure 0.1 shows the cascade of recursive invocations triggered by a single call to fib1(n). Notice that many computations are repeated!

A more sensible scheme would store the intermediate results—the values $F_0, F_1, \ldots, F_{n-1}$—as soon as they become known.

```
function fib2(n)
if n = 0: return 0
create an array f[0...n]
f[0] = 0, f[1] = 1
for i = 2...n:
  f[i] = f[i-1] + f[i-2]
return f[n]
```

Figure 0.1 The proliferation of recursive calls in `fib1`.

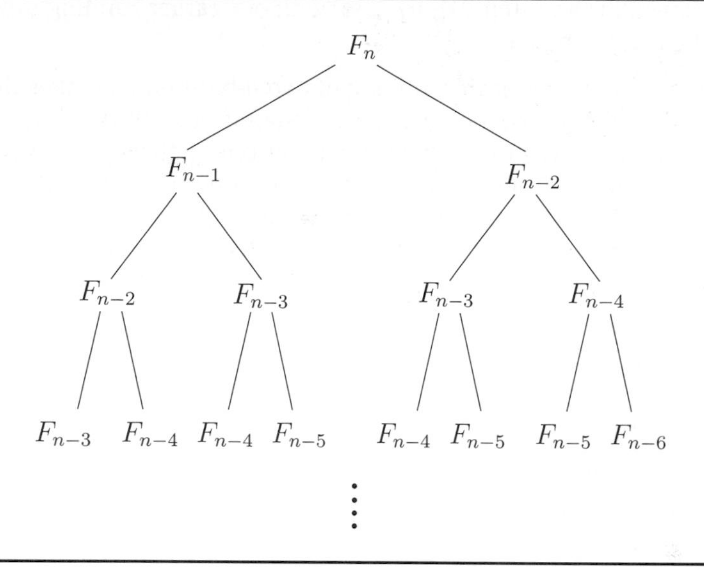

As with `fib1`, the correctness of this algorithm is self-evident because it directly uses the definition of F_n. How long does it take? The inner loop consists of a single computer step and is executed $n - 1$ times. Therefore the number of computer steps used by `fib2` is *linear in n*. From exponential we are down to *polynomial*, a huge breakthrough in running time. It is now perfectly reasonable to compute F_{200} or even $F_{200,000}$.[1]

As we will see repeatedly throughout this book, the right algorithm makes all the difference.

More careful analysis

In our discussion so far, we have been counting the number of *basic computer steps* executed by each algorithm and thinking of these basic steps as taking a constant amount of time. This is a very useful simplification. After all, a processor's instruction set has a variety of basic primitives—branching, storing to memory, comparing numbers, simple arithmetic, and so on—and rather than distinguishing between these elementary operations, it is far more convenient to lump them together into one category.

But looking back at our treatment of Fibonacci algorithms, we have been too liberal with what we consider a basic step. It is reasonable to treat addition as a single computer step if small numbers are being added, 32-bit numbers say. But the nth Fibonacci number is about $0.694n$ bits long, and this can far exceed 32 as n grows.

[1]To better appreciate the importance of this dichotomy between exponential and polynomial algorithms, the reader may want to peek ahead to *the story of Sissa and Moore* in Chapter 8.

Arithmetic operations on arbitrarily large numbers cannot possibly be performed in a single, constant-time step. We need to audit our earlier running time estimates and make them more honest.

We will see in Chapter 1 that the addition of two n-bit numbers takes time roughly proportional to n; this is not too hard to understand if you think back to the grade-school procedure for addition, which works on one digit at a time. Thus `fib1`, which performs about F_n additions, actually uses a number of *basic steps* roughly proportional to nF_n. Likewise, the number of steps taken by `fib2` is proportional to n^2, still polynomial in n and therefore exponentially superior to `fib1`. This correction to the running time analysis does not diminish our breakthrough.

But can we do even better than `fib2`*?* Indeed we can: see Exercise 0.4.

0.3 Big-*O* notation

We've just seen how sloppiness in the analysis of running times can lead to an unacceptable level of inaccuracy in the result. But the opposite danger is also present: it is possible to be *too* precise. An insightful analysis is based on the right simplifications.

Expressing running time in terms of *basic computer steps* is already a simplification. After all, the time taken by one such step depends crucially on the particular processor and even on details such as caching strategy (as a result of which the running time can differ subtly from one execution to the next). Accounting for these architecture-specific minutiae is a nightmarishly complex task and yields a result that does not generalize from one computer to the next. It therefore makes more sense to seek an uncluttered, machine-independent characterization of an algorithm's efficiency. To this end, we will always express running time by counting the number of basic computer steps, as a function of the size of the input.

And this simplification leads to another. Instead of reporting that an algorithm takes, say, $5n^3 + 4n + 3$ steps on an input of size n, it is much simpler to leave out lower-order terms such as $4n$ and 3 (which become insignificant as n grows), and even the detail of the coefficient 5 in the leading term (computers will be five times faster in a few years anyway), and just say that the algorithm takes time $O(n^3)$ (pronounced "big oh of n^3").

It is time to define this notation precisely. In what follows, think of $f(n)$ and $g(n)$ as the running times of two algorithms on inputs of size n.

> *Let $f(n)$ and $g(n)$ be functions from positive integers to positive reals. We say $f = O(g)$ (which means that "f grows no faster than g") if there is a constant $c > 0$ such that $f(n) \leq c \cdot g(n)$.*

Saying $f = O(g)$ is a very loose analog of "$f \leq g$." It differs from the usual notion of $\leq$ because of the constant c, so that for instance $10n = O(n)$. This constant also allows us to disregard what happens for small values of n. For example, suppose we are choosing between two algorithms for a particular computational task. One takes $f_1(n) = n^2$ steps, while the other takes $f_2(n) = 2n + 20$ steps (Figure 0.2). Which is better? Well, this depends on the value of n. For $n \leq 5$, n^2 is smaller; thereafter, $2n + 20$ is the clear winner. In this case, f_2 *scales* much better as n grows, and therefore it is superior.

This superiority is captured by the big-O notation: $f_2 = O(f_1)$, because

$$\frac{f_2(n)}{f_1(n)} = \frac{2n+20}{n^2} \leq 22$$

for all n; on the other hand, $f_1 \neq O(f_2)$, since the ratio $f_1(n)/f_2(n) = n^2/(2n+20)$ can get arbitrarily large, and so no constant c will make the definition work.

Figure 0.2 Which running time is better?

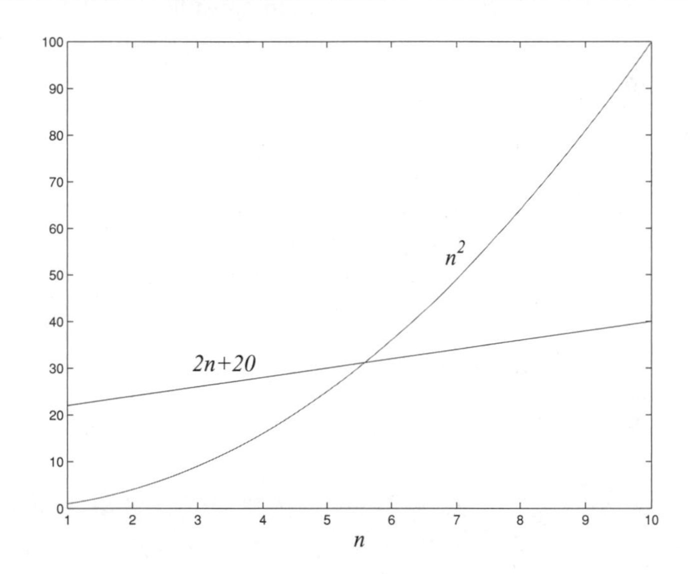

Now another algorithm comes along, one that uses $f_3(n) = n + 1$ steps. Is this better than f_2? Certainly, but only by a constant factor. The discrepancy between $2n + 20$ and $n + 1$ is tiny compared to the huge gap between n^2 and $2n + 20$. In order to stay focused on the big picture, we treat functions as equivalent if they differ only by multiplicative constants.

Returning to the definition of big-O, we see that $f_2 = O(f_3)$:

$$\frac{f_2(n)}{f_3(n)} = \frac{2n+20}{n+1} \leq 20,$$

and of course $f_3 = O(f_2)$, this time with $c = 1$.

Just as $O(\cdot)$ is an analog of $\leq$, we can also define analogs of $\geq$ and $=$ as follows:

$$f = \Omega(g) \text{ means } g = O(f)$$
$$f = \Theta(g) \text{ means } f = O(g) \text{ and } f = \Omega(g).$$

In the preceding example, $f_2 = \Theta(f_3)$ and $f_1 = \Omega(f_3)$.

Big-O notation lets us focus on the big picture. When faced with a complicated function like $3n^2 + 4n + 5$, we just replace it with $O(f(n))$, where $f(n)$ is as simple as possible. In this particular example we'd use $O(n^2)$, because the quadratic portion of the sum dominates the rest. Here are some commonsense rules that help simplify functions by omitting dominated terms:

1. Multiplicative constants can be omitted: $14n^2$ becomes n^2.
2. n^a dominates n^b if $a > b$: for instance, n^2 dominates n.
3. Any exponential dominates any polynomial: 3^n dominates n^5 (it even dominates 2^n).
4. Likewise, any polynomial dominates any logarithm: n dominates $(\log n)^3$. This also means, for example, that n^2 dominates $n \log n$.

Don't misunderstand this cavalier attitude toward constants. Programmers and algorithm developers are *very* interested in constants and would gladly stay up nights in order to make an algorithm run faster by a factor of 2. But understanding algorithms at the level of this book would be impossible without the simplicity afforded by big-O notation.

Exercises

0.1. In each of the following situations, indicate whether $f = O(g)$, or $f = \Omega(g)$, or both (in which case $f = \Theta(g)$).

	$f(n)$	$g(n)$
(a)	$n - 100$	$n - 200$
(b)	$n^{1/2}$	$n^{2/3}$
(c)	$100n + \log n$	$n + (\log n)^2$
(d)	$n \log n$	$10n \log 10n$
(e)	$\log 2n$	$\log 3n$
(f)	$10 \log n$	$\log(n^2)$
(g)	$n^{1.01}$	$n \log^2 n$
(h)	$n^2 / \log n$	$n(\log n)^2$
(i)	$n^{0.1}$	$(\log n)^{10}$
(j)	$(\log n)^{\log n}$	$n / \log n$
(k)	$\sqrt{n}$	$(\log n)^3$
(l)	$n^{1/2}$	$5^{\log_2 n}$
(m)	$n2^n$	3^n

(n)	2^n	2^{n+1}
(o)	$n!$	2^n
(p)	$(\log n)^{\log n}$	$2^{(\log_2 n)^2}$
(q)	$\sum_{i=1}^n i^k$	n^{k+1}

0.2. Show that, if c is a positive real number, then $g(n) = 1 + c + c^2 + \cdots + c^n$ is:

(a) $\Theta(1)$ if $c < 1$.
(b) $\Theta(n)$ if $c = 1$.
(c) $\Theta(c^n)$ if $c > 1$.

The moral: in big-Θ terms, the sum of a geometric series is simply the first term if the series is strictly decreasing, the last term if the series is strictly increasing, or the number of terms if the series is unchanging.

0.3. The Fibonacci numbers $F_0, F_1, F_2, \ldots,$ are defined by the rule

$$F_0 = 0, \;\; F_1 = 1, \;\; F_n = F_{n-1} + F_{n-2}.$$

In this problem we will confirm that this sequence grows exponentially fast and obtain some bounds on its growth.

(a) Use induction to prove that $F_n \geq 2^{0.5n}$ for $n \geq 6$.
(b) Find a constant $c < 1$ such that $F_n \leq 2^{cn}$ for all $n \geq 0$. Show that your answer is correct.
(c) What is the largest c you can find for which $F_n = \Omega(2^{cn})$?

0.4. Is there a faster way to compute the nth Fibonacci number than by `fib2` (page 4)? One idea involves *matrices.*
We start by writing the equations $F_1 = F_1$ and $F_2 = F_0 + F_1$ in matrix notation:

$$\begin{pmatrix} F_1 \\ F_2 \end{pmatrix} = \begin{pmatrix} 0 & 1 \\ 1 & 1 \end{pmatrix} \cdot \begin{pmatrix} F_0 \\ F_1 \end{pmatrix}.$$

Similarly,

$$\begin{pmatrix} F_2 \\ F_3 \end{pmatrix} = \begin{pmatrix} 0 & 1 \\ 1 & 1 \end{pmatrix} \cdot \begin{pmatrix} F_1 \\ F_2 \end{pmatrix} = \begin{pmatrix} 0 & 1 \\ 1 & 1 \end{pmatrix}^2 \cdot \begin{pmatrix} F_0 \\ F_1 \end{pmatrix}$$

and in general

$$\begin{pmatrix} F_n \\ F_{n+1} \end{pmatrix} = \begin{pmatrix} 0 & 1 \\ 1 & 1 \end{pmatrix}^n \cdot \begin{pmatrix} F_0 \\ F_1 \end{pmatrix}.$$

So, in order to compute F_n, it suffices to raise this 2×2 matrix, call it X, to the nth power.

(a) Show that two 2×2 matrices can be multiplied using 4 additions and 8 multiplications.

But how many matrix multiplications does it take to compute X^n?

(b) Show that $O(\log n)$ matrix multiplications suffice for computing X^n. (*Hint:* Think about computing X^8.)

Thus the number of arithmetic operations needed by our matrix-based algorithm, call it `fib3`, is just $O(\log n)$, as compared to $O(n)$ for `fib2`. *Have we broken another exponential barrier?*

The catch is that our new algorithm involves multiplication, not just addition; and multiplications of large numbers are slower than additions. We have already seen that, when the complexity of arithmetic operations is taken into account, the running time of `fib2` becomes $O(n^2)$.

(c) Show that all intermediate results of `fib3` are $O(n)$ bits long.

(d) Let $M(n)$ be the running time of an algorithm for multiplying n-bit numbers, and assume that $M(n) = O(n^2)$ (the school method for multiplication, recalled in Chapter 1, achieves this). Prove that the running time of `fib3` is $O(M(n)\log n)$.

(e) Can you prove that the running time of `fib3` is $O(M(n))$? Assume $M(n) = \Theta(n^a)$ for some $1 \leq a \leq 2$. (*Hint:* The lengths of the numbers being multiplied get doubled with every squaring.)

In conclusion, whether `fib3` is faster than `fib2` depends on whether we can multiply n-bit integers faster than $O(n^2)$. Do you think this is possible? (The answer is in Chapter 2.)

Finally, there is a formula for the Fibonacci numbers:

$$F_n = \frac{1}{\sqrt{5}}\left(\frac{1+\sqrt{5}}{2}\right)^n - \frac{1}{\sqrt{5}}\left(\frac{1-\sqrt{5}}{2}\right)^n.$$

So, it would appear that we only need to raise a couple of numbers to the nth power in order to compute F_n. The problem is that these numbers are irrational, and computing them to sufficient accuracy is nontrivial. In fact, our matrix method `fib3` can be seen as a roundabout way of raising these irrational numbers to the nth power. If you know your linear algebra, you should see why. (*Hint:* What are the eigenvalues of the matrix X?)

Chapter 1
Algorithms with numbers

One of the main themes of this chapter is the dramatic contrast between two ancient problems that at first seem very similar:

> FACTORING: Given a number N, express it as a product of its prime factors.
>
> PRIMALITY: Given a number N, determine whether it is a prime.

Factoring is hard. Despite centuries of effort by some of the world's smartest mathematicians and computer scientists, the fastest methods for factoring a number N take time exponential in the number of bits of N.

On the other hand, we shall soon see that we *can* efficiently test whether N is prime! And (it gets even more interesting) this strange disparity between the two intimately related problems, one very hard and the other very easy, lies at the heart of the technology that enables secure communication in today's global information environment.

En route to these insights, we need to develop algorithms for a variety of computational tasks involving numbers. We begin with basic arithmetic, an especially appropriate starting point because, as we know, the word *algorithms* originally applied only to methods for these problems.

1.1 Basic arithmetic

1.1.1 Addition

We were so young when we learned the standard technique for addition that we would scarcely have thought to ask *why* it works. But let's go back now and take a closer look.

It is a basic property of decimal numbers that

> *The sum of any three single-digit numbers is at most two digits long.*

Quick check: the sum is at most $9 + 9 + 9 = 27$, two digits long. In fact, this rule holds not just in decimal but in *any* base $b \geq 2$ (Exercise 1.1). In binary, for instance, the maximum possible sum of three single-bit numbers is 3, which is a 2-bit number.

Bases and logs

Naturally, there is nothing special about the number 10—we just happen to have 10 fingers, and so 10 was an obvious place to pause and take counting to the next level. The Mayans developed a similar positional system based on the number 20 (no shoes, see?). And of course today computers represent numbers in binary.

How many digits are needed to represent the number $N \geq 0$ in base b? Let's see—with k digits in base b we can express numbers up to $b^k - 1$; for instance, in decimal, three digits get us all the way up to $999 = 10^3 - 1$. By solving for k, we find that $\lceil \log_b(N+1) \rceil$ digits (about $\log_b N$ digits, give or take 1) are needed to write N in base b.

How much does the size of a number change when we change bases? Recall the rule for converting logarithms from base a to base b: $\log_b N = (\log_a N)/(\log_a b)$. So the size of integer N in base a is the same as its size in base b, times a constant factor $\log_a b$. In big-O notation, therefore, the base is irrelevant, and we write the size simply as $O(\log N)$. When we do not specify a base, as we almost never will, we mean $\log_2 N$.

Incidentally, this function $\log N$ appears repeatedly in our subject, in many guises. Here's a sampling:

1. $\log N$ is, of course, the power to which you need to raise 2 in order to obtain N.
2. Going backward, it can also be seen as the number of times you must halve N to get down to 1. (More precisely: $\lceil \log N \rceil$.) This is useful when a number is halved at each iteration of an algorithm, as in several examples later in the chapter.
3. It is the number of bits in the binary representation of N. (More precisely: $\lceil \log(N+1) \rceil$.)
4. It is also the depth of a complete binary tree with N nodes. (More precisely: $\lfloor \log N \rfloor$.)
5. It is even the sum $1 + \frac{1}{2} + \frac{1}{3} + \cdots + \frac{1}{N}$, to within a constant factor (Exercise 1.5).

This simple rule gives us a way to add two numbers in any base: align their right-hand ends, and then perform a single right-to-left pass in which the sum is computed digit by digit, maintaining the overflow as a carry. Since we know each individual sum is a two-digit number, *the carry is always a single digit*, and so at any given step, three single-digit numbers are added. Here's an example showing the addition 53 + 35 in binary.

$$
\begin{array}{rccccccccl}
\text{Carry:} & 1 & & & 1 & 1 & 1 & & \\
 & & 1 & 1 & 0 & 1 & 0 & 1 & (53) \\
 & & 1 & 0 & 0 & 0 & 1 & 1 & (35) \\
\hline
 & 1 & 0 & 1 & 1 & 0 & 0 & 0 & (88)
\end{array}
$$

Ordinarily we would spell out the algorithm in pseudocode, but in this case it is so familiar that we do not repeat it. Instead we move straight to analyzing its efficiency.

Given two binary numbers x and y, how long does our algorithm take to add them? This is the kind of question we shall persistently be asking throughout this book. We want the answer expressed as a function of *the size of the input*: the number of bits of x and y, the number of keystrokes needed to type them in.

Suppose x and y are each n bits long; in this chapter we will consistently use the letter n for the sizes of numbers. Then the sum of x and y is $n+1$ bits at most, and each individual bit of this sum gets computed in a fixed amount of time. The total running time for the addition algorithm is therefore of the form $c_0 + c_1 n$, where c_0 and c_1 are some constants; in other words, it is *linear*. Instead of worrying about the precise values of c_0 and c_1, we will focus on the big picture and denote the running time as $O(n)$.

Now that we have a working algorithm whose running time we know, our thoughts wander inevitably to the question of whether there is something even better.

Is there a faster algorithm? (This is another persistent question.) For addition, the answer is easy: in order to add two n-bit numbers we must at least read them and write down the answer, and even that requires n operations. So the addition algorithm is optimal, up to multiplicative constants!

Some readers may be confused at this point: Why $O(n)$ operations? Isn't binary addition something that computers today perform by just one instruction? There are two answers. First, it is certainly true that in a single instruction we can add integers whose size in bits is within the *word length* of today's computers—32 perhaps. But, as will become apparent later in this chapter, it is often useful and necessary to handle numbers much larger than this, perhaps several thousand bits long. Adding and multiplying such large numbers on real computers is very much like performing the operations bit by bit. Second, when we want to understand algorithms, it makes sense to study even the basic algorithms that are encoded in the hardware of today's computers. In doing so, we shall focus on the *bit complexity* of the algorithm, the number of elementary operations on individual bits—because this accounting reflects the amount of hardware, transistors and wires, necessary for implementing the algorithm.

1.1.2 Multiplication and division

Onward to multiplication! The grade-school algorithm for multiplying two numbers x and y is to create an array of intermediate sums, each representing the product of x by a single digit of y. These values are appropriately left-shifted and then added up. Suppose for instance that we want to multiply 13×11, or in binary notation, $x = 1101$ and $y = 1011$. The multiplication would proceed thus.

```
                1  1  0  1
            ×   1  0  1  1
  ------------------------
                1  1  0  1     (1101 times 1)
             1  1  0  1        (1101 times 1, shifted once)
          0  0  0  0           (1101 times 0, shifted twice)
  +    1  1  0  1              (1101 times 1, shifted thrice)
  ------------------------
    1  0  0  0  1  1  1  1     (binary 143)
```

In binary this is particularly easy since each intermediate row is either zero or x itself, left-shifted an appropriate amount of times. Also notice that left-shifting is just a quick way to multiply by the base, which in this case is 2. (Likewise, the effect of a right shift is to divide by the base, rounding down if needed.)

The correctness of this multiplication procedure is the subject of Exercise 1.6; let's move on and figure out how long it takes. If x and y are both n bits, then there are n intermediate rows, with lengths of up to $2n$ bits (taking the shifting into account). The total time taken to add up these rows, doing two numbers at a time, is

$$\underbrace{O(n) + O(n) + \cdots + O(n)}_{n-1 \text{ times}},$$

which is $O(n^2)$, *quadratic* in the size of the inputs: still polynomial but much slower than addition (as we have all suspected since elementary school).

But Al Khwarizmi knew another way to multiply, a method which is used today in some European countries. To multiply two decimal numbers x and y, write them next to each other, as in the example below. Then repeat the following: divide the first number by 2, rounding down the result (that is, dropping the .5 if the number was odd), and double the second number. Keep going till the first number gets down to 1. Then strike out all the rows in which the first number is even, and add up whatever remains in the second column.

11	13	
5	26	
2	52	(strike out)
1	104	
	143	(answer)

But if we now compare the two algorithms, binary multiplication and multiplication by repeated halvings of the multiplier, we notice that they are doing the same thing! The three numbers added in the second algorithm are precisely the multiples of 13 by powers of 2 that were added in the binary method. Only this time 11 was not given to us explicitly in binary, and so we had to extract its binary representation by looking at the parity of the numbers obtained from it by successive divisions by 2. Al Khwarizmi's second algorithm is a fascinating mixture of decimal and binary!

The same algorithm can thus be repackaged in different ways. For variety we adopt a third formulation, the recursive algorithm of Figure 1.1, which directly implements the rule

$$x \cdot y = \begin{cases} 2(x \cdot \lfloor y/2 \rfloor) & \text{if } y \text{ is even} \\ x + 2(x \cdot \lfloor y/2 \rfloor) & \text{if } y \text{ is odd.} \end{cases}$$

Is this algorithm correct? The preceding recursive rule is transparently correct; so

Figure 1.1 Multiplication à la Français.

```
function multiply(x, y)
Input: Two n−bit integers x and y, where y ≥ 0
Output: Their product

if y = 0: return 0
z = multiply(x, ⌊y/2⌋)
if y is even:
   return 2z
else:
   return x + 2z
```

checking the correctness of the algorithm is merely a matter of verifying that it mimics the rule and that it handles the base case ($y = 0$) properly.

How long does the algorithm take? It must terminate after n recursive calls, because at each call y is halved—that is, its number of bits is decreased by one. And each recursive call requires these operations: a division by 2 (right shift); a test for odd/even (looking up the last bit); a multiplication by 2 (left shift); and possibly one addition, a total of $O(n)$ bit operations. The total time taken is thus $O(n^2)$, just as before.

Can we do better? Intuitively, it seems that multiplication requires adding about n multiples of one of the inputs, and we know that each addition is linear, so it would appear that n^2 bit operations are inevitable. Astonishingly, in Chapter 2 we'll see that we *can* do significantly better!

Division is next. To divide an integer x by another integer $y \neq 0$ means to find a quotient q and a remainder r, where $x = yq + r$ and $r < y$. We show the recursive version of division in Figure 1.2; like multiplication, it takes quadratic time. The analysis of this algorithm is the subject of Exercise 1.8.

Figure 1.2 Division.

```
function divide(x, y)
Input: Two n−bit integers x and y, where y ≥ 1
Output: The quotient and remainder of x divided by y

if x = 0: return (q, r) = (0, 0)
(q, r) = divide(⌊x/2⌋, y)
q = 2 · q,  r = 2 · r
if x is odd: r = r + 1
if r ≥ y: r = r − y,  q = q + 1
return (q, r)
```

1.2 Modular arithmetic

With repeated addition or multiplication, numbers can get cumbersomely large. So it is fortunate that we reset the hour to zero whenever it reaches 24, and the month to January after every stretch of 12 months. Similarly, for the built-in arithmetic operations of computer processors, numbers are restricted to some size, 32 bits say, which is considered generous enough for most purposes.

For the applications we are working toward—primality testing and cryptography—it is necessary to deal with numbers that are significantly larger than 32 bits, but whose range is nonetheless limited.

Modular arithmetic is a system for dealing with restricted ranges of integers. We define x *modulo* N to be the remainder when x is divided by N; that is, if $x = qN + r$ with $0 \leq r < N$, then x modulo N is equal to r. This gives an enhanced notion of equivalence between numbers: x and y are *congruent modulo* N if they differ by a multiple of N, or in symbols,

$$x \equiv y \pmod{N} \iff N \text{ divides } (x - y).$$

For instance, $253 \equiv 13 \pmod{60}$ because $253 - 13$ is a multiple of 60; more familiarly, 253 minutes is 4 hours and 13 minutes. These numbers can also be negative, as in $59 \equiv -1 \pmod{60}$: when it is 59 minutes past the hour, it is also 1 minute short of the next hour.

Figure 1.3 Addition modulo 8.

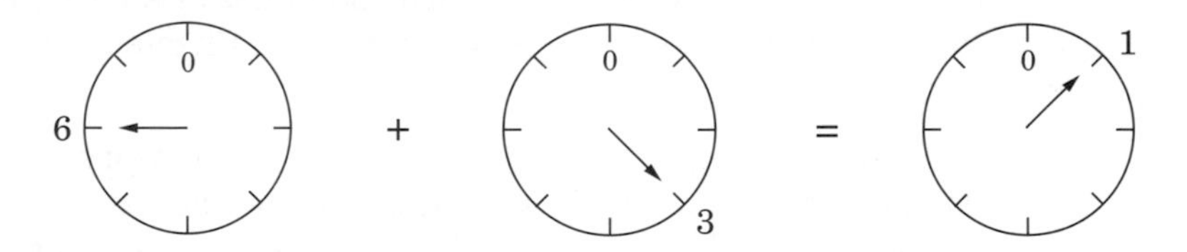

One way to think of modular arithmetic is that it limits numbers to a predefined range $\{0, 1, \ldots, N - 1\}$ and wraps around whenever you try to leave this range—like the hand of a clock (Figure 1.3).

Another interpretation is that modular arithmetic deals with all the integers, but divides them into N *equivalence classes*, each of the form $\{i + kN : k \in \mathbb{Z}\}$ for some i between 0 and $N - 1$. For example, there are three equivalence classes modulo 3:

$$\begin{array}{ccccccccc} \cdots & -9 & -6 & -3 & 0 & 3 & 6 & 9 & \cdots \\ \cdots & -8 & -5 & -2 & 1 & 4 & 7 & 10 & \cdots \\ \cdots & -7 & -4 & -1 & 2 & 5 & 8 & 11 & \cdots \end{array}$$

Any member of an equivalence class is substitutable for any other; when viewed modulo 3, the numbers 5 and 11 are no different. Under such substitutions, addition and multiplication remain well-defined:

Two's complement

Modular arithmetic is nicely illustrated in *two's complement*, the most common format for storing signed integers. It uses n bits to represent numbers in the range $[-2^{n-1}, 2^{n-1}-1]$ and is usually described as follows:

- Positive integers, in the range 0 to $2^{n-1}-1$, are stored in regular binary and have a leading bit of 0.
- Negative integers $-x$, with $1 \le x \le 2^{n-1}$, are stored by first constructing x in binary, then flipping all the bits, and finally adding 1. The leading bit in this case is 1.

(And the usual description of addition and multiplication in this format is even more arcane!)

Here's a much simpler way to think about it: any number in the range -2^{n-1} to $2^{n-1}-1$ is stored modulo 2^n. Negative numbers $-x$ therefore end up as $2^n - x$. Arithmetic operations like addition and subtraction can be performed directly in this format, ignoring any overflow bits that arise.

Substitution rule *If $x \equiv x' \pmod N$ and $y \equiv y' \pmod N$, then:*

$$x + y \equiv x' + y' \pmod N \quad \text{and} \quad xy \equiv x'y' \pmod N.$$

(See Exercise 1.9.) For instance, suppose you watch an entire season of your favorite television show in one sitting, starting at midnight. There are 25 episodes, each lasting 3 hours. At what time of day are you done? Answer: the hour of completion is $(25 \times 3) \bmod 24$, which (since $25 \equiv 1 \bmod 24$) is $1 \times 3 = 3 \bmod 24$, or three o'clock in the morning.

It is not hard to check that in modular arithmetic, the usual associative, commutative, and distributive properties of addition and multiplication continue to apply, for instance:

$$\begin{aligned} x + (y + z) &\equiv (x + y) + z \pmod N && \text{Associativity} \\ xy &\equiv yx \pmod N && \text{Commutativity} \\ x(y + z) &\equiv xy + yz \pmod N && \text{Distributivity} \end{aligned}$$

Taken together with the substitution rule, this implies that while performing a sequence of arithmetic operations, it is legal to reduce intermediate results to their remainders modulo N at any stage. Such simplifications can be a dramatic help in big calculations. Witness, for instance:

$$2^{345} \equiv (2^5)^{69} \equiv 32^{69} \equiv 1^{69} \equiv 1 \pmod{31}.$$

1.2.1 Modular addition and multiplication

To *add* two numbers x and y modulo N, we start with regular addition. Since x and y are each in the range 0 to $N-1$, their sum is between 0 and $2(N-1)$. If the sum exceeds $N-1$, we merely need to subtract off N to bring it back into the required range. The overall computation therefore consists of an addition, and possibly a subtraction, of numbers that never exceed $2N$. Its running time is linear in the sizes of these numbers, in other words $O(n)$, where $n = \lceil \log N \rceil$ is the size of N; as a reminder, our convention is to use the letter n to denote input size.

To *multiply* two mod-N numbers x and y, we again just start with regular multiplication and then reduce the answer modulo N. The product can be as large as $(N-1)^2$, but this is still at most $2n$ bits long since $\log(N-1)^2 = 2\log(N-1) \leq 2n$. To reduce the answer modulo N, we compute the remainder upon dividing it by N, using our quadratic-time division algorithm. Multiplication thus remains a quadratic operation.

Division is not quite so easy. In ordinary arithmetic there is just one tricky case—division by zero. It turns out that in modular arithmetic there are potentially other such cases as well, which we will characterize toward the end of this section. Whenever division is legal, however, it can be managed in cubic time, $O(n^3)$.

To complete the suite of modular arithmetic primitives we need for cryptography, we next turn to *modular exponentiation*, and then to the *greatest common divisor*, which is the key to division. For both tasks, the most obvious procedures take exponentially long, but with some ingenuity polynomial-time solutions can be found. A careful choice of algorithm makes all the difference.

1.2.2 Modular exponentiation

In the cryptosystem we are working toward, it is necessary to compute $x^y \bmod N$ for values of x, y, and N that are several hundred bits long. Can this be done quickly?

The result is some number modulo N and is therefore itself a few hundred bits long. However, the raw value of x^y could be much, much longer than this. Even when x and y are just 20-bit numbers, x^y is at least $(2^{19})^{(2^{19})} = 2^{(19)(524288)}$, about 10 million bits long! Imagine what happens if y is a 500-bit number!

To make sure the numbers we are dealing with never grow too large, we need to perform all intermediate computations modulo N. So here's an idea: calculate $x^y \bmod N$ by repeatedly multiplying by x modulo N. The resulting sequence of intermediate products,

$$x \bmod N \rightarrow x^2 \bmod N \rightarrow x^3 \bmod N \rightarrow \cdots \rightarrow x^y \bmod N,$$

consists of numbers that are smaller than N, and so the individual multiplications do not take too long. But there's a problem: if y is 500 bits long, we need to perform $y - 1 \approx 2^{500}$ multiplications! This algorithm is clearly exponential in the size of y.

Luckily, we *can* do better: starting with x and *squaring repeatedly* modulo N, we get

$$x \bmod N \rightarrow x^2 \bmod N \rightarrow x^4 \bmod N \rightarrow x^8 \bmod N \rightarrow \cdots \rightarrow x^{2^{\lfloor \log y \rfloor}} \bmod N.$$

Each takes just $O(\log^2 N)$ time to compute, and in this case there are only $\log y$ multiplications. To determine $x^y \bmod N$, we simply multiply together an appropriate subset of these powers, those corresponding to 1's in the binary representation of y. For instance,

$$x^{25} = x^{11001_2} = x^{10000_2} \cdot x^{1000_2} \cdot x^{1_2} = x^{16} \cdot x^8 \cdot x^1.$$

A polynomial-time algorithm is finally within reach!

Figure 1.4 Modular exponentiation.

```
function modexp(x, y, N)
Input: Two n-bit integers x and N, an integer exponent y
Output: x^y mod N

if y = 0: return 1
z = modexp(x, ⌊y/2⌋, N)
if y is even:
   return z^2 mod N
else:
   return x · z^2 mod N
```

We can package this idea in a particularly simple form: the recursive algorithm of Figure 1.4, which works by executing, modulo N, the self-evident rule

$$x^y = \begin{cases} \left(x^{\lfloor y/2 \rfloor}\right)^2 & \text{if } y \text{ is even} \\ x \cdot \left(x^{\lfloor y/2 \rfloor}\right)^2 & \text{if } y \text{ is odd.} \end{cases}$$

In doing so, it closely parallels our recursive multiplication algorithm (Figure 1.1). For instance, that algorithm would compute the product $x \cdot 25$ by an analogous decomposition to the one we just saw: $x \cdot 25 = x \cdot 16 + x \cdot 8 + x \cdot 1$. And whereas for multiplication the terms $x \cdot 2^i$ come from repeated *doubling*, for exponentiation the corresponding terms x^{2^i} are generated by repeated squaring.

Let n be the size in bits of x, y, and N (whichever is largest of the three). As with multiplication, the algorithm will halt after at most n recursive calls, and during each call it multiplies n-bit numbers (doing computation modulo N saves us here), for a total running time of $O(n^3)$.

1.2.3 Euclid's algorithm for greatest common divisor

Our next algorithm was discovered well over 2000 years ago by the mathematician Euclid, in ancient Greece. Given two integers a and b, it finds the largest integer that divides both of them, known as their *greatest common divisor* (gcd).

The most obvious approach is to first factor a and b, and then multiply together their common factors. For instance, $1035 = 3^2 \cdot 5 \cdot 23$ and $759 = 3 \cdot 11 \cdot 23$, so their gcd is $3 \cdot 23 = 69$. However, we have no efficient algorithm for factoring. Is there some other way to compute greatest common divisors?

Euclid's algorithm uses the following simple formula.

Euclid of Alexandria
BC 325–265

© Corbis

Euclid's rule *If x and y are positive integers with $x \geq y$, then* $\gcd(x, y) = \gcd(x \bmod y, y)$.

Proof. It is enough to show the slightly simpler rule $\gcd(x, y) = \gcd(x - y, y)$ from which the one stated can be derived by repeatedly subtracting y from x.

Here it goes. Any integer that divides both x and y must also divide $x - y$, so $\gcd(x, y) \leq \gcd(x - y, y)$. Likewise, any integer that divides both $x - y$ and y must also divide both x and y, so $\gcd(x, y) \geq \gcd(x - y, y)$. ▮

Figure 1.5 **Euclid's algorithm for finding the greatest common divisor of two numbers.**

```
function Euclid(a, b)
Input: Two integers a and b with a ≥ b ≥ 0
Output: gcd(a, b)

if b = 0: return a
return Euclid(b, a mod b)
```

Euclid's rule allows us to write down an elegant recursive algorithm (Figure 1.5), and its correctness follows immediately from the rule. In order to figure out its running time, we need to understand how quickly the arguments (a, b) decrease with each successive recursive call. In a single round, arguments (a, b) become $(b, a \bmod b)$: their order is swapped, and the larger of them, a, gets reduced to $a \bmod b$. This is a substantial reduction.

Lemma *If $a \geq b$, then $a \bmod b < a/2$.*

Proof. Witness that either $b \leq a/2$ or $b > a/2$. These two cases are shown in the following figure. If $b \leq a/2$, then we have $a \bmod b < b \leq a/2$; and if $b > a/2$, then $a \bmod b = a - b < a/2$. ∎

This means that after any two consecutive rounds, both arguments, a and b, are at the very least halved in value—the length of each decreases by at least one bit. If they are initially n-bit integers, then the base case will be reached within $2n$ recursive calls. And since each call involves a quadratic-time division, the total time is $O(n^3)$.

1.2.4 An extension of Euclid's algorithm

A small extension to Euclid's algorithm is the key to dividing in the modular world.

To motivate it, suppose someone claims that d is the greatest common divisor of a and b: how can we check this? It is not enough to verify that d divides both a and b, because this only shows d to be a common factor, not necessarily the largest one. Here's a test that can be used if d is of a particular form.

Lemma *If d divides both a and b, and $d = ax + by$ for some integers x and y, then necessarily $d = \gcd(a, b)$.*

Proof. By the first two conditions, d is a common divisor of a and b and so it cannot exceed the greatest common divisor; that is, $d \leq \gcd(a, b)$. On the other hand, since $\gcd(a, b)$ is a common divisor of a and b, it must also divide $ax + by = d$, which implies $\gcd(a, b) \leq d$. Putting these together, $d = \gcd(a, b)$. ∎

So, if we can supply two numbers x and y such that $d = ax + by$, then we can be sure $d = \gcd(a, b)$. For instance, we know $\gcd(13, 4) = 1$ because $13 \cdot 1 + 4 \cdot (-3) = 1$. But when can we find these numbers: under what circumstances can $\gcd(a, b)$ be expressed in this checkable form? It turns out that it *always* can. What is even better, the coefficients x and y can be found by a small extension to Euclid's algorithm; see Figure 1.6.

Figure 1.6 A simple extension of Euclid's algorithm.

```
function extended-Euclid(a, b)
Input: Two positive integers a and b with a ≥ b ≥ 0
Output: Integers x, y, d such that d = gcd(a, b) and ax + by = d

if b = 0: return (1, 0, a)
(x', y', d) = extended-Euclid(b, a mod b)
return (y', x' − ⌊a/b⌋y', d)
```

Lemma *For any positive integers a and b, the extended Euclid algorithm returns integers x, y, and d such that* $\gcd(a, b) = d = ax + by$.

Proof. The first thing to confirm is that if you ignore the x's and y's, the extended algorithm is exactly the same as the original. So, at least we compute $d = \gcd(a, b)$.

For the rest, the recursive nature of the algorithm suggests a proof by induction. The recursion ends when $b = 0$, so it is convenient to do induction on the value of b.

The base case $b = 0$ is easy enough to check directly. Now pick any larger value of b. The algorithm finds $\gcd(a, b)$ by calling $\gcd(b, a \bmod b)$. Since $a \bmod b < b$, we can apply the inductive hypothesis to this recursive call and conclude that the x' and y' it returns are correct:

$$\gcd(b, a \bmod b) = bx' + (a \bmod b)y'.$$

Writing $(a \bmod b)$ as $(a - \lfloor a/b \rfloor b)$, we find

$$\begin{aligned} d &= \gcd(a, b) = \gcd(b, a \bmod b) = bx' + (a \bmod b)y' \\ &= bx' + (a - \lfloor a/b \rfloor b)y' = ay' + b(x' - \lfloor a/b \rfloor y'). \end{aligned}$$

Therefore $d = ax + by$ with $x = y'$ and $y = x' - \lfloor a/b \rfloor y'$, thus validating the algorithm's behavior on input (a, b). ∎

Example. To compute $\gcd(25, 11)$, Euclid's algorithm would proceed as follows:

$$\begin{aligned} \underline{25} &= 2 \cdot \underline{11} + 3 \\ \underline{11} &= 3 \cdot \underline{3} + 2 \\ \underline{3} &= 1 \cdot \underline{2} + 1 \\ \underline{2} &= 2 \cdot \underline{1} + 0 \end{aligned}$$

(at each stage, the gcd computation has been reduced to the underlined numbers). Thus $\gcd(25, 11) = \gcd(11, 3) = \gcd(3, 2) = \gcd(2, 1) = \gcd(1, 0) = 1$.

To find x and y such that $25x + 11y = 1$, we start by expressing 1 in terms of the last pair $(1, 0)$. Then we work backwards and express it in terms of $(2, 1)$, $(3, 2)$, $(11, 3)$, and finally $(25, 11)$. The first step is:

$$1 = \underline{1} - \underline{0}.$$

To rewrite this in terms of $(2, 1)$, we use the substitution $0 = 2 - 2 \cdot 1$ from the last line of the gcd calculation to get:

$$1 = \underline{1} - (\underline{2} - 2 \cdot \underline{1}) = -1 \cdot \underline{2} + 3 \cdot \underline{1}.$$

The second-last line of the gcd calculation tells us that $1 = 3 - 1 \cdot 2$. Substituting:

$$1 = -1 \cdot \underline{2} + 3(\underline{3} - 1 \cdot \underline{2}) = 3 \cdot \underline{3} - 4 \cdot \underline{2}.$$

Continuing in this same way with substitutions $2 = 11 - 3 \cdot 3$ and $3 = 25 - 2 \cdot 11$ gives:

$$1 = 3 \cdot \underline{3} - 4(\underline{11} - 3 \cdot \underline{3}) = -4 \cdot \underline{11} + 15 \cdot \underline{3} = -4 \cdot \underline{11} + 15(\underline{25} - 2 \cdot \underline{11}) = 15 \cdot \underline{25} - 34 \cdot \underline{11}.$$

We're done: $15 \cdot 25 - 34 \cdot 11 = 1$, so $x = 15$ and $y = -34$.

1.2.5 Modular division

In real arithmetic, every number $a \neq 0$ has an inverse, $1/a$, and dividing by a is the same as multiplying by this inverse. In modular arithmetic, we can make a similar definition.

We say x is the multiplicative inverse of a modulo N if $ax \equiv 1 \pmod N$.

There can be at most one such x modulo N (Exercise 1.23), and we shall denote it by a^{-1}. However, this inverse does not always exist! For instance, 2 is not invertible modulo 6: that is, $2x \not\equiv 1 \bmod 6$ for every possible choice of x. In this case, a and N are both even and thus then $a \bmod N$ is always even, since $a \bmod N = a - kN$ for some k. More generally, we can be certain that $\gcd(a, N)$ divides $ax \bmod N$, because this latter quantity can be written in the form $ax + kN$. So if $\gcd(a, N) > 1$, then $ax \not\equiv 1 \bmod N$, no matter what x might be, and therefore a cannot have a multiplicative inverse modulo N.

In fact, this is the only circumstance in which a is not invertible. When $\gcd(a, N) = 1$ (we say a and N are *relatively prime*), the extended Euclid algorithm gives us integers x and y such that $ax + Ny = 1$, which means that $ax \equiv 1 \pmod N$. Thus x is a's sought inverse.

Example. Continuing with our previous example, suppose we wish to compute $11^{-1} \bmod 25$. Using the extended Euclid algorithm, we find that $15 \cdot 25 - 34 \cdot 11 = 1$. Reducing both sides modulo 25, we have $-34 \cdot 11 \equiv 1 \bmod 25$. So $-34 \equiv 16 \bmod 25$ is the inverse of 11 mod 25.

Modular division theorem *For any a mod N, a has a multiplicative inverse modulo N if and only if it is relatively prime to N. When this inverse exists, it can be found in time $O(n^3)$ (where as usual n denotes the number of bits of N) by running the extended Euclid algorithm.*

This resolves the issue of modular division: when working modulo N, we can divide by numbers relatively prime to N—and only by these. And to actually carry out the division, we multiply by the inverse.

1.3 Primality testing

Is there some litmus test that will tell us whether a number is prime without actually trying to factor the number? We place our hopes in a theorem from the year 1640.

Fermat's little theorem *If p is prime, then for every $1 \leq a < p$,*

$$a^{p-1} \equiv 1 \pmod p.$$

Proof. Let S be the nonzero integers modulo p; that is, $S = \{1, 2, \ldots, p-1\}$. Here's the crucial observation: the effect of multiplying these numbers by a (modulo p) is simply to permute them. For instance, here's a picture of the case $a = 3$, $p = 7$:

Is your social security number a prime?

The numbers 7, 17, 19, 71, and 79 are primes, but how about 717-19-7179? Telling whether a reasonably large number is a prime seems tedious because there are far too many candidate factors to try. However, there are some clever tricks to speed up the process. For instance, you can omit even-valued candidates after you have eliminated the number 2. You can actually omit all candidates except those that are themselves primes.

In fact, a little further thought will convince you that you can proclaim N a prime as soon as you have rejected all candidates *up to* $\sqrt{N}$, for if N can indeed be factored as $N = K \cdot L$, then it is impossible for both factors to exceed $\sqrt{N}$.

We seem to be making progress! Perhaps by omitting more and more candidate factors, a truly efficient primality test can be discovered.

Unfortunately, there is no fast primality test down this road. The reason is that we have been trying to tell if a number is a prime *by factoring it.* And factoring is a hard problem!

Modern cryptography, as well as the balance of this chapter, is about the following important idea: *factoring is hard and primality is easy.* We cannot factor large numbers, but we can easily test huge numbers for primality! (Presumably, if a number is composite, such a test will detect this *without finding a factor.*)

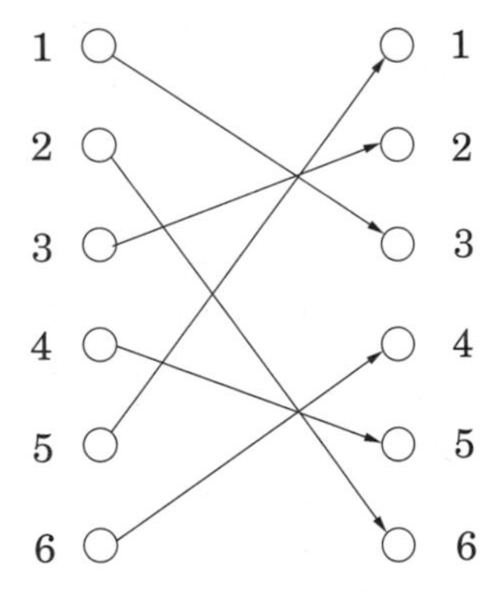

Let's carry this example a bit further. From the picture, we can conclude

$$\{1, 2, \ldots, 6\} = \{3 \cdot 1 \bmod 7,\ 3 \cdot 2 \bmod 7, \ldots,\ 3 \cdot 6 \bmod 7\}.$$

Multiplying all the numbers in each representation then gives $6! \equiv 3^6 \cdot 6! \pmod 7$, and dividing by $6!$ we get $3^6 \equiv 1 \pmod 7$, exactly the result we wanted in the case $a = 3$, $p = 7$.

Now let's generalize this argument to other values of a and p, with $S = \{1, 2, \ldots, p-1\}$. We'll prove that when the elements of S are multiplied by a modulo p, the resulting numbers are all distinct and nonzero. And since they lie in the range $[1, p-1]$, they must simply be a permutation of S.

The numbers $a \cdot i \bmod p$ are distinct because if $a \cdot i \equiv a \cdot j \pmod{p}$, then dividing both sides by a gives $i \equiv j \pmod{p}$. They are nonzero because $a \cdot i \equiv 0$ similarly implies $i \equiv 0$. (And we *can* divide by a, because by assumption it is nonzero and therefore relatively prime to p.)

We now have two ways to write set S:

$$S = \{1, 2, \ldots, p-1\} = \{a \cdot 1 \bmod p,\ a \cdot 2 \bmod p, \ldots,\ a \cdot (p-1) \bmod p\}.$$

We can multiply together its elements in each of these representations to get

$$(p-1)! \equiv a^{p-1} \cdot (p-1)! \pmod{p}.$$

Dividing by $(p-1)!$ (which we can do because it is relatively prime to p, since p is assumed prime) then gives the theorem. ∎

This theorem suggests a "factorless" test for determining whether a number N is prime:

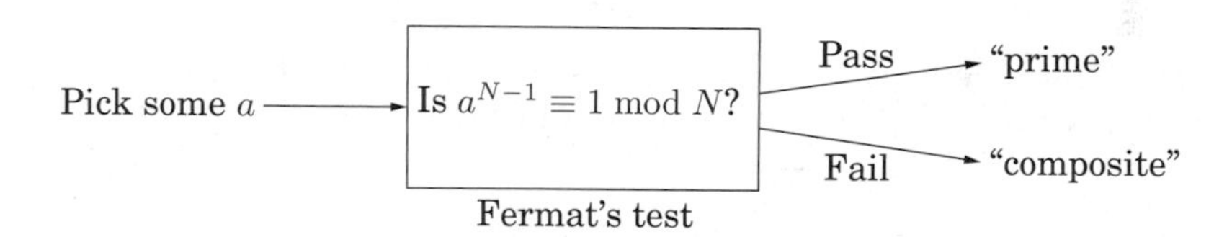

The problem is that Fermat's theorem is not an if-and-only-if condition; it doesn't say what happens when N is *not* prime, so in these cases the preceding diagram is questionable. In fact, it *is* possible for a composite number N to pass Fermat's test (that is, $a^{N-1} \equiv 1 \bmod N$) for certain choices of a. For instance, $341 = 11 \cdot 31$ is not prime, and yet $2^{340} \equiv 1 \bmod 341$. Nonetheless, we might hope that for composite N, *most* values of a will fail the test. This is indeed true, in a sense we will shortly make precise, and motivates the algorithm of Figure 1.7: rather than fixing an arbitrary value of a in advance, we should choose it *randomly* from $\{1, \ldots, N-1\}$.

Figure 1.7 An algorithm for testing primality.

```
function primality(N)
Input: Positive integer N
Output: yes/no

Pick a positive integer a < N at random
if a^(N-1) ≡ 1 (mod N):
   return yes
else:
   return no
```

In analyzing the behavior of this algorithm, we first need to get a minor bad case out of the way. It turns out that certain extremely rare composite numbers N, called *Carmichael numbers*, pass Fermat's test for *all* a relatively prime to N. On such numbers our algorithm will fail; but they are pathologically rare, and we will later see how to deal with them (page 28), so let's ignore these numbers for the time being.

In a Carmichael-free universe, our algorithm works well. Any prime number N will of course pass Fermat's test and produce the right answer. On the other hand, any non-Carmichael composite number N must fail Fermat's test for some value of a; and as we will now show, this implies immediately that N fails Fermat's test for *at least half the possible values of* a!

Lemma *If $a^{N-1} \not\equiv 1 \bmod N$ for some a relatively prime to N, then it must hold for at least half the choices of $a < N$.*

Proof. Fix some value of a for which $a^{N-1} \not\equiv 1 \bmod N$. The key is to notice that every element $b < N$ that passes Fermat's test with respect to N (that is, $b^{N-1} \equiv 1 \bmod N$) has a twin, $a \cdot b$, that fails the test:

$$(a \cdot b)^{N-1} \equiv a^{N-1} \cdot b^{N-1} \equiv a^{N-1} \not\equiv 1 \bmod N.$$

Moreover, all these elements $a \cdot b$, for fixed a but different choices of b, are distinct, for the same reason $a \cdot i \not\equiv a \cdot j$ in the proof of Fermat's test: just divide by a.

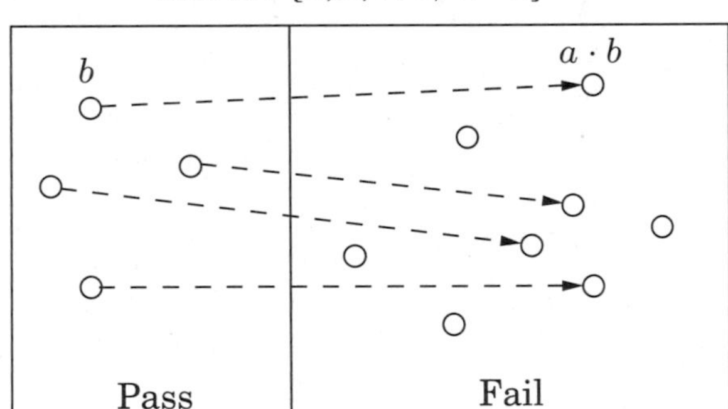

The one-to-one function $b \mapsto a \cdot b$ shows that at least as many elements fail the test as pass it. ∎

We are ignoring Carmichael numbers, so we can now assert

> If N is prime, then $a^{N-1} \equiv 1 \bmod N$ for all $a < N$.
> If N is not prime, then $a^{N-1} \equiv 1 \bmod N$ for at most half the values of $a < N$.

The algorithm of Figure 1.7 therefore has the following probabilistic behavior.

$$\Pr(\text{Algorithm 1.7 returns } \texttt{yes} \text{ when } N \text{ is prime}) = 1$$

$$\Pr(\text{Algorithm 1.7 returns } \texttt{yes} \text{ when } N \text{ is not prime}) \leq \frac{1}{2}$$

Hey, that was group theory!

For any integer N, the set of all numbers mod N that are relatively prime to N constitute what mathematicians call a *group*:

- There is a multiplication operation defined on this set.
- The set contains a neutral element (namely 1: any number multiplied by this remains unchanged).
- All elements have a well-defined inverse.

This particular group is called the *multiplicative group of* N, usually denoted $\mathbb{Z}_N^*$.

Group theory is a very well developed branch of mathematics. One of its key concepts is that a group can contain a *subgroup*—a subset that is a group in and of itself. And an important fact about a subgroup is that its size must divide the size of the whole group.

Consider now the set $B = \{b : b^{N-1} \equiv 1 \bmod N\}$. It is not hard to see that it is a subgroup of $\mathbb{Z}_N^*$ (just check that B is closed under multiplication and inverses). Thus the size of B must divide that of $\mathbb{Z}_N^*$. Which means that if B doesn't contain all of $\mathbb{Z}_N^*$, the next largest size it can have is $|\mathbb{Z}_N^*|/2$.

We can reduce this *one-sided error* by repeating the procedure many times, by randomly picking several values of a and testing them all (Figure 1.8).

$$\Pr(\text{Algorithm 1.8 returns } \texttt{yes} \text{ when } N \text{ is not prime}) \leq \frac{1}{2^k}$$

This probability of error drops exponentially fast, and can be driven arbitrarily low by choosing k large enough. Testing $k = 100$ values of a makes the probability of failure at most 2^{-100}, which is miniscule: far less, for instance, than the probability that a random cosmic ray will sabotage the computer during the computation!

Figure 1.8 An algorithm for testing primality, with low error probability.

```
function primality2(N)
Input: Positive integer N
Output: yes/no

Pick positive integers a_1, a_2, ..., a_k < N at random
if a_i^(N-1) ≡ 1 (mod N) for all i = 1, 2, ..., k:
   return yes
else:
   return no
```

Carmichael numbers

The smallest Carmichael number is 561. It is not a prime: $561 = 3 \cdot 11 \cdot 17$; yet it fools the Fermat test, because $a^{560} \equiv 1 \pmod{561}$ for all values of a relatively prime to 561. For a long time it was thought that there might be only finitely many numbers of this type; now we know they are infinite, but exceedingly rare.

There *is* a way around Carmichael numbers, using a slightly more refined primality test due to Rabin and Miller. Write $N - 1$ in the form $2^t u$. As before we'll choose a random base a and check the value of $a^{N-1} \bmod N$. Perform this computation by first determining $a^u \bmod N$ and then repeatedly squaring, to get the sequence:

$$a^u \bmod N,\ a^{2u} \bmod N,\ \ldots,\ a^{2^t u} = a^{N-1} \bmod N.$$

If $a^{N-1} \not\equiv 1 \bmod N$, then N is composite by Fermat's little theorem, and we're done. But if $a^{N-1} \equiv 1 \bmod N$, we conduct a little follow-up test: somewhere in the preceding sequence, we ran into a 1 for the first time. If this happened after the first position (that is, if $a^u \bmod N \neq 1$), and if the preceding value in the list is not $-1 \bmod N$, then we declare N composite.

In the latter case, we have found a *nontrivial square root* of 1 modulo N: a number that is not $\pm 1 \bmod N$ but that when squared is equal to $1 \bmod N$. Such a number can only exist if N is composite (Exercise 1.40). It turns out that if we combine this square-root check with our earlier Fermat test, then at least three-fourths of the possible values of a between 1 and $N - 1$ will reveal a composite N, even if it is a Carmichael number.

1.3.1 Generating random primes

We are now close to having all the tools we need for cryptographic applications. The final piece of the puzzle is a fast algorithm for choosing random primes that are a few hundred bits long. What makes this task quite easy is that primes are abundant—a random n-bit number has roughly a one-in-n chance of being prime (actually about $1/(\ln 2^n) \approx 1.44/n$). For instance, *about 1 in 20 social security numbers is prime!*

Lagrange's prime number theorem *Let $\pi(x)$ be the number of primes $\leq x$. Then $\pi(x) \approx x/(\ln x)$, or more precisely,*

$$\lim_{x \to \infty} \frac{\pi(x)}{(x/\ln x)} = 1.$$

Such abundance makes it simple to generate a random n-bit prime:

- Pick a random n-bit number N.
- Run a primality test on N.
- If it passes the test, output N; else repeat the process.

How fast is this algorithm? If the randomly chosen N is truly prime, which happens with probability at least $1/n$, then it will certainly pass the test. So on each iteration,

Randomized algorithms: a virtual chapter

Surprisingly—almost paradoxically—some of the fastest and most clever algorithms we have rely on *chance*: at specified steps they proceed according to the outcomes of random coin tosses. These *randomized algorithms* are often very simple and elegant, and their output is allowed to be incorrect *with small probability*. This bound on the failure probability holds for every input; it only depends on the random choices made by the algorithm itself, and can easily be made as small as one likes.

Instead of devoting a special chapter to this topic, in this book we intersperse randomized algorithms at the chapters and sections where they arise most naturally. Furthermore, no specialized knowledge of probability is necessary to follow what is happening. You just need to be familiar with the concept of probability, expected value, the expected number of times we must flip a coin before getting heads, and the property known as "linearity of expectation."

Here are pointers to the major randomized algorithms in this book: One of the earliest and most dramatic examples of a randomized algorithm is the probabilistic primality test of Figure 1.8. Although a deterministic primality test was recently discovered, the randomized test is much faster and therefore remains the algorithm of choice. Later in this chapter, in Section 1.5 (page 35), we discuss hashing, a general randomized data structure that supports inserts, deletes, and lookups. Again, in practice it leads to faster data access than deterministic schemes like binary search trees.

There are two varieties of randomized algorithms. *Monte Carlo* algorithms always run fast but their output has a small chance of being incorrect; the primality test is an example. *Las Vegas* algorithms, on the other hand, always output the correct answer but guarantee a short running time with high probability. Examples of this are the randomized algorithms for sorting and median finding described in Chapter 2 (on pages 50 and 53, respectively).

The fastest known algorithm for the minimum cut problem is a randomized Monte Carlo algorithm, described in the box on page 139. Randomization plays an important role in heuristics as well; these are described in Section 9.3. And finally the quantum algorithm for factoring (Section 10.7) works very much like a randomized algorithm, its output being correct with high probability—except that it draws its randomness not from coin tosses, but from the superposition principle in quantum mechanics.

Virtual exercises: 1.29, 1.34, 1.46, 2.24, 2.33, 5.35, 9.8, 10.8.

this procedure has at least a $1/n$ chance of halting. Therefore on average it will halt within $O(n)$ rounds (Exercise 1.34).

Next, exactly which primality test should be used? In this application, since the numbers we are testing for primality are chosen at random rather than by an adversary, it is sufficient to perform the Fermat test with base $a = 2$ (or to be really safe, $a = 2, 3, 5$), because for random numbers the Fermat test has a much smaller failure probability than the worst-case 1/2 bound that we proved earlier. Numbers that pass this test have been jokingly referred to as "industrial grade primes." The resulting algorithm is quite fast, generating primes that are hundreds of bits long in a fraction of a second on a PC.

The important question that remains is: what is the probability that the output of the algorithm is really prime? To answer this we must first understand how discerning the Fermat test is. As a concrete example, suppose we perform the test with base $a = 2$ for all numbers $N \leq 25 \times 10^9$. In this range, there are about 10^9 primes, and about 20,000 composites that pass the test (see the following figure). Thus the chance of erroneously outputting a composite is approximately $20{,}000/10^9 = 2 \times 10^{-5}$. This chance of error decreases rapidly as the length of the numbers involved is increased (to the few hundred digits we expect in our applications).

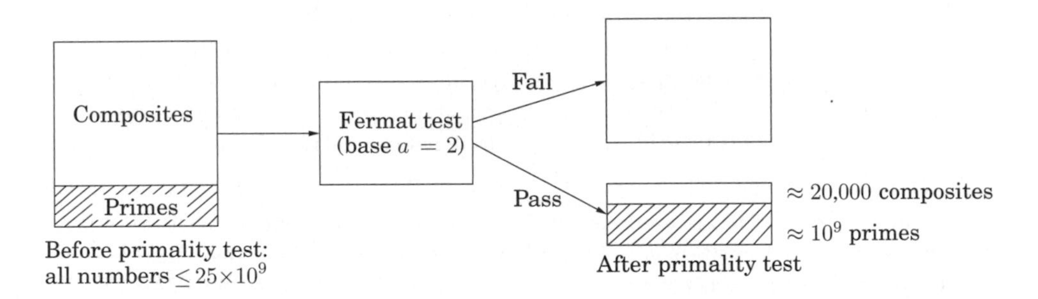

1.4 Cryptography

Our next topic, the Rivest-Shamir-Adleman (RSA) cryptosystem, uses all the ideas we have introduced in this chapter! It derives very strong guarantees of security by ingeniously exploiting the wide gulf between the polynomial-time computability of certain number-theoretic tasks (modular exponentiation, greatest common divisor, primality testing) and the intractability of others (factoring).

The typical setting for cryptography can be described via a cast of three characters: Alice and Bob, who wish to communicate in private, and Eve, an eavesdropper who will go to great lengths to find out what they are saying. For concreteness, let's say Alice wants to send a specific message x, written in binary (why not), to her friend Bob. She encodes it as $e(x)$, sends it over, and then Bob applies his decryption function $d(\cdot)$ to decode it: $d(e(x)) = x$. Here $e(\cdot)$ and $d(\cdot)$ are appropriate transformations of the messages.

Alice and Bob are worried that the eavesdropper, Eve, will intercept $e(x)$: for instance, she might be a sniffer on the network. But ideally the encryption function

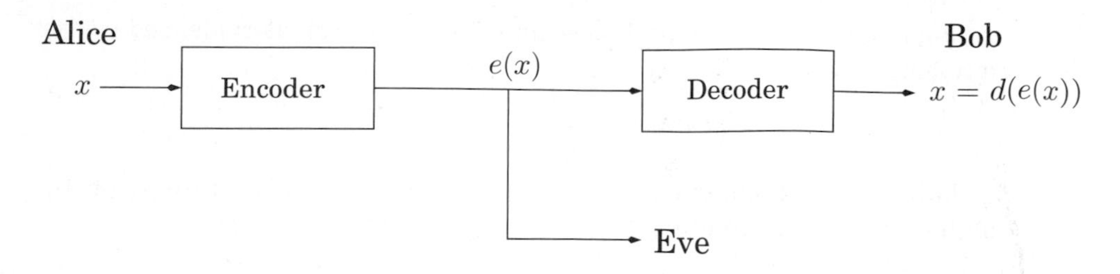

$e(\cdot)$ is so chosen that without knowing $d(\cdot)$, Eve cannot do anything with the information she has picked up. In other words, knowing $e(x)$ tells her little or nothing about what x might be.

For centuries, cryptography was based on what we now call *private-key protocols*. In such a scheme, Alice and Bob meet beforehand and together choose a secret codebook, with which they encrypt all future correspondence between them. Eve's only hope, then, is to collect some encoded messages and use them to at least partially figure out the codebook.

Public-key schemes such as RSA are significantly more subtle and tricky: they allow Alice to send a message to Bob without their ever having met before. Bob's encryption function e(·) is publicly available, and Alice can encrypt her message with this function, thereby *digitally locking* it.Only Bob knows the key to quickly unlocking this digital lock: the decryption function d(·). The point is that Alice and Bob need only perform simple calculations to lock and unlock the message respectively—operations that any pocket computing device could handle. By contrast, to unlock the message without the key, Eve must perform operations like factoring large numbers, which requires more computational power than would be afforded by the world's most powerful computers combined. This compelling guarantee enables secure Web commerce, such as sending credit card numbers to companies over the Internet.

1.4.1 Private-key schemes: one-time pad and AES

If Alice wants to transmit an important private message to Bob, it would be wise of her to scramble it with an encryption function,

$$e : \langle \text{messages} \rangle \rightarrow \langle \text{encoded messages} \rangle.$$

Of course, this function must be invertible—for decoding to be possible—and is therefore a bijection. Its inverse is the decryption function $d(\cdot)$.

In the *one-time pad*, Alice and Bob meet beforehand and secretly choose a bi nary string r of the same length—say, n bits—as the important message x that Alice will later send. Alice's encryption function is then a *bitwise exclusive-or*, $e_r(x) = x \oplus r$: each position in the encoded message is the exclusive-or of the corresponding positions in x and r. For instance, if $r = 01110010$, then the message 11110000 is scrambled thus:

$$e_r(11110000) = 11110000 \oplus 01110010 = 10000010.$$

This function e_r is a bijection from n-bit strings to n-bit strings, as evidenced by the fact that it is its own inverse!

$$e_r(e_r(x)) = (x \oplus r) \oplus r = x \oplus (r \oplus r) = x \oplus \overline{0} = x,$$

where $\overline{0}$ is the string of all zeros. Thus Bob can decode Alice's transmission by applying the same encryption function a second time: $d_r(y) = y \oplus r$.

How should Alice and Bob choose r for this scheme to be secure? Simple: they should pick r *at random*, flipping a coin for each bit, so that the resulting string is equally likely to be any element of $\{0, 1\}^n$. This will ensure that if Eve intercepts the encoded message $y = e_r(x)$, she gets no information about x. Suppose, for example, that Eve finds out $y = 10$; what can she deduce? She doesn't know r, and the possible values it can take all correspond to different original messages x:

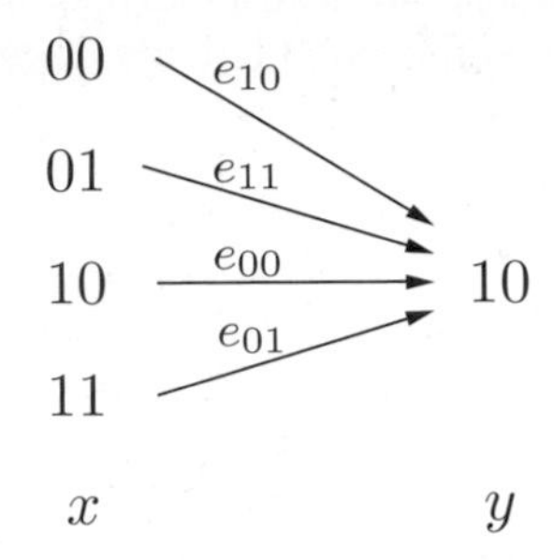

So given what Eve knows, all possibilities for x are equally likely!

The downside of the one-time pad is that it has to be discarded after use, hence the name. A second message encoded with the same pad would not be secure, because if Eve knew $x \oplus r$ and $z \oplus r$ for two messages x and z, then she could take the exclusive-or to get $x \oplus z$, which might be important information—for example, (1) it reveals whether the two messages begin or end the same, and (2) if one message contains a long sequence of zeros (as could easily be the case if the message is an image), then the corresponding part of the other message will be exposed. Therefore the random string that Alice and Bob share has to be the combined length of all the messages they will need to exchange.

The one-time pad is a toy cryptographic scheme whose behavior and theoretical properties are completely clear. At the other end of the spectrum lies the *advanced encryption standard* (AES), a very widely used cryptographic protocol that was approved by the U.S. National Institute of Standards and Technologies in 2001. AES is once again private-key: Alice and Bob have to agree on a shared random string r. But this time the string is of a small fixed size, 128 to be precise (variants with 192 or 256 bits also exist), and specifies a bijection e_r from 128-bit strings to 128-bit strings. The crucial difference is that this function can be used repeatedly, so for instance a long message can be encoded by splitting it into segments of 128 bits and applying e_r to each segment.

The security of AES has not been rigorously established, but certainly at present the general public does not know how to break the code—to recover x from $e_r(x)$—except using techniques that are not very much better than the brute-force approach of trying all possibilities for the shared string r.

1.4.2 RSA

Unlike the previous two protocols, the RSA scheme is an example of *public-key cryptography*: anybody can send a message to anybody else using publicly available information, rather like addresses or phone numbers. Each person has a public key known to the whole world and a secret key known only to him- or herself. When Alice wants to send message x to Bob, she encodes it using his public key. He decrypts it using his secret key, to retrieve x. Eve is welcome to see as many encrypted messages for Bob as she likes, but she will not be able to decode them, under certain simple assumptions.

The RSA scheme is based heavily upon number theory. Think of messages from Alice to Bob as numbers modulo N; messages larger than N can be broken into smaller pieces. The encryption function will then be a bijection on $\{0, 1, \ldots, N-1\}$, and the decryption function will be its inverse. What values of N are appropriate, and what bijection should be used?

Property *Pick any two primes p and q and let $N = pq$. For any e relatively prime to $(p-1)(q-1)$:*

1. *The mapping $x \mapsto x^e \bmod N$ is a bijection on $\{0, 1, \ldots, N-1\}$.*
2. *Moreover, the inverse mapping is easily realized: let d be the inverse of e modulo $(p-1)(q-1)$. Then for all $x \in \{0, \ldots, N-1\}$,*

$$(x^e)^d \equiv x \bmod N.$$

The first property tells us that the mapping $x \mapsto x^e \bmod N$ is a reasonable way to encode messages x; no information is lost. So, if Bob publishes (N, e) as his *public key*, everyone else can use it to send him encrypted messages. The second property then tells us how decryption can be achieved. Bob should retain the value d as his *secret key*, with which he can decode all messages that come to him by simply raising them to the dth power modulo N.

Example. Let $N = 55 = 5 \cdot 11$. Choose encryption exponent $e = 3$, which satisfies the condition $\gcd(e, (p-1)(q-1)) = \gcd(3, 40) = 1$. The decryption exponent is then $d = 3^{-1} \bmod 40 = 27$. Now for any message $x \bmod 55$, the encryption of x is $y = x^3 \bmod 55$, and the decryption of y is $x = y^{27} \bmod 55$. So, for example, if $x = 13$, then $y = 13^3 = 52 \bmod 55$ and $13 = 52^{27} \bmod 55$.

Let's prove the assertion above and then examine the security of the scheme.

Proof. If the mapping $x \mapsto x^e \bmod N$ is invertible, it must be a bijection; hence statement 2 implies statement 1. To prove statement 2, we start by observing that e is invertible modulo $(p-1)(q-1)$ because it is relatively prime to this number. To see that $(x^e)^d \equiv x \bmod N$, we examine the exponent: since $ed \equiv 1 \bmod (p-1)(q-1)$, we can write ed in the form $1 + k(p-1)(q-1)$ for some k. Now we need to show that the difference

$$x^{ed} - x = x^{1+k(p-1)(q-1)} - x$$

is always 0 modulo N. The second form of the expression is convenient because it can be simplified using Fermat's little theorem. It is divisible by p (since $x^{p-1} \equiv 1 \bmod p$) and likewise by q. Since p and q are primes, this expression must also be divisible by their product N. Hence $x^{ed} - x = x^{1+k(p-1)(q-1)} - x \equiv 0 \pmod{N}$, exactly as we need. ∎

Figure 1.9 RSA.

Bob chooses his public and secret keys.

- He starts by picking two large (n-bit) random primes p and q.
- His public key is (N, e) where $N = pq$ and e is a $2n$-bit number relatively prime to $(p-1)(q-1)$. A common choice is $e = 3$ because it permits fast encoding.
- His secret key is d, the inverse of e modulo $(p-1)(q-1)$, computed using the extended Euclid algorithm.

Alice wishes to send message x to Bob.

- She looks up his public key (N, e) and sends him $y = (x^e \bmod N)$, computed using an efficient modular exponentiation algorithm.
- He decodes the message by computing $y^d \bmod N$.

The RSA protocol is summarized in Figure 1.9. It is certainly convenient: the computations it requires of Alice and Bob are elementary. But how secure is it against Eve?

The security of RSA hinges upon a simple assumption:

> *Given N, e, and $y = x^e \bmod N$, it is computationally intractable to determine x.*

This assumption is quite plausible. How might Eve try to guess x? She could experiment with all possible values of x, each time checking whether $x^e \equiv y \bmod N$, but this would take exponential time. Or she could try to factor N to retrieve p and q, and then figure out d by inverting e modulo $(p-1)(q-1)$, but we believe factoring to be hard. Intractability is normally a source of dismay; the insight of RSA lies in using it to advantage.

1.5 Universal hashing

We end this chapter with an application of number theory to the design of *hash functions*. Hashing is a very useful method of storing data items in a table so as to support insertions, deletions, and lookups.

Suppose, for instance, that we need to maintain an ever-changing list of about 250 IP (Internet protocol) addresses, perhaps the addresses of the currently active customers of a Web service. (Recall that an IP address consists of 32 bits encoding the location of a computer on the Internet, usually shown broken down into four 8-bit fields, for example, 128.32.168.80.) We could obtain fast lookup times if we maintained the records in an array indexed by IP address. But this would be very wasteful of memory: the array would have $2^{32} \approx 4 \times 10^9$ entries, the vast majority of them blank. Or alternatively, we could use a linked list of just the 250 records. But then accessing records would be very slow, taking time proportional to 250, the total number of customers. Is there a way to get the best of both worlds, to use an amount of memory that is proportional to the number of customers and yet also achieve fast lookup times? This is exactly where hashing comes in.

1.5.1 Hash tables

Here's a high-level view of hashing. We will give a short "nickname" to each of the 2^{32} possible IP addresses. You can think of this short name as just a number between 1 and 250 (we will later adjust this range very slightly). Thus many IP addresses will inevitably have the same nickname; however, we hope that most of the 250 IP addresses of our particular customers are assigned distinct names, and we will store their records in an array of size 250 indexed by these names. What if there is more than one record associated with the same name? Easy: each entry of the array points to a linked list containing all records with that name. So the total amount of storage is proportional to 250, the number of customers, and is independent of the total number of possible IP addresses. Moreover, if not too many customer IP addresses are assigned the same name, lookup is fast, because the average size of the linked list we have to scan through is small.

But how do we assign a short name to each IP address? This is the role of a *hash function*: in our example, a function h that maps IP addresses to positions in a table of length about 250 (the expected number of data items). The name assigned to an

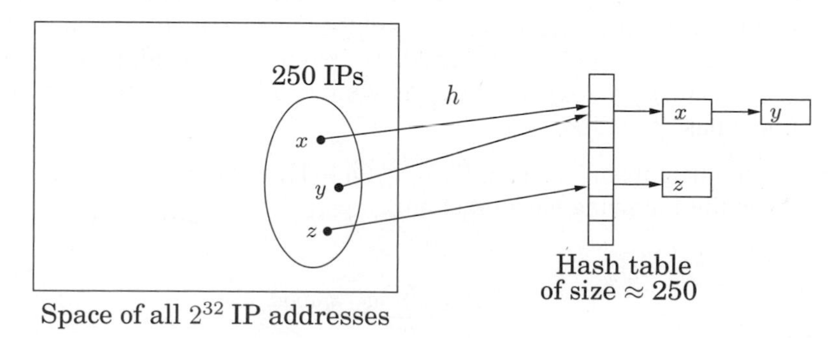

IP address x is thus $h(x)$, and the record for x is stored in position $h(x)$ of the table. As described before, each position of the table is in fact a *bucket*, a linked list that contains all current IP addresses that map to it. Hopefully, there will be very few buckets that contain more than a handful of IP addresses.

1.5.2 Families of hash functions

Designing hash functions is tricky. A hash function must in some sense be "random" (so that it scatters data items around), but it should also be a function and therefore "consistent" (so that we get the same result every time we apply it). And the statistics of the data items may work against us. In our example, one possible hash function would map an IP address to the 8-bit number that is its last segment: $h(128.32.168.80) = 80$. A table of $n = 256$ buckets would then be required. But is this a good hash function? Not if, for example, the last segment of an IP address tends to be a small (single- or double-digit) number; then low-numbered buckets would be crowded. Taking the first segment of the IP address also invites disaster—for example, if most of our customers come from Asia.

There is nothing inherently wrong with these two functions. If our 250 IP addresses were uniformly drawn from among all $N = 2^{32}$ possibilities, then these functions would behave well. The problem is we have no guarantee that the distribution of IP addresses is uniform.

Conversely, there is no single hash function, no matter how sophisticated, that behaves well on all sets of data. Since a hash function maps 2^{32} IP addresses to just 250 names, there must be a collection of at least $2^{32}/250 \approx 2^{24} \approx 16{,}000{,}000$ IP addresses that are assigned the same name (or, in hashing terminology, "collide"). If many of these show up in our customer set, we're in trouble.

Obviously, we need some kind of randomization. Here's an idea: let us pick a hash function *at random* from some class of functions. We will then show that, no matter what set of 250 IP addresses we actually care about, most choices of the hash function will give very few collisions among these addresses.

To this end, we need to define a class of hash functions from which we can pick at random; and this is where we turn to number theory. Let us take the number of buckets to be not 250 but $n = 257$—*a prime number*! And we consider every IP address x as a quadruple $x = (x_1, \ldots, x_4)$ of integers modulo n—recall that it is in fact a quadruple of integers between 0 and 255, so there is no harm in this. We can define a function h from IP addresses to a number mod n as follows: fix any four numbers mod $n = 257$, say 87, 23, 125, and 4. Now map the IP address $(x_1, \ldots, x_4)$ to $h(x_1, \ldots, x_4) = (87x_1 + 23x_2 + 125x_3 + 4x_4) \bmod 257$. Indeed, any four numbers mod n define a hash function.

For any four coefficients $a_1, \ldots, a_4 \in \{0, 1, \ldots, n-1\}$, write $a = (a_1, a_2, a_3, a_4)$ and define h_a to be the following hash function:

$$h_a(x_1, \ldots, x_4) = \sum_{i=1}^{4} a_i \cdot x_i \bmod n.$$

We will show that if we pick these coefficients a at random, then h_a is very likely to be good in the following sense.

Property *Consider any pair of distinct IP addresses* $x = (x_1, \ldots, x_4)$ *and* $y = (y_1, \ldots, y_4)$. *If the coefficients* $a = (a_1, a_2, a_3, a_4)$ *are chosen uniformly at random from* $\{0, 1, \ldots, n-1\}$, *then*

$$\Pr\{h_a(x_1, \ldots, x_4) = h_a(y_1, \ldots, y_4)\} = \frac{1}{n}.$$

In other words, the chance that x and y collide under h_a is the same as it would be if each were assigned nicknames randomly and independently. This condition guarantees that the expected lookup time for any item is small. Here's why. If we wish to look up x in our hash table, the time required is dominated by the size of its bucket, that is, the number of items that are assigned the same name as x. But there are only 250 items in the hash table, and the probability that any one item gets the same name as x is $1/n = 1/257$. Therefore the expected number of items that are assigned the same name as x by a randomly chosen hash function h_a is $250/257 \approx 1$, which means the expected size of x's bucket is less than 2.[1]

Let us now prove the preceding property.

Proof. Since $x = (x_1, \ldots, x_4)$ and $y = (y_1, \ldots, y_4)$ are distinct, these quadruples must differ in some component; without loss of generality let us assume that $x_4 \neq y_4$. We wish to compute the probability $\Pr[h_a(x_1, \ldots, x_4) = h_a(y_1, \ldots, y_4)]$, that is, the probability that $\sum_{i=1}^{4} a_i \cdot x_i \equiv \sum_{i=1}^{4} a_i \cdot y_i \bmod n$. This last equation can be rewritten as

$$\sum_{i=1}^{3} a_i \cdot (x_i - y_i) \equiv a_4 \cdot (y_4 - x_4) \bmod n. \tag{1}$$

Suppose that we draw a random hash function h_a by picking $a = (a_1, a_2, a_3, a_4)$ at random. We start by drawing a_1, a_2, and a_3, and then we pause and think: What is the probability that the last drawn number a_4 is such that equation (1) holds? So far the left-hand side of equation (1) evaluates to some number, call it c. And since n is prime and $x_4 \neq y_4$, $(y_4 - x_4)$ has a unique inverse modulo n. Thus for equation (1) to hold, the last number a_4 must be precisely $c \cdot (y_4 - x_4)^{-1} \bmod n$, out of its n possible values. The probability of this happening is $1/n$, and the proof is complete. ∎

Let us step back and see what we just achieved. Since we have no control over the set of data items, we decided instead to select a hash function h uniformly at

[1]When a hash function h_a is chosen at random, let the random variable Y_i (for $i = 1, \ldots, 250$) be 1 if item i gets the same name as x and 0 otherwise. So the expected value of Y_i is $1/n$. Now, $Y = Y_1 + Y_2 + \cdots + Y_{250}$ is the number of items which get the same name as x, and by linearity of expectation, the expected value of Y is simply the sum of the expected values of Y_1 through Y_{250}. It is thus $250/n = 250/257$.

random from among a family $\mathcal{H}$ of hash functions. In our example,

$$\mathcal{H} = \{h_a : a \in \{0, \ldots, n-1\}^4\}.$$

To draw a hash function uniformly at random from this family, we just draw four numbers $a_1, \ldots, a_4$ modulo n. (Incidentally, notice that the two simple hash functions we considered earlier, namely, taking the last or the first 8-bit segment, belong to this class. They are $h_{(0,0,0,1)}$ and $h_{(1,0,0,0)}$, respectively.) And we insisted that the family have the following property:

> *For any two distinct data items x and y, exactly $|\mathcal{H}|/n$ of all the hash functions in $\mathcal{H}$ map x and y to the same bucket, where n is the number of buckets.*

A family of hash functions with this property is called *universal*. In other words, for any two data items, the probability these items collide is $1/n$ if the hash function is randomly drawn from a universal family. This is also the collision probability if we map x and y to buckets uniformly at random—in some sense the gold standard of hashing. We then showed that this property implies that hash table operations have good performance *in expectation*.

This idea, motivated as it was by the hypothetical IP address application, can of course be applied more generally. Start by choosing the table size n to be some prime number that is a little larger than the number of items expected in the table (there is usually a prime number close to any number we start with; actually, to ensure that hash table operations have good performance, it is better to have the size of the hash table be about twice as large as the number of items). Next assume that the size of the domain of all data items is $N = n^k$, a power of n (if we need to overestimate the true number of data items, so be it). Then each data item can be considered as a k-tuple of integers modulo n, and $\mathcal{H} = \{h_a : a \in \{0, \ldots, n-1\}^k\}$ is a universal family of hash functions.

Exercises

1.1. Show that in any base $b \geq 2$, the sum of any three single-digit numbers is at most two digits long.

1.2. Show that any binary integer is at most four times as long as the corresponding decimal integer. For very large numbers, what is the ratio of these two lengths, approximately?

1.3. A d-ary tree is a rooted tree in which each node has at most d children. Show that any d-ary tree with n nodes must have a depth of $\Omega(\log n/\log d)$. Can you give a precise formula for the minimum depth it could possibly have?

1.4. Show that

$$\log(n!) = \Theta(n \log n).$$

(*Hint:* To show an upper bound, compare $n!$ with n^n. To show a lower bound, compare it with $(n/2)^{n/2}$.)

1.5. Unlike a decreasing geometric series, the sum of the *harmonic series* 1, 1/2, 1/3, 1/4, 1/5, ... diverges; that is,

$$\sum_{i=1}^{\infty} \frac{1}{i} = \infty.$$

It turns out that, for large n, the sum of the first n terms of this series can be well approximated as

$$\sum_{i=1}^{n} \frac{1}{i} \approx \ln n + \gamma,$$

where ln is natural logarithm (log base $e = 2.718\ldots$) and γ is a particular constant $0.57721\ldots$. Show that

$$\sum_{i=1}^{n} \frac{1}{i} = \Theta(\log n).$$

(*Hint:* To show an upper bound, decrease each denominator to the next power of two. For a lower bound, increase each denominator to the next power of 2.)

1.6. Prove that the grade-school multiplication algorithm (page 13), when applied to binary numbers, always gives the right answer.

1.7. How long does the recursive multiplication algorithm (page 15) take to multiply an n-bit number by an m-bit number? Justify your answer.

1.8. Justify the correctness of the recursive division algorithm given in page 15, and show that it takes time $O(n^2)$ on n-bit inputs.

1.9. Starting from the definition of $x \equiv y \bmod N$ (namely, that N divides $x - y$), justify the substitution rule

$$x \equiv x' \bmod N,\ y \equiv y' \bmod N \ \Rightarrow\ x + y \equiv x' + y' \bmod N,$$

and also the corresponding rule for multiplication.

1.10. Show that if $a \equiv b \pmod N$ and if M divides N then $a \equiv b \pmod M$.

1.11. Is $4^{1536} - 9^{4824}$ divisible by 35?

1.12. What is $2^{2^{2006}} \pmod 3$?

1.13. Is the difference of $5^{30,000}$ and $6^{123,456}$ a multiple of 31?

1.14. Suppose you want to compute the nth Fibonacci number F_n, modulo an integer p. Can you find an efficient way to do this? (*Hint:* Recall Exercise 0.4.)

1.15. Determine necessary and sufficient conditions on x and c so that the following holds: for any a, b, if $ax \equiv bx \bmod c$, then $a \equiv b \bmod c$.

1.16. The algorithm for computing $a^b \bmod c$ by repeated squaring does not necessarily lead to the minimum number of multiplications. Give an example of $b > 10$ where the exponentiation can be performed using fewer multiplications, by some other method.

1.17. Consider the problem of computing x^y for given integers x and y: we want the *whole* answer, not modulo a third integer. We know two algorithms for doing this: the iterative algorithm which performs $y-1$ multiplications by x; and the recursive algorithm based on the binary expansion of y.

Compare the time requirements of these two algorithms, assuming that the time to multiply an n-bit number by an m-bit number is $O(mn)$.

1.18. Compute gcd(210, 588) two different ways: by finding the factorization of each number, and by using Euclid's algorithm.

1.19. The *Fibonacci numbers* $F_0, F_1, \ldots$ are given by the recurrence $F_{n+1} = F_n + F_{n-1}$, $F_0 = 0$, $F_1 = 1$. Show that for any $n \geq 1$, $\gcd(F_{n+1}, F_n) = 1$.

1.20. Find the inverse of: 20 mod 79, 3 mod 62, 21 mod 91, 5 mod 23.

1.21. How many integers modulo 11^3 have inverses? (Note: $11^3 = 1331$.)

1.22. Prove or disprove: If a has an inverse modulo b, then b has an inverse modulo a.

1.23. Show that if a has a multiplicative inverse modulo N, then this inverse is unique (modulo N).

1.24. If p is prime, how many elements of $\{0, 1, \ldots, p^n - 1\}$ have an inverse modulo p^n?

1.25. Calculate $2^{125} \bmod 127$ using any method you choose. (*Hint:* 127 is prime.)

1.26. What is the least significant decimal digit of $17^{17^{17}}$? (*Hint:* For distinct primes p, q, and any a relatively prime to pq, we proved the formula $a^{(p-1)(q-1)} \equiv 1 \pmod{pq}$ in Section 1.4.2.)

1.27. Consider an RSA key set with $p = 17, q = 23, N = 391$, and $e = 3$ (as in Figure 1.9). What value of d should be used for the secret key? What is the encryption of the message $M = 41$?

1.28. In an RSA cryptosystem, $p = 7$ and $q = 11$ (as in Figure 1.9). Find appropriate exponents d and e.

1.29. Let $[m]$ denote the set $\{0, 1, \ldots, m-1\}$. For each of the following families of hash functions, say whether or not it is universal, and determine how many random bits are needed to choose a function from the family.

(a) $H = \{h_{a_1,a_2} : a_1, a_2 \in [m]\}$, where m is a fixed prime and

$$h_{a_1,a_2}(x_1, x_2) = a_1 x_1 + a_2 x_2 \bmod m.$$

Notice that each of these functions has signature $h_{a_1,a_2} : [m]^2 \to [m]$, that is, it maps a pair of integers in $[m]$ to a single integer in $[m]$.

(b) H is as before, except that now $m = 2^k$ is some fixed power of 2.

(c) H is the set of all functions $f : [m] \to [m-1]$.

1.30. The grade-school algorithm for multiplying two n-bit binary numbers x and y consists of adding together n copies of x, each appropriately left-shifted. Each copy, when shifted, is at most $2n$ bits long.

In this problem, we will examine a scheme for adding n binary numbers, each m bits long, using a *circuit* or a *parallel architecture*. The main parameter of interest in this question is therefore the depth of the circuit or the longest path from the input to the output of the circuit. This determines the total time taken for computing the function.

To add two m-bit binary numbers naively, we must wait for the carry bit from position $i-1$ before we can figure out the ith bit of the answer. This leads to a circuit of depth $O(m)$. However carry lookahead circuits (see wikipedia.com if you want to know more about this) can add in $O(\log m)$ depth.

(a) Assuming you have carry lookahead circuits for addition, show how to add n numbers each m bits long using a circuit of depth $O((\log n)(\log m))$.

(b) When adding *three* m-bit binary numbers $x+y+z$, there is a trick we can use to parallelize the process. Instead of carrying out the addition completely, we can re-express the result as the sum of just *two* binary numbers $r+s$, such that the ith bits of r and s can be computed independently of the other bits. Show how this can be done. (*Hint:* One of the numbers represents carry bits.)

(c) Show how to use the trick from the previous part to design a circuit of depth $O(\log n)$ for multiplying two n-bit numbers.

1.31. Consider the problem of computing $N! = 1 \cdot 2 \cdot 3 \cdots N$.

(a) If N is an n-bit number, how many bits long is $N!$, approximately (in $\Theta(\cdot)$ form)?

(b) Give an algorithm to compute $N!$ and analyze its running time.

1.32. A positive integer N is a *power* if it is of the form q^k, where q, k are positive integers and $k > 1$.

(a) Give an efficient algorithm that takes as input a number N and determines whether it is a square, that is, whether it can be written as q^2 for some positive integer q. What is the running time of your algorithm?

(b) Show that if $N = q^k$ (with N, q, and k all positive integers), then either $k \le \log N$ or $N = 1$.

(c) Give an efficient algorithm for determining whether a positive integer N is a power. Analyze its running time.

1.33. Give an efficient algorithm to compute the *least common multiple* of two n-bit numbers x and y, that is, the smallest number divisible by both x and y. What is the running time of your algorithm as a function of n?

1.34. On page 29, we claimed that since about a $1/n$ fraction of n-bit numbers are prime, on average it is sufficient to draw $O(n)$ random n-bit numbers before hitting a prime. We now justify this rigorously.

Suppose a particular coin has a probability p of coming up heads. How many times must you toss it, on average, before it comes up heads? (*Hint:* Method 1: start by showing that the correct expression is $\sum_{i=1}^{\infty} i(1-p)^{i-1}p$. Method 2: if E is the average number of coin tosses, show that $E = 1 + (1-p)E$.)

1.35. Wilson's theorem says that a number N is prime if and only if

$$(N-1)! \equiv -1 \pmod{N}.$$

(a) If p is prime, then we know every number $1 \leq x < p$ is invertible modulo p. Which of these numbers are their own inverse?

(b) By pairing up multiplicative inverses, show that $(p-1)! \equiv -1 \pmod{p}$ for prime p.

(c) Show that if N is *not* prime, then $(N-1)! \not\equiv -1 \pmod{N}$. (*Hint:* Consider $d = \gcd(N, (N-1)!)$.)

(d) Unlike Fermat's Little theorem, Wilson's theorem is an if-and-only-if condition for primality. Why can't we immediately base a primality test on this rule?

1.36. *Square roots.* In this problem, we'll see that it is easy to compute square roots modulo a prime p with $p \equiv 3 \pmod{4}$.

(a) Suppose $p \equiv 3 \pmod{4}$. Show that $(p+1)/4$ is an integer.

(b) We say x is a *square root* of a modulo p if $a \equiv x^2 \pmod{p}$. Show that if $p \equiv 3 \pmod{4}$ and if a has a square root modulo p, then $a^{(p+1)/4}$ is such a square root.

1.37. *The Chinese remainder theorem.*

(a) Make a table with three columns. The first column is all numbers from 0 to 14. The second is the residues of these numbers modulo 3; the third column is the residues modulo 5.

(b) Prove that if p and q are distinct primes, then for every pair (j, k) with $0 \leq j < p$ and $0 \leq k < q$, there is a unique integer $0 \leq i < pq$ such that $i \equiv j \bmod p$ and $i \equiv k \bmod q$. (*Hint:* Prove that no two different i's in this range can have the same (j, k), and then count.)

(c) In this one-to-one correspondence between integers and pairs, it is easy to go from i to (j, k). Prove that the following formula takes you the other way:

$$i = \{j \cdot q \cdot (q^{-1} \bmod p) + k \cdot p \cdot (p^{-1} \bmod q)\} \bmod pq.$$

(d) Can you generalize parts (b) and (c) to more than two primes?

1.38. To see if a number, say 562437487, is divisible by 3, you just add up the digits of its decimal representation, and see if the result is divisible by 3. $(5+6+2+4+3+7+4+8+7=46$, so it is not divisible by 3.)

To see if the same number is divisible by 11, you can do this: subdivide the number into pairs of digits, from the right-hand end $(87, 74, 43, 62, 5)$, add these numbers, and see if the sum is divisible by 11 (if it's too big, repeat).

How about 37? To see if the number is divisible by 37, subdivide it into triples from the end $(487, 437, 562)$ add these up, and see if the sum is divisible by 37.

This is true for any prime p other than 2 and 5. That is, for any prime $p \neq 2, 5$, there is an integer r such that in order to see if p divides a decimal number n, we break n into r-tuples of decimal digits (starting from the right-hand end), add up these r-tuples, and check if the sum is divisible by p.

(a) What is the smallest such r for $p = 13$? For $p = 17$?

(b) Show that r is a divisor of $p - 1$.

1.39. Give a polynomial-time algorithm for computing $a^{b^c} \bmod p$, given a, b, c, and prime p.

1.40. Show that if x is a nontrivial square root of 1 modulo N, that is, if $x^2 \equiv 1 \bmod N$ but $x \not\equiv \pm 1 \bmod N$, then N must be composite. (For instance, $4^2 \equiv 1 \bmod 15$ but $4 \not\equiv \pm 1 \bmod 15$; thus 4 is a nontrivial square root of 1 modulo 15.)

1.41. *Quadratic residues.* Fix a positive integer N. We say that a is a *quadratic residue* modulo N if there exists x such that $a \equiv x^2 \bmod N$.

(a) Let N be an odd prime and a be a non-zero quadratic residue modulo N. Show that there are exactly two values in $\{0, 1, \ldots, N - 1\}$ satisfying $x^2 \equiv a \bmod N$.

(b) Show that if N is an odd prime, there are exactly $(N + 1)/2$ quadratic residues in $\{0, 1, \ldots, N - 1\}$.

(c) Give an example of positive integers a and N such that $x^2 \equiv a \bmod N$ has more than two solutions in $\{0, 1, \ldots, N - 1\}$.

1.42. Suppose that instead of using a composite $N = pq$ in the RSA cryptosystem (Figure 1.9), we simply use a prime modulus p. As in RSA, we would have an encryption exponent e, and the encryption of a message $m \bmod p$ would be $m^e \bmod p$. Prove that this new cryptosystem is not secure, by giving an efficient algorithm to decrypt: that is, an algorithm that given p, e, and $m^e \bmod p$ as input, computes $m \bmod p$. Justify the correctness and analyze the running time of your decryption algorithm.

1.43. In the RSA cryptosystem, Alice's public key (N, e) is available to everyone. Suppose that her private key d is compromised and becomes known to Eve. Show that if $e = 3$ (a common choice) then Eve can efficiently factor N.

1.44. Alice and her three friends are all users of the RSA cryptosystem. Her friends have public keys $(N_i, e_i = 3)$, $i = 1, 2, 3$, where as always, $N_i = p_i q_i$ for randomly chosen n-bit primes p_i, q_i. Show that if Alice sends the same n-bit message M (encrypted using RSA) to each of her friends, then anyone who intercepts all three encrypted messages will be able to efficiently recover M. (*Hint:* It helps to have solved problem 1.37 first.)

1.45. *RSA and digital signatures.* Recall that in the RSA public-key cryptosystem, each user has a public key $P = (N, e)$ and a secret key d. In a *digital signature scheme*, there are two algorithms, `sign` and `verify`. The `sign` procedure takes a message and a secret key, then outputs a signature σ. The `verify` procedure takes a public key (N, e), a signature σ, and a message M, then returns "true" if σ could have been created by `sign` (when called with message M and the secret key corresponding to the public key (N, e)); "false" otherwise.

(a) Why would we want digital signatures?

(b) An RSA signature consists of `sign`$(M, d) = M^d \pmod N$, where d is a secret key and N is part of the public key. Show that anyone who knows the public key (N, e) can perform `verify`$((N, e), M^d, M)$, i.e., they can check that a signature really was created by the private key. Give an implementation and prove its correctness.

(c) Generate your own RSA modulus $N = pq$, public key e, and private key d (you don't need to use a computer). Pick p and q so you have a 4-digit modulus and work by hand. Now sign your name using the private exponent of this RSA modulus. To do this you will need to specify some one-to-one mapping from strings to integers in $[0, N-1]$. Specify any mapping you like. Give the mapping from your name to numbers $m_1, m_2, \ldots m_k$, then sign the first number by giving the value $m_1^d \pmod N$, and finally show that $(m_1^d)^e = m_1 \pmod N$.

(d) Alice wants to write a message that looks like it was digitally signed by Bob. She notices that Bob's public RSA key is $(17, 391)$. To what exponent should she raise her message?

1.46. *Digital signatures, continued.* Consider the signature scheme of Exercise 1.45.

(a) Signing involves decryption, and is therefore risky. Show that if Bob agrees to sign anything he is asked to, Eve can take advantage of this and decrypt any message sent by Alice to Bob.

(b) Suppose that Bob is more careful, and refuses to sign messages if their signatures look suspiciously like text. (We assume that a randomly chosen message—that is, a random number in the range $\{1, \ldots, N-1\}$—is very unlikely to look like text.) Describe a way in which Eve can nevertheless still decrypt messages from Alice to Bob, by getting Bob to sign messages whose signatures look random.

Chapter 2

Divide-and-conquer algorithms

The *divide-and-conquer* strategy solves a problem by:

1. Breaking it into *subproblems* that are themselves smaller instances of the same type of problem
2. Recursively solving these subproblems
3. Appropriately combining their answers

The real work is done piecemeal, in three different places: in the partitioning of problems into subproblems; at the very tail end of the recursion, when the subproblems are so small that they are solved outright; and in the gluing together of partial answers. These are held together and coordinated by the algorithm's core recursive structure.

As an introductory example, we'll see how this technique yields a new algorithm for multiplying numbers, one that is much more efficient than the method we all learned in elementary school!

2.1 Multiplication

The mathematician Carl Friedrich Gauss (1777–1855) once noticed that although the product of two complex numbers

$$(a + bi)(c + di) = ac - bd + (bc + ad)i$$

seems to involve *four* real-number multiplications, it can in fact be done with just *three*: ac, bd, and $(a + b)(c + d)$, since

$$bc + ad = (a + b)(c + d) - ac - bd.$$

In our big-O way of thinking, reducing the number of multiplications from four to three seems wasted ingenuity. But this modest improvement becomes very significant *when applied recursively*.

Let's move away from complex numbers and see how this helps with regular multiplication. Suppose x and y are two n-bit integers, and assume for convenience that n is a power of 2 (the more general case is hardly any different). As a first step toward multiplying x and y, split each of them into their left and right halves, which

Carl Friedrich Gauss
1777–1855

are $n/2$ bits long:

$$x = \boxed{x_L}\boxed{x_R} = 2^{n/2}x_L + x_R$$
$$y = \boxed{y_L}\boxed{y_R} = 2^{n/2}y_L + y_R.$$

For instance, if $x = 10110110_2$ (the subscript 2 means "binary") then $x_L = 1011_2$, $x_R = 0110_2$, and $x = 1011_2 \times 2^4 + 0110_2$. The product of x and y can then be rewritten as

$$xy \;=\; (2^{n/2}x_L + x_R)(2^{n/2}y_L + y_R) \;=\; 2^n\, x_L y_L \;+\; 2^{n/2}\,(x_L y_R + x_R y_L) \;+\; x_R y_R.$$

We will compute xy via the expression on the right. The additions take linear time, as do the multiplications by powers of 2 (which are merely left-shifts). The significant operations are the four $n/2$-bit multiplications, $x_L y_L$, $x_L y_R$, $x_R y_L$, $x_R y_R$; these we can handle by four recursive calls. Thus our method for multiplying n-bit numbers starts by making recursive calls to multiply these four pairs of $n/2$-bit numbers (four subproblems of half the size), and then evaluates the preceding expression in $O(n)$ time. Writing $T(n)$ for the overall running time on n-bit inputs, we get the *recurrence relation*

$$T(n) \;=\; 4T(n/2) + O(n).$$

We will soon see general strategies for solving such equations. In the meantime, this particular one works out to $O(n^2)$, the same running time as the traditional grade-school multiplication technique. So we have a radically new algorithm, but we haven't yet made any progress in efficiency. How can our method be sped up?

This is where Gauss's trick comes to mind. Although the expression for xy seems to demand four $n/2$-bit multiplications, as before just three will do: $x_L y_L$, $x_R y_R$, and $(x_L + x_R)(y_L + y_R)$, since $x_L y_R + x_R y_L = (x_L + x_R)(y_L + y_R) - x_L y_L - x_R y_R$. The resulting

Figure 2.1 A divide-and-conquer algorithm for integer multiplication.

```
function multiply(x, y)
Input: n-bit positive integers x and y
Output: Their product

if n=1: return xy

x_L, x_R = leftmost ⌈n/2⌉, rightmost ⌊n/2⌋ bits of x
y_L, y_R = leftmost ⌈n/2⌉, rightmost ⌊n/2⌋ bits of y

P_1 = multiply(x_L, y_L)
P_2 = multiply(x_R, y_R)
P_3 = multiply(x_L + x_R, y_L + y_R)
return P_1 × 2^n + (P_3 − P_1 − P_2) × 2^(n/2) + P_2
```

algorithm, shown in Figure 2.1, has an improved running time of[1]

$$T(n) = 3T(n/2) + O(n).$$

The point is that now the constant factor improvement, from 4 to 3, occurs *at every level of the recursion*, and this compounding effect leads to a dramatically lower time bound of $O(n^{1.59})$.

This running time can be derived by looking at the algorithm's pattern of recursive calls, which form a tree structure, as in Figure 2.2. Let's try to understand the shape of this tree. At each successive level of recursion the subproblems get halved in size. At the $(\log_2 n)^{\text{th}}$ level, the subproblems get down to size 1, and so the recursion ends. Therefore, the height of the tree is $\log_2 n$. The branching factor is 3—each problem recursively produces three smaller ones—with the result that at depth k in the tree there are 3^k subproblems, each of size $n/2^k$.

For each subproblem, a linear amount of work is done in identifying further subproblems and combining their answers. Therefore the total time spent at depth k in the tree is

$$3^k \times O\left(\frac{n}{2^k}\right) = \left(\frac{3}{2}\right)^k \times O(n).$$

[1] Actually, the recurrence should read

$$T(n) \leq 3T(n/2 + 1) + O(n)$$

since the numbers $(x_L + x_R)$ and $(y_L + y_R)$ could be $n/2 + 1$ bits long. The one we're using is simpler to deal with and can be seen to imply exactly the same big-O running time.

Figure 2.2 Divide-and-conquer integer multiplication. (a) Each problem is divided into three subproblems. (b) The levels of recursion.

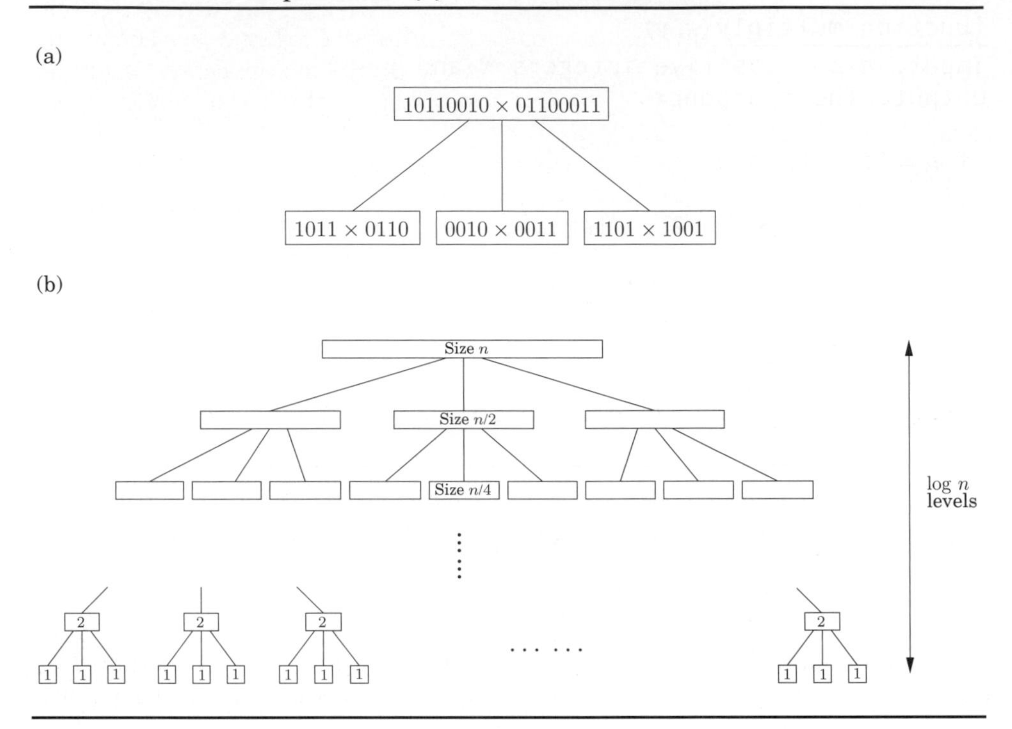

At the very top level, when $k = 0$, this works out to $O(n)$. At the bottom, when $k = \log_2 n$, it is $O(3^{\log_2 n})$, which can be rewritten as $O(n^{\log_2 3})$ (do you see why?). Between these two endpoints, the work done increases *geometrically* from $O(n)$ to $O(n^{\log_2 3})$, by a factor of $3/2$ per level. The sum of any increasing geometric series is, within a constant factor, simply the last term of the series: such is the rapidity of the increase (Exercise 0.2). Therefore the overall running time is $O(n^{\log_2 3})$, which is about $O(n^{1.59})$.

In the absence of Gauss's trick, the recursion tree would have the same height, but the branching factor would be 4. There would be $4^{\log_2 n} = n^2$ leaves, and therefore the running time would be at least this much. In divide-and-conquer algorithms, the number of subproblems translates into the branching factor of the recursion tree; small changes in this coefficient can have a big impact on running time.

A practical note: it generally does not make sense to recurse all the way down to 1 bit. For most processors, 16- or 32-bit multiplication is a single operation, so by the time the numbers get into this range they should be handed over to the built-in procedure.

Finally, the eternal question: *Can we do better?* It turns out that even faster algorithms for multiplying numbers exist, based on another important divide-and-conquer algorithm: the fast Fourier transform, to be explained in Section 2.6.

2.2 Recurrence relations

Divide-and-conquer algorithms often follow a generic pattern: they tackle a problem of size n by recursively solving, say, a subproblems of size n/b and then combining these answers in $O(n^d)$ time, for some $a, b, d > 0$ (in the multiplication algorithm, $a = 3$, $b = 2$, and $d = 1$). Their running time can therefore be captured by the equation $T(n) = aT(\lceil n/b \rceil) + O(n^d)$. We next derive a closed-form solution to this general recurrence so that we no longer have to solve it explicitly in each new instance.

Master theorem[2] *If $T(n) = aT(\lceil n/b \rceil) + O(n^d)$ for some constants $a > 0$, $b > 1$, and $d \geq 0$, then*

$$T(n) = \begin{cases} O(n^d) & \text{if } d > \log_b a \\ O(n^d \log n) & \text{if } d = \log_b a \\ O(n^{\log_b a}) & \text{if } d < \log_b a. \end{cases}$$

This single theorem tells us the running times of most of the divide-and-conquer procedures we are likely to use.

Proof. To prove the claim, let's start by assuming for the sake of convenience that n is a power of b. This will not influence the final bound in any important way—after all, n is at most a multiplicative factor of b away from some power of b (Exercise 2.2)—and it will allow us to ignore the rounding effect in $\lceil n/b \rceil$.

Next, notice that the size of the subproblems decreases by a factor of b with each level of recursion, and therefore reaches the base case after $\log_b n$ levels. This is the height of the recursion tree. Its branching factor is a, so the kth level of the tree is made up of a^k subproblems, each of size n/b^k (Figure 2.3). The total work done at this level is

$$a^k \times O\left(\frac{n}{b^k}\right)^d = O(n^d) \times \left(\frac{a}{b^d}\right)^k.$$

As k goes from 0 (the root) to $\log_b n$ (the leaves), these numbers form a geometric series with ratio a/b^d. Finding the sum of such a series in big-O notation is easy (Exercise 0.2), and comes down to three cases.

1. The ratio is less than 1.
   Then the series is decreasing, and its sum is just given by its first term, $O(n^d)$.
2. The ratio is greater than 1.
   The series is increasing and its sum is given by its last term, $O(n^{\log_b a})$:

$$n^d \left(\frac{a}{b^d}\right)^{\log_b n} = n^d \left(\frac{a^{\log_b n}}{(b^{\log_b n})^d}\right) = a^{\log_b n} = a^{(\log_a n)(\log_b a)} = n^{\log_b a}.$$

3. The ratio is exactly 1.
   In this case all $O(\log n)$ terms of the series are equal to $O(n^d)$.

These cases translate directly into the three contingencies in the theorem statement. ▮

[2]There are even more general results of this type, but we will not be needing them.

Binary search

The ultimate divide-and-conquer algorithm is, of course, *binary search:* to find a key k in a large file containing keys $z[0, 1, \ldots, n-1]$ in sorted order, we first compare k with $z[n/2]$, and depending on the result we recurse either on the first half of the file, $z[0, \ldots, n/2-1]$, or on the second half, $z[n/2, \ldots, n-1]$. The recurrence now is $T(n) = T(\lceil n/2 \rceil) + O(1)$, which is the case $a = 1$, $b = 2$, $d = 0$. Plugging into our master theorem we get the familiar solution: a running time of just $O(\log n)$.

Figure 2.3 Each problem of size n is divided into a subproblems of size n/b.

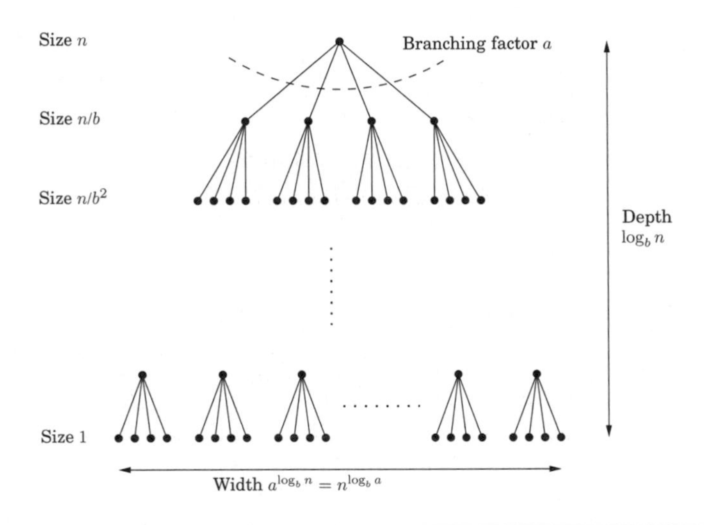

2.3 Mergesort

The problem of sorting a list of numbers lends itself immediately to a divide-and-conquer strategy: split the list into two halves, recursively sort each half, and then *merge* the two sorted sublists.

```
function mergesort(a[1...n])
Input: An array of numbers a[1...n]
Output: A sorted version of this array

if n > 1:
   return merge(mergesort(a[1...⌊n/2⌋]),
                mergesort(a[⌊n/2⌋+1...n]))
else:
   return a
```

The correctness of this algorithm is self-evident, as long as a correct merge subroutine is specified. If we are given two sorted arrays $x[1\ldots k]$ and $y[1\ldots l]$, how do we efficiently merge them into a single sorted array $z[1\ldots k+l]$? Well, the very first element of z is either $x[1]$ or $y[1]$, whichever is smaller. The rest of $z[\cdot]$ can then be constructed recursively.

```
function merge(x[1...k], y[1...l])
if k=0: return y[1...l]
if l=0: return x[1...k]
if x[1] ≤ y[1]:
   return x[1] ∘ merge(x[2...k], y[1...l])
else:
   return y[1] ∘ merge(x[1...k], y[2...l])
```

Here $\circ$ denotes concatenation. This merge procedure does a constant amount of work per recursive call (provided the required array space is allocated in advance), for a total running time of $O(k+l)$. Thus merge's are linear, and the overall time taken by mergesort is

$$T(n) = 2T(n/2) + O(n),$$

or $O(n \log n)$.

Looking back at the mergesort algorithm, we see that all the real work is done in merging, which doesn't start until the recursion gets down to singleton arrays. The singletons are merged in pairs, to yield arrays with two elements. Then pairs of these 2-tuples are merged, producing 4-tuples, and so on. Figure 2.4 shows an example.

This viewpoint also suggests how mergesort might be made iterative. At any given moment, there is a set of "active" arrays—initially, the singletons—which are merged in pairs to give the next batch of active arrays. These arrays can be organized in a queue, and processed by repeatedly removing two arrays from the front of the queue, merging them, and putting the result at the end of the queue.

In the following pseudocode, the primitive operation inject adds an element to the end of the queue while eject removes and returns the element at the front of the queue.

```
function iterative-mergesort(a[1...n])
Input: elements a_1, a_2, ..., a_n to be sorted

Q = [ ] (empty queue)
for i = 1 to n:
   inject(Q, [a_i])
while |Q| > 1:
   inject(Q, merge(eject(Q), eject(Q)))
return eject(Q)
```

Figure 2.4 The sequence of merge operations in `mergesort`.

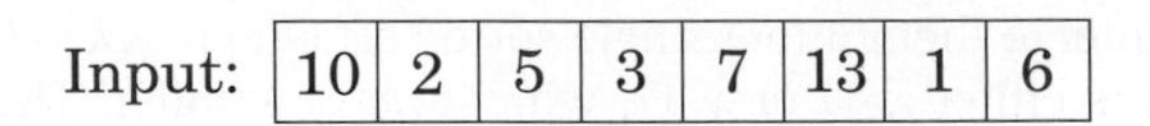

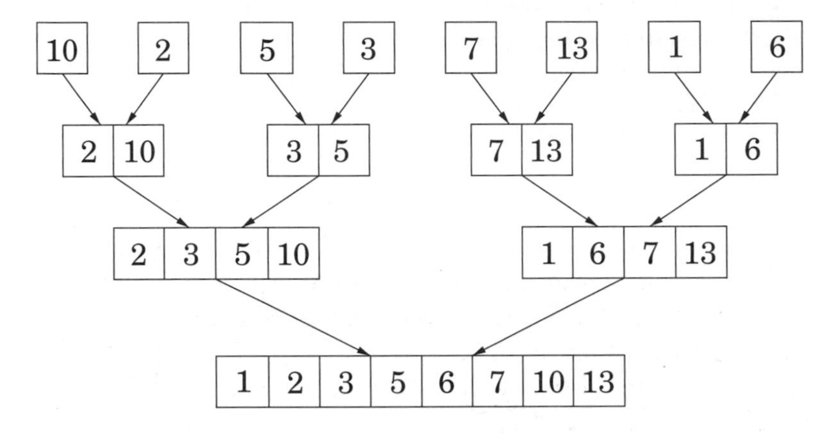

An $n \log n$ lower bound for sorting

Sorting algorithms can be depicted as trees. The one in the following figure sorts an array of three elements, a_1, a_2, a_3. It starts by comparing a_1 to a_2 and, if the first is larger, compares it with a_3; otherwise it compares a_2 and a_3. And so on. Eventually we end up at a leaf, and this leaf is labeled with the true order of the three elements as a permutation of 1, 2, 3. For example, if $a_2 < a_1 < a_3$, we get the leaf labeled "2 1 3."

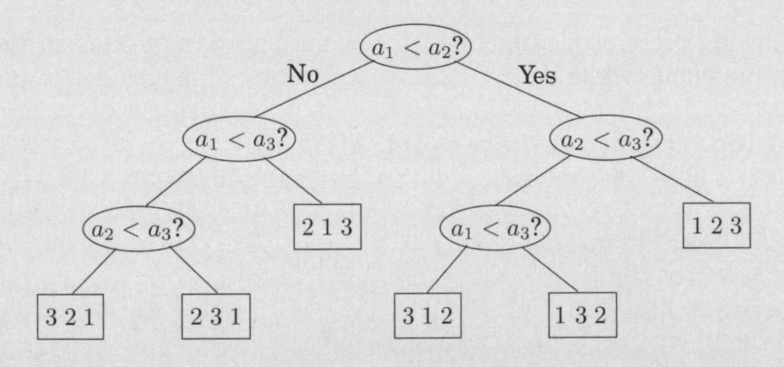

The *depth* of the tree—the number of comparisons on the longest path from root to leaf, in this case 3—is exactly the worst-case time complexity of the algorithm.

An $n \log n$ lower bound for sorting (*continued*)

This way of looking at sorting algorithms is useful because it allows one to argue that *mergesort is optimal*, in the sense that $\Omega(n \log n)$ comparisons are necessary for sorting n elements.

Here is the argument: Consider any such tree that sorts an array of n elements. Each of its leaves is labeled by a permutation of $\{1, 2, \ldots, n\}$. In fact, *every* permutation must appear as the label of a leaf. The reason is simple: if a particular permutation is missing, what happens if we feed the algorithm an input ordered according to this same permutation? And since there are $n!$ permutations of n elements, it follows that the tree has at least $n!$ leaves.

We are almost done: This is a binary tree, and we argued that it has at least $n!$ leaves. Recall now that a binary tree of depth d has at most 2^d leaves (proof: an easy induction on d). So, the depth of our tree—and the complexity of our algorithm—must be at least $\log(n!)$.

And it is well known that $\log(n!) \geq c \cdot n \log n$ for some $c > 0$. There are many ways to see this. The easiest is to notice that $n! \geq (n/2)^{(n/2)}$ because $n! = 1 \cdot 2 \cdot \cdots \cdot n$ contains at least $n/2$ factors larger than $n/2$; and to then take logs of both sides. Another is to recall Stirling's formula

$$n! \approx \sqrt{\pi \left(2n + \frac{1}{3}\right)} \cdot n^n \cdot e^{-n}.$$

Either way, we have established that any comparison tree that sorts n elements must make, in the worst case, $\Omega(n \log n)$ comparisons, and hence mergesort is optimal!
Well, there is some fine print: this neat argument applies only to *algorithms that use comparisons*. Is it conceivable that there are alternative sorting strategies, perhaps using sophisticated numerical manipulations, that work in linear time? The answer is *yes*, under certain exceptional circumstances: the canonical such example is when the elements to be sorted are integers that lie in a small range (Exercise 2.20).

2.4 Medians

The *median* of a list of numbers is its 50th percentile: half the numbers are bigger than it, and half are smaller. For instance, the median of [45, 1, 10, 30, 25] is 25, since this is the middle element when the numbers are arranged in order. If the list has even length, there are two choices for what the middle element could be, in which case we pick the smaller of the two, say.

The purpose of the median is to summarize a set of numbers by a single, typical value. The *mean*, or average, is also very commonly used for this, but the median is in a sense more typical of the data: it is always one of the data values, unlike the mean, and it is less sensitive to outliers. For instance, the median of a list of a hundred 1's is (rightly) 1, as is the mean. However, if just one of these numbers gets accidentally corrupted to 10,000, the mean shoots up above 100, while the median is unaffected.

Computing the median of n numbers is easy: just sort them. The drawback is that this takes $O(n \log n)$ time, whereas we would ideally like something linear. We have reason to be hopeful, because sorting is doing far more work than we really need—we just want the middle element and don't care about the relative ordering of the rest of them.

When looking for a recursive solution, it is paradoxically often easier to work with a *more general* version of the problem—for the simple reason that this gives a more powerful step to recurse upon. In our case, the generalization we will consider is *selection*.

Selection

Input: A list of numbers S; an integer k

Output: The kth smallest element of S

For instance, if $k = 1$, the minimum of S is sought, whereas if $k = \lfloor |S|/2 \rfloor$, it is the median.

A randomized divide-and-conquer algorithm for selection

Here's a divide-and-conquer approach to selection. For any number v, imagine splitting list S into three categories: elements smaller than v, those equal to v (there might be duplicates), and those greater than v. Call these S_L, S_v, and S_R respectively. For instance, if the array

S :	2	36	5	21	8	13	11	20	5	4	1

is split on $v = 5$, the three subarrays generated are

S_L :	2	4	1

S_v :	5	5

S_R :	36	21	8	13	11	20

The search can instantly be narrowed down to one of these sublists. If we want, say, the *eighth*-smallest element of S, we know it must be the *third*-smallest element of S_R since $|S_L| + |S_v| = 5$. That is, selection$(S, 8)$ = selection$(S_R, 3)$. More generally, by checking k against the sizes of the subarrays, we can quickly determine which of them holds the desired element:

$$\text{selection}(S, k) = \begin{cases} \text{selection}(S_L, k) & \text{if } k \leq |S_L| \\ v & \text{if } |S_L| < k \leq |S_L| + |S_v| \\ \text{selection}(S_R, k - |S_L| - |S_v|) & \text{if } k > |S_L| + |S_v|. \end{cases}$$

The three sublists S_L, S_v, and S_R can be computed from S in linear time; in fact, this computation can even be done *in place*, that is, without allocating new memory (Exercise 2.15). We then recurse on the appropriate sublist. The effect of the split is thus to shrink the number of elements from $|S|$ to at most $\max\{|S_L|, |S_R|\}$.

Our divide-and-conquer algorithm for selection is now fully specified, except for the crucial detail of how to choose v. It should be picked quickly, and it should shrink the array substantially, the ideal situation being $|S_L|, |S_R| \approx \frac{1}{2}|S|$. If we could always guarantee this situation, we would get a running time of

$$T(n) = T(n/2) + O(n),$$

which is linear as desired. But this requires picking v to be the median, which is our ultimate goal! Instead, we follow a much simpler alternative: *we pick v randomly from S.*

Efficiency analysis

Naturally, the running time of our algorithm depends on the random choices of v. It is possible that due to persistent bad luck we keep picking v to be the largest element of the array (or the smallest element), and thereby shrink the array by only one element each time. In the earlier example, we might first pick $v = 36$, then $v = 21$, and so on. This *worst-case* scenario would force our selection algorithm to perform

$$n + (n-1) + (n-2) + \cdots + \frac{n}{2} = \Theta(n^2)$$

operations (when computing the median), but it is extremely unlikely to occur. Equally unlikely is the *best* possible case we discussed before, in which each randomly chosen v just happens to split the array perfectly in half, resulting in a running time of $O(n)$. Where, in this spectrum from $O(n)$ to $\Theta(n^2)$, does the *average* running time lie? Fortunately, it lies very close to the best-case time.

To distinguish between lucky and unlucky choices of v, we will call v *good* if it lies within the 25th to 75th percentile of the array that it is chosen from. We like these choices of v because they ensure that the sublists S_L and S_R have size at most three-fourths that of S (do you see why?), so that the array shrinks substantially. Fortunately, good v's are abundant: half the elements of any list must fall between the 25th to 75th percentile!

Given that a randomly chosen v has a 50% chance of being good, how many v's do we need to pick on average before getting a good one? Here's a more familiar reformulation (see also Exercise 1.34):

Lemma *On average a fair coin needs to be tossed two times before a "heads" is seen.*

Proof. Let E be the expected number of tosses before a heads is seen. We certainly need at least one toss, and if it's heads, we're done. If it's tails (which occurs with probability 1/2), we need to repeat. Hence $E = 1 + \frac{1}{2}E$, which works out to $E = 2$. ∎

Therefore, after two split operations on average, the array will shrink to at most three-fourths of its size. Letting $T(n)$ be the *expected* running time on an array of size n, we get

$$T(n) \leq T(3n/4) + O(n).$$

This follows by taking expected values of both sides of the following statement:

Time taken on an array of size n

$\leq$ (time taken on an array of size $3n/4$) + (time to reduce array size to $\leq 3n/4$),

and, for the right-hand side, using the familiar property that *the expectation of the sum is the sum of the expectations.*

The unix `sort` command

Comparing the algorithms for sorting and median-finding we notice that, beyond the common divide-and-conquer philosophy and structure, they are exact opposites. Mergesort splits the array in two in the most convenient way (first half, second half), without any regard to the magnitudes of the elements in each half; but then it works hard to put the sorted subarrays together. In contrast, the median algorithm is careful about its splitting (smaller numbers first, then the larger ones), but its work ends with the recursive call.

Quicksort is a sorting algorithm that splits the array in exactly the same way as the median algorithm; and once the subarrays are sorted, by two recursive calls, there is nothing more to do. Its worst-case performance is $\Theta(n^2)$, like that of median-finding. But it can be proved (Exercise 2.24) that its average case is $O(n \log n)$; furthermore, empirically it outperforms other sorting algorithms. This has made quicksort a favorite in many applications—for instance, it is the basis of the code by which really enormous files are sorted.

From this recurrence we conclude that $T(n) = O(n)$: on *any* input, our algorithm returns the correct answer after a linear number of steps, on the average.

2.5 Matrix multiplication

The product of two $n \times n$ matrices X and Y is a third $n \times n$ matrix $Z = XY$, with (i, j)th entry

$$Z_{ij} = \sum_{k=1}^{n} X_{ik} Y_{kj}.$$

To make it more visual, Z_{ij} is the dot product of the ith row of X with the jth column of Y:

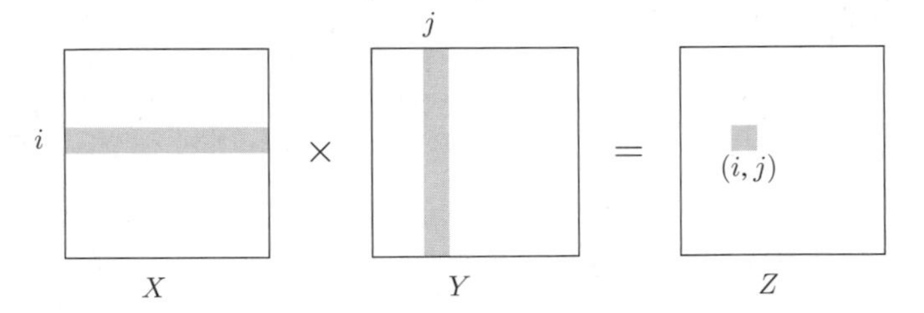

In general, XY is not the same as YX; matrix multiplication is not commutative.

The preceding formula implies an $O(n^3)$ algorithm for matrix multiplication: there are n^2 entries to be computed, and each takes $O(n)$ time. For quite a while, this was widely believed to be the best running time possible, and it was even proved that in certain models of computation no algorithm could do better. It was therefore a source of great excitement when in 1969, the German mathematician Volker Strassen announced a significantly more efficient algorithm, based upon divide-and-conquer.

Matrix multiplication is particularly easy to break into subproblems, because it can be performed *blockwise*. To see what this means, carve X into four $n/2 \times n/2$ blocks, and also Y:

$$X = \begin{bmatrix} A & B \\ C & D \end{bmatrix}, \quad Y = \begin{bmatrix} E & F \\ G & H \end{bmatrix}.$$

Then their product can be expressed in terms of these blocks and is exactly as if the blocks were single elements (Exercise 2.11).

$$XY = \begin{bmatrix} A & B \\ C & D \end{bmatrix} \begin{bmatrix} E & F \\ G & H \end{bmatrix} = \begin{bmatrix} AE + BG & AF + BH \\ CE + DG & CF + DH \end{bmatrix}$$

We now have a divide-and-conquer strategy: to compute the size-n product XY, recursively compute eight size-$n/2$ products $AE, BG, AF, BH, CE, DG, CF, DH$, and then do a few $O(n^2)$-time additions. The total running time is described by the recurrence relation

$$T(n) = 8T(n/2) + O(n^2).$$

This comes out to an unimpressive $O(n^3)$, the same as for the default algorithm. But the efficiency *can* be further improved, and as with integer multiplication, the key is some clever algebra. It turns out XY can be computed from just *seven* $n/2 \times n/2$ subproblems, via a decomposition so tricky and intricate that one wonders how Strassen was ever able to discover it!

$$XY = \begin{bmatrix} P_5 + P_4 - P_2 + P_6 & P_1 + P_2 \\ P_3 + P_4 & P_1 + P_5 - P_3 - P_7 \end{bmatrix}$$

where

$$\begin{aligned} P_1 &= A(F - H) & P_5 &= (A + D)(E + H) \\ P_2 &= (A + B)H & P_6 &= (B - D)(G + H) \\ P_3 &= (C + D)E & P_7 &= (A - C)(E + F) \\ P_4 &= D(G - E) \end{aligned}$$

The new running time is

$$T(n) = 7T(n/2) + O(n^2),$$

which by the master theorem works out to $O(n^{\log_2 7}) \approx O(n^{2.81})$.

2.6 The fast Fourier transform

We have so far seen how divide-and-conquer gives fast algorithms for multiplying integers and matrices; our next target is *polynomials*. The product of two degree-d polynomials is a polynomial of degree $2d$, for example:

$$(1 + 2x + 3x^2) \cdot (2 + x + 4x^2) = 2 + 5x + 12x^2 + 11x^3 + 12x^4.$$

More generally, if $A(x) = a_0 + a_1x + \cdots + a_dx^d$ and $B(x) = b_0 + b_1x + \cdots + b_dx^d$, their product $C(x) = A(x) \cdot B(x) = c_0 + c_1x + \cdots + c_{2d}x^{2d}$ has coefficients

$$c_k = a_0b_k + a_1b_{k-1} + \cdots + a_kb_0 = \sum_{i=0}^{k} a_ib_{k-i}$$

(for $i > d$, take a_i and b_i to be zero). Computing c_k from this formula takes $O(k)$ steps, and finding all $2d+1$ coefficients would therefore seem to require $\Theta(d^2)$ time. *Can we possibly multiply polynomials faster than this?*

The solution we will develop, the fast Fourier transform, has revolutionized—indeed, defined—the field of signal processing (see the following box). Because of its huge importance, and its wealth of insights from different fields of study, we will approach it a little more leisurely than usual. The reader who wants just the core algorithm can skip directly to Section 2.6.4.

2.6.1 An alternative representation of polynomials

To arrive at a fast algorithm for polynomial multiplication we take inspiration from an important property of polynomials.

Fact *A degree-d polynomial is uniquely characterized by its values at any $d+1$ distinct points.*

A familiar instance of this is that "any two points determine a line." We will later see why the more general statement is true (page 64), but for the time being it gives us an *alternative representation* of polynomials. Fix any distinct points $x_0, \ldots, x_d$. We can specify a degree-d polynomial $A(x) = a_0 + a_1x + \cdots + a_dx^d$ by either one of the following:

1. Its coefficients $a_0, a_1, \ldots, a_d$
2. The values $A(x_0), A(x_1), \ldots, A(x_d)$

Of these two representations, the second is the more attractive for polynomial multiplication. Since the product $C(x)$ has degree $2d$, it is completely determined by its value at any $2d+1$ points. And its value at any given point z is easy enough to figure out, just $A(z)$ times $B(z)$. Thus *polynomial multiplication takes linear time in the value representation.*

The problem is that we expect the input polynomials, and also their product, to be specified by coefficients. So we need to first translate from coefficients to values—which is just a matter of *evaluating* the polynomial at the chosen points—then multiply in the value representation, and finally translate back to coefficients, a process called *interpolation*.

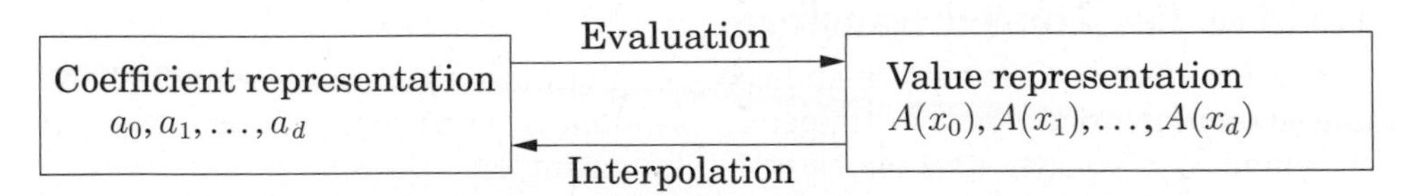

Figure 2.5 presents the resulting algorithm.

Why multiply polynomials?

For one thing, it turns out that the fastest algorithms we have for multiplying integers rely heavily on polynomial multiplication; after all, polynomials and binary integers are quite similar—just replace the variable x by the base 2, and watch out for carries. But perhaps more importantly, multiplying polynomials is crucial for *signal processing*.

A *signal* is any quantity that is a function of time (as in Figure (a)) or of position. It might, for instance, capture a human voice by measuring fluctuations in air pressure close to the speaker's mouth, or alternatively, the pattern of stars in the night sky, by measuring brightness as a function of angle.

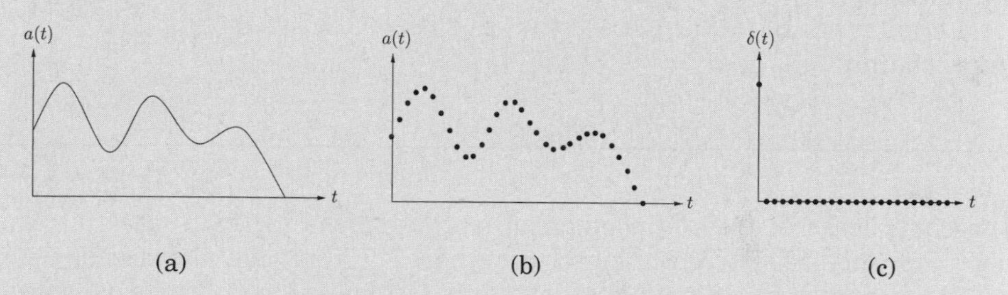

(a) (b) (c)

In order to extract information from a signal, we need to first *digitize* it by sampling (Figure (b))—and, then, to put it through a *system* that will transform it in some way. The output is called the *response* of the system:

$$\text{signal} \longrightarrow \boxed{\text{SYSTEM}} \longrightarrow \text{response.}$$

An important class of systems are those that are *linear*—the response to the sum of two signals is just the sum of their individual responses—and *time invariant*—shifting the input signal by time t produces the same output, also shifted by t. Any system with these properties is completely characterized by its response to the simplest possible input signal: the *unit impulse* $\delta(t)$, consisting solely of a "jerk" at $t = 0$ (Figure (c)). To see this, first consider the close relative $\delta(t - i)$, a shifted impulse in which the jerk occurs at time i. Any signal $a(t)$ can be expressed as a linear combination of these, letting $\delta(t - i)$ pick out its behavior at time i,

$$a(t) = \sum_{i=0}^{T-1} a(i)\delta(t - i)$$

(if the signal consists of T samples). By linearity, the system response to input $a(t)$ is determined by the responses to the various $\delta(t - i)$. And by time invariance, these are in turn just shifted copies of the *impulse response* $b(t)$, the response to $\delta(t)$.

In other words, the output of the system at time k is

$$c(k) = \sum_{i=0}^{k} a(i)b(k - i),$$

exactly the formula for polynomial multiplication!

Figure 2.5 Polynomial multiplication

Input: Coefficients of two polynomials, $A(x)$ and $B(x)$, of degree d
Output: Their product $C = A \cdot B$

Selection
Pick some points $x_0, x_1, \ldots, x_{n-1}$, where $n \geq 2d+1$
Evaluation
Compute $A(x_0), A(x_1), \ldots, A(x_{n-1})$ and $B(x_0), B(x_1), \ldots, B(x_{n-1})$
Multiplication
Compute $C(x_k) = A(x_k)B(x_k)$ for all $k = 0, \ldots, n-1$
Interpolation
Recover $C(x) = c_0 + c_1 x + \cdots + c_{2d}x^{2d}$

The equivalence of the two polynomial representations makes it clear that this high-level approach is correct, but how efficient is it? Certainly the selection step and the n multiplications are no trouble at all, just linear time.[3] But (leaving aside interpolation, about which we know even less) how about evaluation? Evaluating a polynomial of degree $d \leq n$ at a single point takes $O(n)$ steps (Exercise 2.29), and so the baseline for n points is $\Theta(n^2)$. We'll now see that the fast Fourier transform (FFT) does it in just $O(n \log n)$ time, for a particularly clever choice of $x_0, \ldots, x_{n-1}$ in which the computations required by the individual points overlap with one another and can be shared.

2.6.2 Evaluation by divide-and-conquer

Here's an idea for how to pick the n points at which to evaluate a polynomial $A(x)$ of degree $\leq n-1$. If we choose them to be positive-negative pairs, that is,

$$\pm x_0, \pm x_1, \ldots, \pm x_{n/2-1},$$

then the computations required for each $A(x_i)$ and $A(-x_i)$ overlap a lot, because the even powers of x_i coincide with those of $-x_i$.

To investigate this, we need to split $A(x)$ into its odd and even powers, for instance

$$3 + 4x + 6x^2 + 2x^3 + x^4 + 10x^5 \;=\; (3 + 6x^2 + x^4) + x(4 + 2x^2 + 10x^4).$$

Notice that the terms in parentheses are polynomials in x^2. More generally,

$$A(x) \;=\; A_e(x^2) + x A_o(x^2),$$

---

[3]In a typical setting for polynomial multiplication, the coefficients of the polynomials are real numbers and, moreover, are small enough that basic arithmetic operations (adding and multiplying) take unit time. We will assume this to be the case without any great loss of generality; in particular, the time bounds we obtain are easily adjustable to situations with larger numbers.

where $A_e(\cdot)$, with the even-numbered coefficients, and $A_o(\cdot)$, with the odd-numbered coefficients, are polynomials of degree $\le n/2 - 1$ (assume for convenience that n is even). Given *paired* points $\pm x_i$, the calculations needed for $A(x_i)$ can be recycled toward computing $A(-x_i)$:

$$A(x_i) = A_e(x_i^2) + x_i A_o(x_i^2)$$
$$A(-x_i) = A_e(x_i^2) - x_i A_o(x_i^2).$$

In other words, evaluating $A(x)$ at n paired points $\pm x_0, \ldots, \pm x_{n/2-1}$ reduces to evaluating $A_e(x)$ and $A_o(x)$ (which each have half the degree of $A(x)$) at just $n/2$ points, $x_0^2, \ldots, x_{n/2-1}^2$.

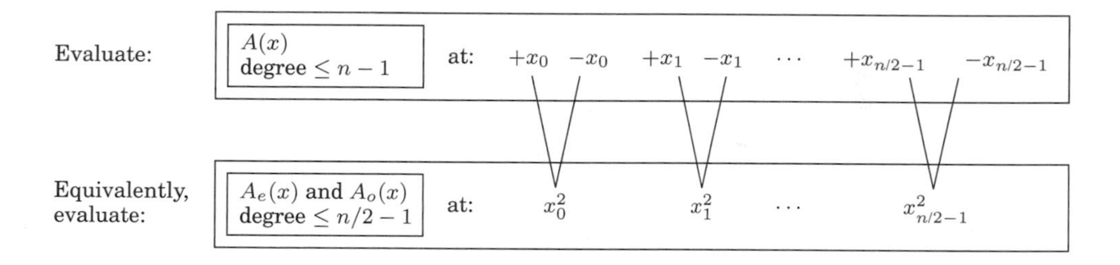

The original problem of size n is in this way recast as two subproblems of size $n/2$, followed by some linear-time arithmetic. If we could recurse, we would get a divide-and-conquer procedure with running time

$$T(n) = 2T(n/2) + O(n),$$

which is $O(n \log n)$, exactly what we want.

But we have a problem: The plus-minus trick only works at the top level of the recursion. To recurse at the next level, we need the $n/2$ evaluation points $x_0^2, x_1^2, \ldots, x_{n/2-1}^2$ to be *themselves* plus-minus pairs. But how can a square be negative? The task seems impossible! *Unless, of course, we use complex numbers.*

Fine, but which complex numbers? To figure this out, let us "reverse engineer" the process. At the very bottom of the recursion, we have a single point. This point might as well be 1, in which case the level above it must consist of its square roots, $\pm\sqrt{1} = \pm 1$.

The next level up then has $\pm\sqrt{+1} = \pm 1$ as well as the *complex* numbers $\pm\sqrt{-1} = \pm i$, where i is the imaginary unit. By continuing in this manner, we eventually reach the initial set of n points. Perhaps you have already guessed what

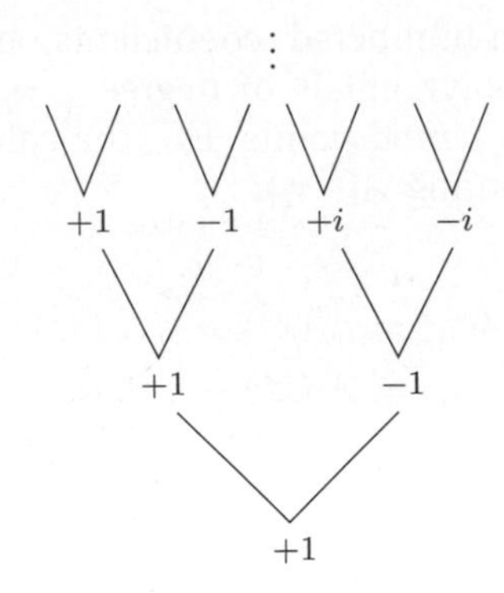

they are: the *complex nth roots of unity*, that is, the n complex solutions to the equation $z^n = 1$.

Figure 2.6 is a pictorial review of some basic facts about complex numbers. The third panel of this figure introduces the nth roots of unity: the complex numbers $1, \omega, \omega^2, \ldots, \omega^{n-1}$, where $\omega = e^{2\pi i/n}$. If n is even,

1. The nth roots are plus-minus paired, $\omega^{n/2+j} = -\omega^j$.
2. Squaring them produces the $(n/2)$nd roots of unity.

Therefore, if we start with these numbers for some n that is a power of 2, then at successive levels of recursion we will have the $(n/2^k)$th roots of unity, for $k = 0, 1, 2, 3, \ldots$. All these sets of numbers are plus-minus paired, and so our divide-and-conquer, as shown in the last panel, works perfectly. The resulting algorithm is the fast Fourier transform (Figure 2.7).

2.6.3 Interpolation

Let's take stock of where we are. We first developed a high-level scheme for multiplying polynomials (Figure 2.5), based on the observation that polynomials can be represented in two ways, in terms of their *coefficients* or in terms of their *values* at a selected set of points.

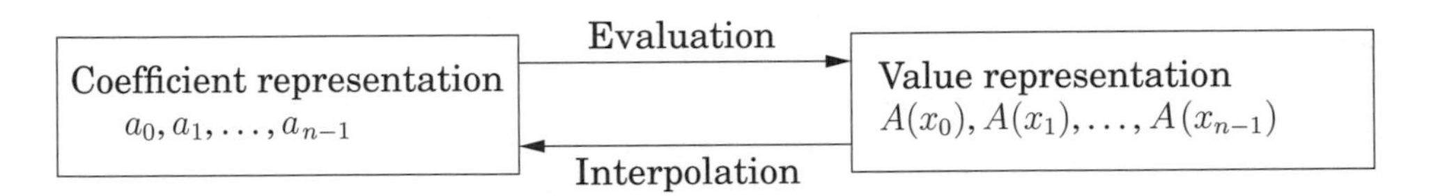

The value representation makes it trivial to multiply polynomials, but we cannot ignore the coefficient representation since it is the form in which the input and output of our overall algorithm are specified.

So we designed the FFT, a way to move from coefficients to values in time just $O(n \log n)$, when the points $\{x_i\}$ are complex nth roots of unity $(1, \omega, \omega^2, \ldots, \omega^{n-1})$.

$$\langle \text{values} \rangle = \text{FFT}(\langle \text{coefficients} \rangle, \omega).$$

Figure 2.6 The complex roots of unity are ideal for our divide-and-conquer scheme.

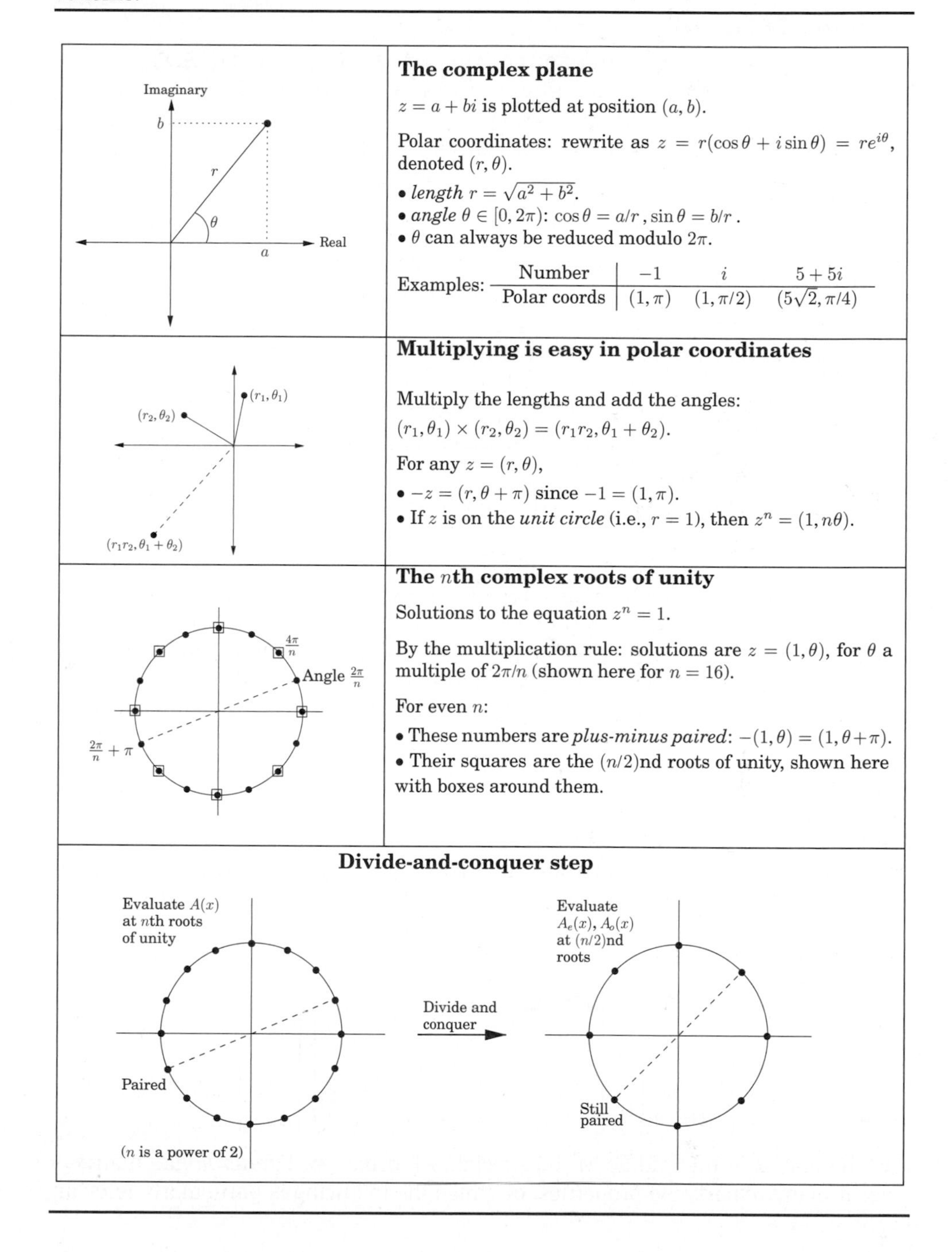

Figure 2.7 The fast Fourier transform (polynomial formulation)

```
function FFT(A, ω)
Input: Coefficient representation of a polynomial A(x)
       of degree ≤ n−1, where n is a power of 2
       ω, an nth root of unity
Output: Value representation A(ω^0), ..., A(ω^(n−1))

if ω = 1: return A(1)
express A(x) in the form A_e(x^2) + x A_o(x^2)
call FFT(A_e, ω^2) to evaluate A_e at even powers of ω
call FFT(A_o, ω^2) to evaluate A_o at even powers of ω
for j = 0 to n−1:
   compute A(ω^j) = A_e(ω^(2j)) + ω^j A_o(ω^(2j))

return A(ω^0), ..., A(ω^(n−1))
```

The last remaining piece of the puzzle is the inverse operation, interpolation. It will turn out, amazingly, that

$$\langle\text{coefficients}\rangle = \frac{1}{n}\,\text{FFT}(\langle\text{values}\rangle, \omega^{-1}).$$

Interpolation is thus solved in the most simple and elegant way we could possibly have hoped for—using the same FFT algorithm, but called with ω^{-1} in place of ω! This might seem like a miraculous coincidence, but it will make a lot more sense when we recast our polynomial operations in the language of linear algebra. Meanwhile, our $O(n \log n)$ polynomial multiplication algorithm (Figure 2.5) is now fully specified.

A matrix reformulation

To get a clearer view of interpolation, let's explicitly set down the relationship between our two representations for a polynomial $A(x)$ of degree $\leq n - 1$. They are both vectors of n numbers, and one is a linear transformation of the other:

$$\begin{bmatrix} A(x_0) \\ A(x_1) \\ \vdots \\ A(x_{n-1}) \end{bmatrix} = \begin{bmatrix} 1 & x_0 & x_0^2 & \cdots & x_0^{n-1} \\ 1 & x_1 & x_1^2 & \cdots & x_1^{n-1} \\ & & \vdots & & \\ 1 & x_{n-1} & x_{n-1}^2 & \cdots & x_{n-1}^{n-1} \end{bmatrix} \begin{bmatrix} a_0 \\ a_1 \\ \vdots \\ a_{n-1} \end{bmatrix}.$$

Call the matrix in the middle M. Its specialized format—a *Vandermonde* matrix—gives it many remarkable properties, of which the following is particularly relevant to us.

If $x_0, \ldots, x_{n-1}$ are distinct numbers, then M is invertible.

The existence of M^{-1} allows us to invert the preceding matrix equation so as to express coefficients in terms of values. In brief,

Evaluation is multiplication by M, while interpolation is multiplication by M^{-1}.

This reformulation of our polynomial operations reveals their essential nature more clearly. Among other things, it finally justifies an assumption we have been making throughout, that $A(x)$ is uniquely characterized by its values at any n points—in fact, we now have an explicit formula that will give us the coefficients of $A(x)$ in this situation. Vandermonde matrices also have the distinction of being quicker to invert than more general matrices, in $O(n^2)$ time instead of $O(n^3)$. However, using this for interpolation would still not be fast enough for us, so once again we turn to our special choice of points—the complex roots of unity.

Interpolation resolved

In linear algebra terms, the FFT multiplies an arbitrary n-dimensional vector—which we have been calling the *coefficient representation*—by the $n \times n$ matrix

$$M_n(\omega) = \begin{bmatrix} 1 & 1 & 1 & \cdots & 1 \\ 1 & \omega & \omega^2 & \cdots & \omega^{n-1} \\ 1 & \omega^2 & \omega^4 & \cdots & \omega^{2(n-1)} \\ & & \vdots & & \\ 1 & \omega^j & \omega^{2j} & \cdots & \omega^{(n-1)j} \\ & & \vdots & & \\ 1 & \omega^{(n-1)} & \omega^{2(n-1)} & \cdots & \omega^{(n-1)(n-1)} \end{bmatrix} \begin{matrix} \longleftarrow & \text{row for } \omega^0 = 1 \\ \longleftarrow & \omega \\ \longleftarrow & \omega^2 \\ & \vdots \\ \longleftarrow & \omega^j \\ & \vdots \\ \longleftarrow & \omega^{n-1} \end{matrix}$$

where ω is a complex nth root of unity, and n is a power of 2. Notice how simple this matrix is to describe: its (j, k)th entry (starting row- and column-count at zero) is ω^{jk}.

Multiplication by $M = M_n(\omega)$ maps the kth coordinate axis (the vector with all zeros except for a 1 at position k) onto the kth column of M. Now here's the crucial observation, which we'll prove shortly: *the columns of M are orthogonal (at right angles) to each other.* Therefore they can be thought of as the axes of an alternative coordinate system, which is often called the *Fourier basis*. The effect of multiplying a vector by M is to rotate it from the standard basis, with the usual set of axes, into the Fourier basis, which is defined by the columns of M (Figure 2.8). The FFT is thus a change of basis, a *rigid rotation*. The inverse of M is the opposite rotation, from the Fourier basis back into the standard basis. When we write out the orthogonality condition precisely, we will be able to read off this inverse transformation with ease:

Inversion formula $\quad M_n(\omega)^{-1} = \frac{1}{n} M_n(\omega^{-1})$.

But ω^{-1} is also an nth root of unity, and so interpolation—or equivalently, multiplication by $M_n(\omega)^{-1}$—is itself just an FFT operation, but with ω replaced by ω^{-1}.

Now let's get into the details. Take ω to be $e^{2\pi i/n}$ for convenience, and think of the columns of M as vectors in $\mathbb{C}^n$. Recall that the *angle* between two vectors $u = (u_0, \ldots, u_{n-1})$ and $v = (v_0, \ldots, v_{n-1})$ in $\mathbb{C}^n$ is just a scaling factor times their

Figure 2.8 The FFT takes points in the standard coordinate system, whose axes are shown here as x_1, x_2, x_3, and rotates them into the Fourier basis, whose axes are the columns of $M_n(\omega)$, shown here as f_1, f_2, f_3. For instance, points in direction x_1 get mapped into direction f_1.

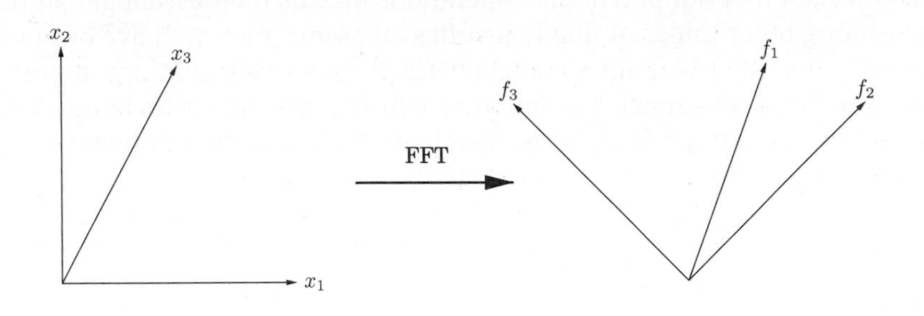

inner product

$$u \cdot v^* = u_0 v_0^* + u_1 v_1^* + \cdots + u_{n-1} v_{n-1}^*,$$

where z^* denotes the complex conjugate[4] of z. This quantity is maximized when the vectors lie in the same direction and is zero when the vectors are orthogonal to each other.

The fundamental observation we need is the following.

Lemma *The columns of matrix M are orthogonal to each other.*

Proof. Take the inner product of any columns j and k of matrix M,

$$1 + \omega^{j-k} + \omega^{2(j-k)} + \cdots + \omega^{(n-1)(j-k)}.$$

This is a geometric series with first term 1, last term $\omega^{(n-1)(j-k)}$, and ratio $\omega^{(j-k)}$. Therefore it evaluates to $(1 - \omega^{n(j-k)})/(1 - \omega^{(j-k)})$, which is 0—except when $j = k$, in which case all terms are 1 and the sum is n. ∎

The orthogonality property can be summarized in the single equation

$$MM^* = nI,$$

since $(MM^*)_{ij}$ is the inner product of the ith and jth columns of M (do you see why?). This immediately implies $M^{-1} = (1/n)M^*$: we have an inversion formula! But is it the same formula we earlier claimed? Let's see—the (j, k)th entry of M^* is the complex conjugate of the corresponding entry of M, in other words ω^{-jk}. Whereupon $M^* = M_n(\omega^{-1})$, and we're done.

And now we can finally step back and view the whole affair geometrically. The task we need to perform, polynomial multiplication, is a lot easier in the Fourier

[4]The *complex conjugate* of a complex number $z = re^{i\theta}$ is $z^* = re^{-i\theta}$. The complex conjugate of a vector (or matrix) is obtained by taking the complex conjugates of all its entries.

basis than in the standard basis. Therefore, we first rotate vectors into the Fourier basis (*evaluation*), then perform the task (*multiplication*), and finally rotate back (*interpolation*). The initial vectors are *coefficient representations*, while their rotated counterparts are *value representations*. To efficiently switch between these, back and forth, is the province of the FFT.

2.6.4 A closer look at the fast Fourier transform

Now that our efficient scheme for polynomial multiplication is fully realized, let's hone in more closely on the core subroutine that makes it all possible, the fast Fourier transform.

The definitive FFT algorithm

The FFT takes as input a vector $a = (a_0, \ldots, a_{n-1})$ and a complex number ω whose powers $1, \omega, \omega^2, \ldots, \omega^{n-1}$ are the complex nth roots of unity. It multiplies vector a by the $n \times n$ matrix $M_n(\omega)$, which has (j, k)th entry (starting row- and column-count at zero) ω^{jk}. The potential for using divide-and-conquer in this matrix-vector multiplication becomes apparent when M's columns are segregated into evens and odds:

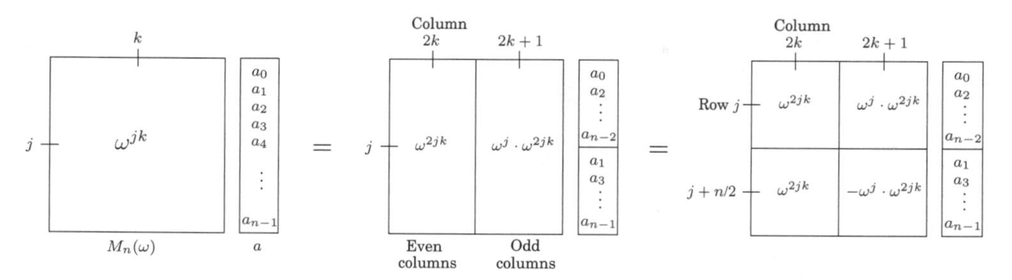

In the second step, we have simplified entries in the bottom half of the matrix using $\omega^{n/2} = -1$ and $\omega^n = 1$. Notice that the top left $n/2 \times n/2$ submatrix is $M_{n/2}(\omega^2)$, as is the one on the bottom left. And the top and bottom right submatrices are almost the same as $M_{n/2}(\omega^2)$, but with their jth rows multiplied through by ω^j and $-\omega^j$, respectively. Therefore the final product is the vector.

$$\begin{array}{r} \text{Row } j \\ \\ j + n/2 \end{array} \left[\begin{array}{c} M_{n/2} \begin{bmatrix} a_0 \\ a_2 \\ \vdots \\ a_{n-2} \end{bmatrix} + \omega^j \, M_{n/2} \begin{bmatrix} a_1 \\ a_3 \\ \vdots \\ a_{n-1} \end{bmatrix} \\ \\ M_{n/2} \begin{bmatrix} a_0 \\ a_2 \\ \vdots \\ a_{n-2} \end{bmatrix} - \omega^j \, M_{n/2} \begin{bmatrix} a_1 \\ a_3 \\ \vdots \\ a_{n-1} \end{bmatrix} \end{array} \right]$$

In short, the product of $M_n(\omega)$ with vector $(a_0, \ldots, a_{n-1})$, a size-n problem, can be expressed in terms of two size-$n/2$ problems: the product of $M_{n/2}(\omega^2)$ with $(a_0, a_2, \ldots, a_{n-2})$ and with $(a_1, a_3, \ldots, a_{n-1})$. This divide-and-conquer strategy leads to the definitive FFT algorithm of Figure 2.9, whose running time is $T(n) = 2T(n/2) + O(n) = O(n \log n)$.

Figure 2.9 The fast Fourier transform

```
function FFT(a, ω)
Input: An array a = (a_0, a_1, ..., a_{n-1}), for n a power of 2
   A primitive nth root of unity, ω
Output: M_n(ω) a

if ω = 1: return a
(s_0, s_1, ..., s_{n/2-1}) = FFT((a_0, a_2, ..., a_{n-2}), ω^2)
(s'_0, s'_1, ..., s'_{n/2-1}) = FFT((a_1, a_3, ..., a_{n-1}), ω^2)
for j = 0 to n/2 - 1:
   r_j = s_j + ω^j s'_j
   r_{j+n/2} = s_j - ω^j s'_j
return (r_0, r_1, ..., r_{n-1})
```

The fast Fourier transform unraveled

Throughout all our discussions so far, the fast Fourier transform has remained tightly cocooned within a divide-and-conquer formalism. To fully expose its structure, we now unravel the recursion.

The divide-and-conquer step of the FFT can be drawn as a very simple circuit. Here is how a problem of size n is reduced to two subproblems of size $n/2$ (for clarity, one pair of outputs $(j, j + n/2)$ is singled out):

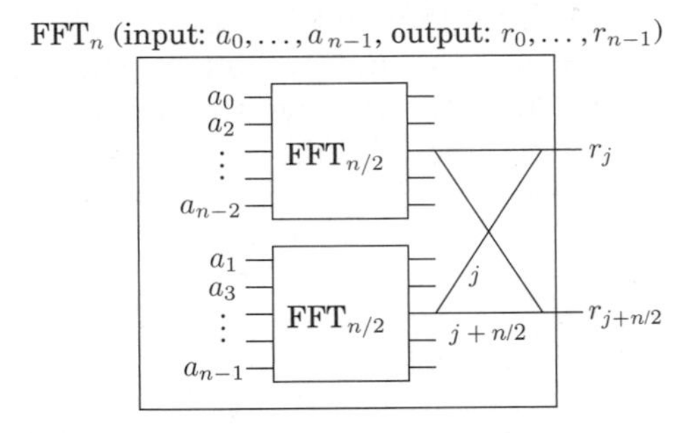

We're using a particular shorthand: the edges are wires carrying complex numbers from left to right. A weight of j means "multiply the number on this wire by ω^j." And when two wires come into a junction from the left, the numbers they are carrying get added up. So the two outputs depicted are executing the commands

$$r_j = s_j + \omega^j s'_j$$
$$r_{j+n/2} = s_j - \omega^j s'_j$$

from the FFT algorithm (Figure 2.9), via a pattern of wires known as a *butterfly*: ⧖.

Unraveling the FFT circuit completely for $n = 8$ elements, we get Figure 2.10. Notice the following.

1. For n inputs there are $\log_2 n$ levels, each with n nodes, for a total of $n \log n$ operations.
2. The inputs are arranged in a peculiar order: $0, 4, 2, 6, 1, 5, 3, 7$.

Why? Recall that at the top level of recursion, we first bring up the even coefficients of the input and then move on to the odd ones. Then at the next level, the even coefficients of this first group (which therefore are multiples of 4, or equivalently, have zero as their two least significant bits) are brought up, and so on. To put it otherwise, the inputs are arranged by increasing *last* bit of the binary representation of their index, resolving ties by looking at the next more significant bit(s). The resulting order in binary, 000, 100, 010, 110, 001, 101, 011, 111, is the same as the natural one, 000, 001, 010, 011, 100, 101, 110, 111 *except the bits are mirrored!*

3. There is a unique path between each input a_j and each output $A(\omega^k)$.

This path is most easily described using the binary representations of j and k (shown in Figure 2.10). There are two edges out of each node, one going up (the 0-edge) and

Figure 2.10 The fast Fourier transform circuit.

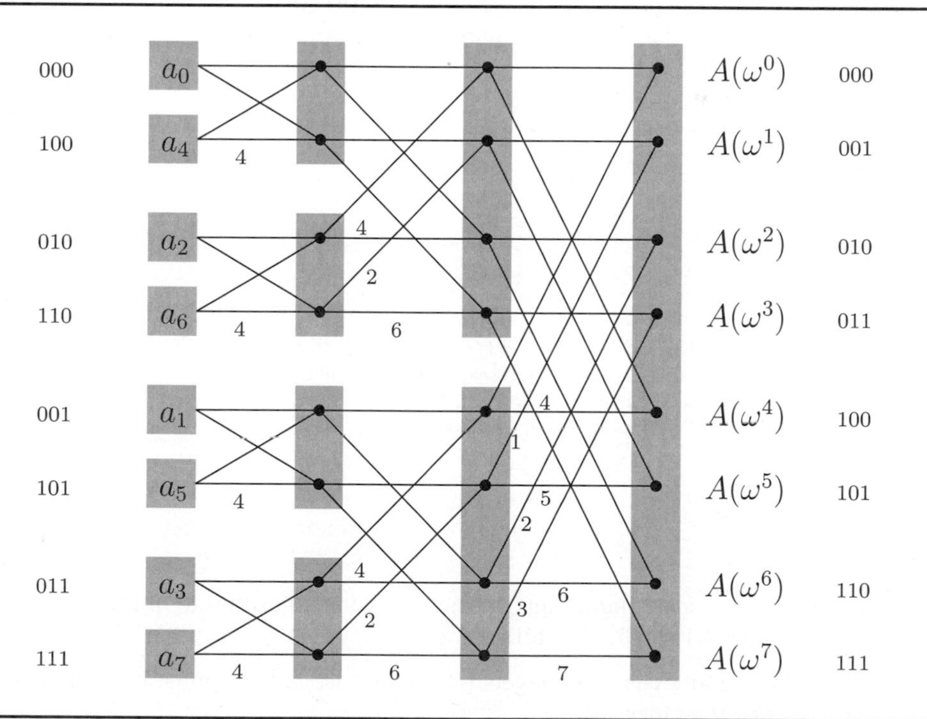

one going down (the 1-edge). To get to $A(\omega^k)$ from any input node, simply follow the edges specified in the bit representation of k, starting from the rightmost bit. (Can you similarly specify the path in the reverse direction?)

4. On the path between a_j and $A(\omega^k)$, the labels add up to $jk \bmod 8$.

Since $\omega^8 = 1$, this means that the contribution of input a_j to output $A(\omega^k)$ is $a_j\omega^{jk}$, and therefore the circuit computes correctly the values of polynomial $A(x)$.

5. And finally, notice that the FFT circuit is a natural for parallel computation and direct implementation in hardware.

The slow spread of a fast algorithm

In 1963, during a meeting of President Kennedy's scientific advisors, John Tukey, a mathematician from Princeton, explained to IBM's Dick Garwin a fast method for computing Fourier transforms. Garwin listened carefully, because he was at the time working on ways to detect nuclear explosions from seismographic data, and Fourier transforms were the bottleneck of his method. When he went back to IBM, he asked John Cooley to implement Tukey's algorithm; they decided that a paper should be published so that the idea could not be patented.

Tukey was not very keen to write a paper on the subject, so Cooley took the initiative. And this is how one of the most famous and most cited scientific papers was published in 1965, co-authored by Cooley and Tukey. The reason Tukey was reluctant to publish the FFT was not secretiveness or pursuit of profit via patents. He just felt that this was a simple observation that was probably already known. This was typical of the period: back then (and for some time later) algorithms were considered second-class mathematical objects, devoid of depth and elegance, and unworthy of serious attention.

But Tukey was right about one thing: it was later discovered that British engineers had used the FFT for hand calculations during the late 1930s. And—to end this chapter with the same great mathematician who started it—a paper by Gauss in the early 1800s on (what else?) interpolation contained essentially the same idea in it! Gauss's paper had remained a secret for so long because it was protected by an old-fashioned cryptographic technique: like most scientific papers of its era, it was written in Latin.

Exercises

2.1. Use the divide-and-conquer integer multiplication algorithm to multiply the two binary integers 10011011 and 10111010.

2.2. Show that for any positive integers n and any base b, there must be some power of b lying in the range $[n, bn]$.

2.3. Section 2.2 describes a method for solving recurrence relations which is based on analyzing the recursion tree and deriving a formula for the work done at each level. Another (closely related) method is to expand out the recurrence a few times, until a pattern emerges. For instance, let's start with the familiar $T(n) = 2T(n/2) + O(n)$. Think of $O(n)$ as being $\leq cn$ for some constant c, so: $T(n) \leq 2T(n/2) + cn$. By repeatedly applying this rule, we can bound $T(n)$ in terms of $T(n/2)$, then $T(n/4)$, then $T(n/8)$, and so on, at each step getting closer to the value of $T(\cdot)$ we do know, namely $T(1) = O(1)$.

$$\begin{aligned} T(n) &\leq 2T(n/2) + cn \\ &\leq 2[2T(n/4) + cn/2] + cn = 4T(n/4) + 2cn \\ &\leq 4[2T(n/8) + cn/4] + 2cn = 8T(n/8) + 3cn \\ &\leq 8[2T(n/16) + cn/8] + 3cn = 16T(n/16) + 4cn \\ &\ \vdots \end{aligned}$$

A pattern is emerging... the general term is

$$T(n) \leq 2^k T(n/2^k) + kcn.$$

Plugging in $k = \log_2 n$, we get $T(n) \leq nT(1) + cn\log_2 n = O(n \log n)$.

(a) Do the same thing for the recurrence $T(n) = 3T(n/2) + O(n)$. What is the general kth term in this case? And what value of k should be plugged in to get the answer?

(b) Now try the recurrence $T(n) = T(n-1) + O(1)$, a case which is not covered by the master theorem. Can you solve this too?

2.4. Suppose you are choosing between the following three algorithms:

- Algorithm A solves problems by dividing them into five subproblems of half the size, recursively solving each subproblem, and then combining the solutions in linear time.
- Algorithm B solves problems of size n by recursively solving two subproblems of size $n-1$ and then combining the solutions in constant time.
- Algorithm C solves problems of size n by dividing them into nine subproblems of size $n/3$, recursively solving each subproblem, and then combining the solutions in $O(n^2)$ time.

What are the running times of each of these algorithms (in big-O notation), and which would you choose?

2.5. Solve the following recurrence relations and give a Θ bound for each of them.

(a) $T(n) = 2T(n/3) + 1$

(b) $T(n) = 5T(n/4) + n$

(c) $T(n) = 7T(n/7) + n$

(d) $T(n) = 9T(n/3) + n^2$

(e) $T(n) = 8T(n/2) + n^3$
(f) $T(n) = 49T(n/25) + n^{3/2} \log n$
(g) $T(n) = T(n-1) + 2$
(h) $T(n) = T(n-1) + n^c$, where $c \geq 1$ is a constant
(i) $T(n) = T(n-1) + c^n$, where $c > 1$ is some constant
(j) $T(n) = 2T(n-1) + 1$
(k) $T(n) = T(\sqrt{n}) + 1$

2.6. A linear, time-invariant system has the following impulse response:

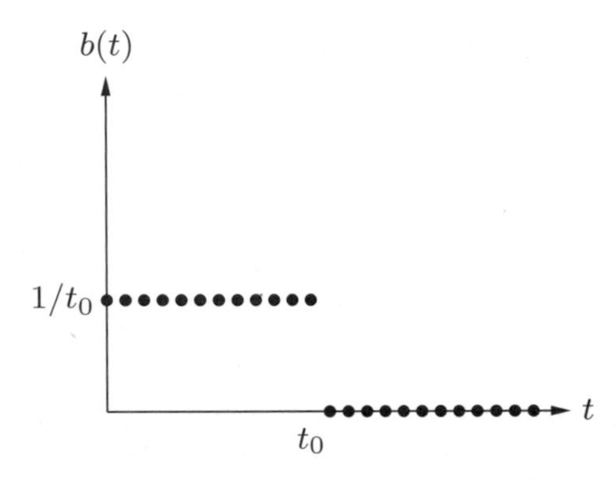

(a) Describe in words the effect of this system.
(b) What is the corresponding polynomial?

2.7. What is the sum of the nth roots of unity? What is their product if n is odd? If n is even?

2.8. Practice with the fast Fourier transform.

(a) What is the FFT of $(1, 0, 0, 0)$? What is the appropriate value of ω in this case? And of which sequence is $(1, 0, 0, 0)$ the FFT?
(b) Repeat for $(1, 0, 1, -1)$.

2.9. Practice with polynomial multiplication by FFT.

(a) Suppose that you want to multiply the two polynomials $x + 1$ and $x^2 + 1$ using the FFT. Choose an appropriate power of two, find the FFT of the two sequences, multiply the results componentwise, and compute the inverse FFT to get the final result.
(b) Repeat for the pair of polynomials $1 + x + 2x^2$ and $2 + 3x$.

2.10. Find the unique polynomial of degree 4 that takes on values $p(1) = 2$, $p(2) = 1$, $p(3) = 0$, $p(4) = 4$, and $p(5) = 0$. Write your answer in the coefficient representation.

2.11. In justifying our matrix multiplication algorithm (Section 2.5), we claimed the following blockwise property: if X and Y are $n \times n$ matrices, and

$$X = \begin{bmatrix} A & B \\ C & D \end{bmatrix}, \quad Y = \begin{bmatrix} E & F \\ G & H \end{bmatrix},$$

where A, B, C, D, E, F, G, and H are $n/2 \times n/2$ submatrices, then the product XY can be expressed in terms of these blocks:

$$XY = \begin{bmatrix} A & B \\ C & D \end{bmatrix} \begin{bmatrix} E & F \\ G & H \end{bmatrix} = \begin{bmatrix} AE + BG & AF + BH \\ CE + DG & CF + DH \end{bmatrix}.$$

Prove this property.

2.12. How many lines, as a function of n (in $\Theta(\cdot)$ form), does the following program print? Write a recurrence and solve it. You may assume n is a power of 2.

```
function f(n)
  if n > 1:
    print_line("still going")
    f(n/2)
    f(n/2)
```

2.13. A binary tree is *full* if all of its vertices have either zero or two children. Let B_n denote the number of full binary trees with n vertices.

(a) By drawing out all full binary trees with 3, 5, or 7 vertices, determine the exact values of B_3, B_5, and B_7. Why have we left out even numbers of vertices, like B_4?

(b) For general n, derive a recurrence relation for B_n.

(c) Show by induction that B_n is $2^{\Omega(n)}$.

2.14. You are given an array of n elements, and you notice that some of the elements are duplicates; that is, they appear more than once in the array. Show how to remove all duplicates from the array in time $O(n \log n)$.

2.15. In our median-finding algorithm (Section 2.4), a basic primitive is the `split` operation, which takes as input an array S and a value v and then divides S into three sets: the elements less than v, the elements equal to v, and the elements greater than v. Show how to implement this `split` operation *in place*, that is, without allocating new memory.

2.16. You are given an infinite array $A[\cdot]$ in which the first n cells contain integers in sorted order and the rest of the cells are filled with ∞. You are *not* given the value of n. Describe an algorithm that takes an integer x as input and finds a position in the array containing x, if such a position exists, in $O(\log n)$ time. (If you are disturbed by the fact that the array A has infinite length, assume instead that it is of length n, but that you don't know this length, and that the implementation of the array data type in your programming language returns the error message ∞ whenever elements $A[i]$ with $i > n$ are accessed.)

2.17. Given a sorted array of distinct integers $A[1, \ldots, n]$, you want to find out whether there is an index i for which $A[i] = i$. Give a divide-and-conquer algorithm that runs in time $O(\log n)$.

2.18. Consider the task of searching a sorted array $A[1 \ldots n]$ for a given element x: a task we usually perform by binary search in time $O(\log n)$. Show that any algorithm that accesses the array only via comparisons (that is, by asking questions of the form "is $A[i] \leq z$?"), must take $\Omega(\log n)$ steps.

2.19. *A k-way merge operation.* Suppose you have k sorted arrays, each with n elements, and you want to combine them into a single sorted array of kn elements.

(a) Here's one strategy: Using the `merge` procedure from Section 2.3, merge the first two arrays, then merge in the third, then merge in the fourth, and so on. What is the time complexity of this algorithm, in terms of k and n?

(b) Give a more efficient solution to this problem, using divide-and-conquer.

2.20. Show that any array of integers $x[1 \ldots n]$ can be sorted in $O(n + M)$ time, where

$$M = \max_i x_i - \min_i x_i.$$

For small M, this is linear time: why doesn't the $\Omega(n \log n)$ lower bound apply in this case?

2.21. *Mean and median.* One of the most basic tasks in statistics is to summarize a set of observations $\{x_1, x_2, \ldots, x_n\} \subseteq \mathbb{R}$ by a single number. Two popular choices for this summary statistic are:

- The median, which we'll call μ_1
- The mean, which we'll call μ_2

(a) Show that the median is the value of μ that minimizes the function

$$\sum_i |x_i - \mu|.$$

You can assume for simplicity that n is odd. (*Hint:* Show that for any $\mu \neq \mu_1$, the function decreases if you move μ either slightly to the left or slightly to the right.)

(b) Show that the mean is the value of μ that minimizes the function

$$\sum_i (x_i - \mu)^2.$$

One way to do this is by calculus. Another method is to prove that for any $\mu \in \mathbb{R}$,

$$\sum_i (x_i - \mu)^2 = \sum_i (x_i - \mu_2)^2 + n(\mu - \mu_2)^2.$$

Notice how the function for μ_2 penalizes points that are far from μ much more heavily than the function for μ_1. Thus μ_2 tries much harder to be close to *all* the observations. This might sound like a good thing at some level, but it is statistically undesirable because just a few outliers can severely throw off the estimate of μ_2. It is therefore sometimes said that μ_1 is a more robust estimator than μ_2. Worse than either of them, however, is μ_∞, the value of μ that minimizes the function

$$\max_i |x_i - \mu|.$$

(c) Show that μ_∞ can be computed in $O(n)$ time (assuming the numbers x_i are small enough that basic arithmetic operations on them take unit time).

2.22. You are given two sorted lists of size m and n. Give an $O(\log m + \log n)$ time algorithm for computing the kth smallest element in the union of the two lists.

2.23. An array $A[1 \ldots n]$ is said to have a *majority element* if more than half of its entries are the same. Given an array, the task is to design an efficient algorithm to tell whether the array has a majority element, and, if so, to find that element. The elements of the array are not necessarily from some ordered domain like the integers, and so there can be no comparisons of the form "is $A[i] > A[j]$?". (Think of the array elements as GIF files, say.) However you *can* answer questions of the form: "is $A[i] = A[j]$?" in constant time.

(a) Show how to solve this problem in $O(n \log n)$ time. (*Hint:* Split the array A into two arrays A_1 and A_2 of half the size. Does knowing the majority elements of A_1 and A_2 help you figure out the majority element of A? If so, you can use a divide-and-conquer approach.)

(b) Can you give a linear-time algorithm? (*Hint:* Here's another divide-and-conquer approach:

- Pair up the elements of A arbitrarily, to get $n/2$ pairs
- Look at each pair: if the two elements are different, discard both of them; if they are the same, keep just one of them

Show that after this procedure there are at most $n/2$ elements left, and that they have a majority element if A does.)

2.24. On page 56 there is a high-level description of the quicksort algorithm.

(a) Write down the pseudocode for quicksort.

(b) Show that its *worst-case* running time on an array of size n is $\Theta(n^2)$.

(c) Show that its *expected* running time satisfies the recurrence relation

$$T(n) \le O(n) + \frac{1}{n}\sum_{i=0}^{n-1}(T(i) + T(n-i)).$$

Then, show that the solution to this recurrence is $O(n \log n)$.

2.25. In Section 2.1 we described an algorithm that multiplies two n-bit binary integers x and y in time n^a, where $a = \log_2 3$. Call this procedure `fastmultiply` (x, y).

(a) We want to convert the decimal integer 10^n (a 1 followed by n zeros) into binary. Here is the algorithm (assume n is a power of 2):

```
function pwr2bin(n)
  if n=1: return 1010₂
  else:
    z=???
    return fastmultiply(z,z)
```

Fill in the missing details. Then give a recurrence relation for the running time of the algorithm, and solve the recurrence.

(b) Next, we want to convert any decimal integer x with n digits (where n is a power of 2) into binary. The algorithm is the following:

```
function dec2bin(x)
  if n=1: return binary[x]
  else:
    split x into two decimal numbers x_L, x_R with n/2
      digits each
    return ???
```

Here `binary[·]` is a vector that contains the binary representation of all one-digit integers. That is, `binary`$[0] = 0_2$, `binary`$[1] = 1_2$, up to `binary`$[9] = 1001_2$. Assume that a lookup in `binary` takes $O(1)$ time. Fill in the missing details. Once again, give a recurrence for the running time of the algorithm, and solve it.

2.26. Professor F. Lake tells his class that it is asymptotically faster to square an n-bit integer than to multiply two n-bit integers. Should they believe him?

2.27. The *square* of a matrix A is its product with itself, AA.

(a) Show that five multiplications are sufficient to compute the square of a 2×2 matrix.

(b) What is wrong with the following algorithm for computing the square of an $n \times n$ matrix?

> "Use a divide-and-conquer approach as in Strassen's algorithm, except that instead of getting 7 subproblems of size $n/2$, we now get 5 subproblems of size $n/2$ thanks to part (a). Using the same analysis as in Strassen's algorithm, we can conclude that the algorithm runs in time $O(n^{\log_2 5})$."

(c) In fact, squaring matrices is no easier than matrix multiplication. Show that if $n \times n$ matrices can be squared in time $O(n^c)$, then any two $n \times n$ matrices can be multiplied in time $O(n^c)$.

2.28. The *Hadamard matrices* $H_0, H_1, H_2, \ldots$ are defined as follows:

- H_0 is the 1×1 matrix $[1]$

- For $k > 0$, H_k is the $2^k \times 2^k$ matrix

$$H_k = \left[\begin{array}{c|c} H_{k-1} & H_{k-1} \\ \hline H_{k-1} & -H_{k-1} \end{array}\right].$$

Show that if v is a column vector of length $n = 2^k$, then the matrix-vector product $H_k v$ can be calculated using $O(n \log n)$ operations. Assume that all the numbers involved are small enough that basic arithmetic operations like addition and multiplication take unit time.

2.29. Suppose we want to evaluate the polynomial $p(x) = a_0 + a_1 x + a_2 x^2 + \cdots + a_n x^n$ at point x.

(a) Show that the following simple routine, known as *Horner's rule*, does the job and leaves the answer in z.

```
z = a_n
for i = n − 1 downto 0:
    z = zx + a_i
```

(b) How many additions and multiplications does this routine use, as a function of n? Can you find a polynomial for which an alternative method is substantially better?

2.30. This problem illustrates how to do the Fourier Transform (FT) in modular arithmetic, for example, modulo 7.

(a) There is a number ω such that all the powers $\omega, \omega^2, \ldots, \omega^6$ are distinct (modulo 7). Find this ω, and show that $\omega + \omega^2 + \cdots + \omega^6 = 0$. (Interestingly, for any prime modulus there is such a number.)

(b) Using the matrix form of the FT, produce the transform of the sequence $(0, 1, 1, 1, 5, 2)$ modulo 7; that is, multiply this vector by the matrix $M_6(\omega)$, for the value of ω you found earlier. In the matrix multiplication, all calculations should be performed modulo 7.

(c) Write down the matrix necessary to perform the inverse FT. Show that multiplying by this matrix returns the original sequence. (Again all arithmetic should be performed modulo 7.)

(d) Now show how to multiply the polynomials $x^2 + x + 1$ and $x^3 + 2x - 1$ using the FT modulo 7.

2.31. In Section 1.2.3, we studied Euclid's algorithm for computing the *greatest common divisor* (gcd) of two positive integers: the largest integer which divides them both. Here we will look at an alternative algorithm based on divide-and-conquer.

(a) Show that the following rule is true.

$$\gcd(a, b) = \begin{cases} 2\gcd(a/2, b/2) & \text{if } a, b \text{ are even} \\ \gcd(a, b/2) & \text{if } a \text{ is odd, } b \text{ is even} \\ \gcd((a-b)/2, b) & \text{if } a, b \text{ are odd} \end{cases}$$

(b) Give an efficient divide-and-conquer algorithm for greatest common divisor.

(c) How does the efficiency of your algorithm compare to Euclid's algorithm if a and b are n-bit integers? (In particular, since n might be large you cannot assume that basic arithmetic operations like addition take constant time.)

2.32. In this problem we will develop a divide-and-conquer algorithm for the following geometric task.

CLOSEST PAIR
Input: A set of points in the plane, $\{p_1 = (x_1, y_1), \ldots, p_n = (x_n, y_n)\}$
Output: The closest pair of points: that is, the pair $p_i \neq p_j$ for which the distance between p_i and p_j, that is,

$$\sqrt{(x_i - x_j)^2 + (y_i - y_j)^2},$$

is minimized.

For simplicity, assume that n is a power of two, and that all the x-coordinates x_i are distinct, as are the y-coordinates.

Here's a high-level overview of the algorithm:

- Find a value x for which exactly half the points have $x_i < x$, and half have $x_i > x$. On this basis, split the points into two groups, L and R.
- Recursively find the closest pair in L and in R. Say these pairs are $p_L, q_L \in L$ and $p_R, q_R \in R$, with distances d_L and d_R respectively. Let d be the smaller of these two distances.
- It remains to be seen whether there is a point in L and a point in R that are less than distance d apart from each other. To this end, discard all points with $x_i < x - d$ or $x_i > x + d$ and sort the remaining points by y-coordinate.
- Now, go through this sorted list, and for each point, compute its distance to the *seven* subsequent points in the list. Let p_M, q_M be the closest pair found in this way.
- The answer is one of the three pairs $\{p_L, q_L\}, \{p_R, q_R\}, \{p_M, q_M\}$, whichever is closest.

(a) In order to prove the correctness of this algorithm, start by showing the following property: any square of size $d \times d$ in the plane contains at most four points of L.

(b) Now show that the algorithm is correct. The only case which needs careful consideration is when the closest pair is split between L and R.

(c) Write down the pseudocode for the algorithm, and show that its running time is given by the recurrence:

$$T(n) = 2T(n/2) + O(n \log n).$$

Show that the solution to this recurrence is $O(n \log^2 n)$.

(d) Can you bring the running time down to $O(n \log n)$?

2.33. Suppose you are given $n \times n$ matrices A, B, C and you wish to check whether $AB = C$. You can do this in $O(n^{\log_2 7})$ steps using Strassen's algorithm. In this question we will explore a much faster, $O(n^2)$ randomized test.

(a) Let $\mathbf{v}$ be an n-dimensional vector whose entries are randomly and independently chosen to be 0 or 1 (each with probablity 1/2). Prove that if M is a non-zero $n \times n$ matrix, then $\Pr[M\mathbf{v} = 0] \leq 1/2$.

(b) Show that $\Pr[AB\mathbf{v} = C\mathbf{v}] \leq 1/2$ if $AB \neq C$. Why does this give an $O(n^2)$ randomized test for checking whether $AB = C$?

2.34. *Linear 3SAT*. The 3SAT problem is defined in Section 8.1. Briefly, the input is a Boolean formula—expressed as a set of clauses—over some set of variables and the goal is to determine whether there is an assignment (of `true`/`false` values) to these variables that makes the entire formula evaluate to `true`.

Consider a 3SAT instance with the following special locality property. Suppose there are n variables in the Boolean formula, and that they are numbered $1,2,\ldots,n$ in such a way that each clause involves variables whose numbers are within ± 10 of each other. Give a linear-time algorithm for solving such an instance of 3SAT.

Chapter 3

Decompositions of graphs

3.1 Why graphs?

A wide range of problems can be expressed with clarity and precision in the concise pictorial language of graphs. For instance, consider the task of coloring a political map. What is the minimum number of colors needed, with the obvious restriction that neighboring countries should have different colors? One of the difficulties in attacking this problem is that the map itself, even a stripped-down version like Figure 3.1(a), is usually cluttered with irrelevant information: intricate boundaries, border posts where three or more countries meet, open seas, and meandering rivers. Such distractions are absent from the mathematical object of Figure 3.1(b), a graph with one *vertex* for each country (1 is Brazil, 11 is Argentina) and *edges* between neighbors. It contains exactly the information needed for coloring, and nothing more. The precise goal is now to assign a color to each vertex so that no edge has endpoints of the same color.

Graph coloring is not the exclusive domain of map designers. Suppose a university needs to schedule examinations for all its classes and wants to use the fewest time slots possible. The only constraint is that two exams cannot be scheduled concurrently if some student will be taking both of them. To express this problem as a graph, use one vertex for each exam and put an edge between two vertices if there is a conflict, that is, if there is somebody taking both endpoint exams. Think of each time slot as having its own color. Then, assigning time slots is exactly the same as coloring this graph!

Some basic operations on graphs arise with such frequency, and in such a diversity of contexts, that a lot of effort has gone into finding efficient procedures for them. This chapter is devoted to some of the most fundamental of these algorithms—those that uncover the basic connectivity structure of a graph.

Formally, a graph is specified by a set of vertices (also called *nodes*) V and by edges E between select pairs of vertices. In the map example, $V = \{1, 2, 3, \ldots, 13\}$ and E includes, among many other edges, $\{1, 2\}$, $\{9, 11\}$, and $\{7, 13\}$. Here an edge between x and y specifically means "x shares a border with y." This is a symmetric

Figure 3.1 (a) A map and (b) its graph.

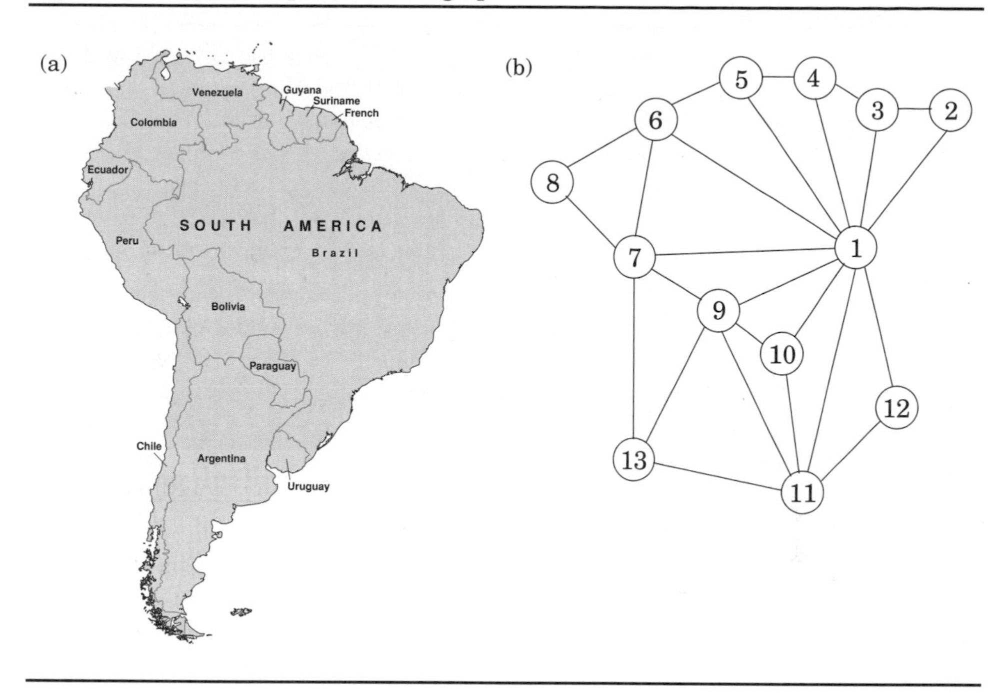

relation—it implies also that y shares a border with x—and we denote it using set notation, $e = \{x, y\}$. Such edges are *undirected* and are part of an *undirected graph*.

Sometimes graphs depict relations that do not have this reciprocity, in which case it is necessary to use edges with directions on them. There can be *directed edges e from x to y* (written $e = (x, y)$), or from y to x (written (y, x)), or both. A particularly enormous example of a *directed graph* is the graph of all links in the World Wide Web. It has a vertex for each site on the Internet, and a directed edge (u, v) whenever site u has a link to site v: in total, billions of nodes and edges! Understanding even the most basic connectivity properties of the Web is of great economic and social interest. Although the size of this problem is daunting, we will soon see that a lot of valuable information about the structure of a graph can, happily, be determined in just linear time.

3.1.1 How is a graph represented?

We can represent a graph by an *adjacency matrix*; if there are $n = |V|$ vertices $v_1, \ldots, v_n$, this is an $n \times n$ array whose (i, j)th entry is

$$a_{ij} = \begin{cases} 1 & \text{if there is an edge from } v_i \text{ to } v_j \\ 0 & \text{otherwise.} \end{cases}$$

For undirected graphs, the matrix is symmetric since an edge $\{u, v\}$ can be taken in either direction.

How big is your graph?

Which of the two representations, adjacency matrix or adjacency list, is better? Well, it depends on the relationship between $|V|$, the number of nodes in the graph, and $|E|$, the number of edges. $|E|$ can be as small as $|V|$ (if it gets much smaller, then the graph degenerates— for example, has isolated vertices), or as large as $|V|^2$ (when all possible edges are present). When $|E|$ is close to the upper limit of this range, we call the graph *dense*. At the other extreme, if $|E|$ is close to $|V|$, the graph is *sparse*. As we shall see in this chapter and the next two chapters, *exactly where $|E|$ lies in this range is usually a crucial factor in selecting the right graph algorithm.*

Or, for that matter, in selecting the graph representation. If it is the World Wide Web graph that we wish to store in computer memory, we should think twice before using an adjacency matrix: at the time of writing, search engines know of about eight billion vertices of this graph, and hence the adjacency matrix would take up *dozens of millions of terabits*. Again at the time we write these lines, it is not clear that there is enough computer memory in the whole world to achieve this. (And waiting a few years until there *is* enough memory is unwise: the Web will grow too and will probably grow faster.)

With adjacency lists, representing the World Wide Web becomes feasible: there are only a few dozen billion hyperlinks in the Web, and each will occupy a few bytes in the adjacency list. You can carry a device that stores the result, a terabyte or two, in your pocket (it may soon fit in your earring, but by that time the Web will have grown too).

The reason why adjacency lists are so much more effective in the case of the World Wide Web is that the Web is very sparse: the average Web page has hyperlinks to only about half a dozen other pages, out of the billions of possibilities.

The biggest convenience of this format is that the presence of a particular edge can be checked in constant time, with just one memory access. On the other hand the matrix takes up $O(n^2)$ space, which is wasteful if the graph does not have very many edges.

An alternative representation, with size proportional to the number of edges, is the *adjacency list*. It consists of $|V|$ linked lists, one per vertex. The linked list for vertex u holds the names of vertices to which u has an outgoing edge—that is, vertices v for which $(u, v) \in E$. Therefore, each edge appears in exactly one of the linked lists if the graph is directed or two of the lists if the graph is undirected. Either way, the total size of the data structure is $O(|E|)$. Checking for a particular edge (u, v) is no longer constant time, because it requires sifting through u's adjacency list. But it is easy to iterate through all neighbors of a vertex (by running down the corresponding linked list), and, as we shall soon see, this turns out to be a very useful operation in graph algorithms. Again, for undirected graphs, this representation has a symmetry of sorts: v is in u's adjacency list if and only if u is in v's adjacency list.

Figure 3.2 Exploring a graph is rather like navigating a maze.

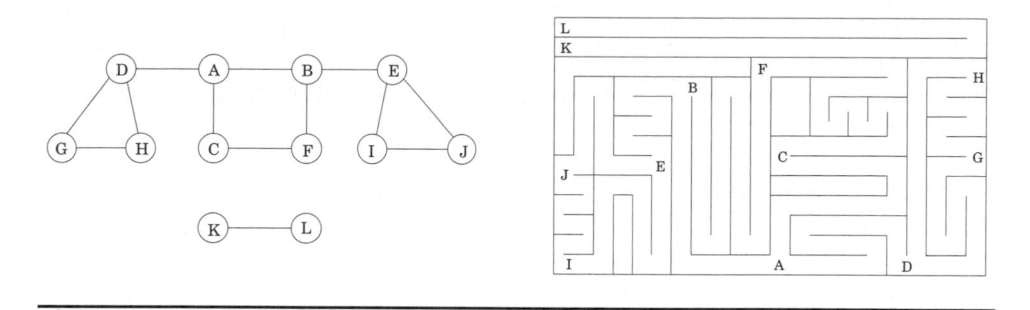

3.2 Depth-first search in undirected graphs

3.2.1 Exploring mazes

Depth-first search is a surprisingly versatile linear-time procedure that reveals a wealth of information about a graph. The most basic question it addresses is,

> *What parts of the graph are reachable from a given vertex?*

To understand this task, try putting yourself in the position of a computer that has just been given a new graph, say in the form of an adjacency list. This representation offers just one basic operation: finding the neighbors of a vertex. With only this primitive, the reachability problem is rather like exploring a labyrinth (Figure 3.2). You start walking from a fixed place and whenever you arrive at any junction (vertex) there are a variety of passages (edges) you can follow. A careless choice of passages might lead you around in circles or might cause you to overlook some accessible part of the maze. Clearly, you need to record some intermediate information during exploration.

This classic challenge has amused people for centuries. Everybody knows that all you need to explore a labyrinth is a ball of string and a piece of chalk. The chalk prevents looping, by marking the junctions you have already visited. The string always takes you back to the starting place, enabling you to return to passages that you previously saw but did not yet investigate.

How can we simulate these two primitives, chalk and string, on a computer? The chalk marks are easy: for each vertex, maintain a Boolean variable indicating whether it has been visited already. As for the ball of string, the correct cyberanalog is a *stack*. After all, the exact role of the string is to offer two primitive operations—*unwind* to get to a new junction (the stack equivalent is to *push* the new vertex) and *rewind* to return to the previous junction (*pop* the stack).

Instead of explicitly maintaining a stack, we will do so implicitly via recursion (which is implemented using a stack of activation records). The resulting algorithm

Figure 3.3 Finding all nodes reachable from a particular node.

```
procedure explore(G, v)
Input: G = (V, E) is a graph; v ∈ V
Output: visited(u) is set to true for all nodes u reachable
   from v

visited(v) = true
previsit(v)
for each edge (v, u) ∈ E:
   if not visited(u): explore(u)
postvisit(v)
```

is shown in Figure 3.3.[1] The `previsit` and `postvisit` procedures are optional, meant for performing operations on a vertex when it is first discovered and also when it is being left for the last time. We will soon see some creative uses for them.

More immediately, we need to confirm that `explore` always works correctly. It certainly does not venture too far, because it only moves from nodes to their neighbors and can therefore never jump to a region that is not reachable from v. But does it find *all* vertices reachable from v? Well, if there is some u that it misses, choose any path from v to u, and look at the last vertex on that path that the procedure actually visited. Call this node z, and let w be the node immediately after it on the same path.

So z was visited but w was not. This is a contradiction: while the `explore` procedure was at node z, it would have noticed w and moved on to it.

Incidentally, this pattern of reasoning arises often in the study of graphs and is in essence a streamlined induction. A more formal inductive proof would start by framing a hypothesis, such as "for any $k \geq 0$, all nodes within k hops from v get visited." The base case is as usual trivial, since v is certainly visited. And the general case—showing that if all nodes k hops away are visited, then so are all nodes $k+1$ hops away—is precisely the same point we just argued.

Figure 3.4 shows the result of running `explore` on our earlier example graph, starting at node A, and breaking ties in alphabetical order whenever there is a choice of nodes to visit. The solid edges are those that were actually traversed, each of which was elicited by a call to `explore` and led to the discovery of a new vertex.

[1] As with many of our graph algorithms, this one applies to both undirected and directed graphs. In such cases, we adopt the *directed* notation for edges, (x, y). If the graph is undirected, then each of its edges should be thought of as existing in both directions: (x, y) and (y, x).

Figure 3.4 The result of explore(A) on the graph of Figure 3.2.

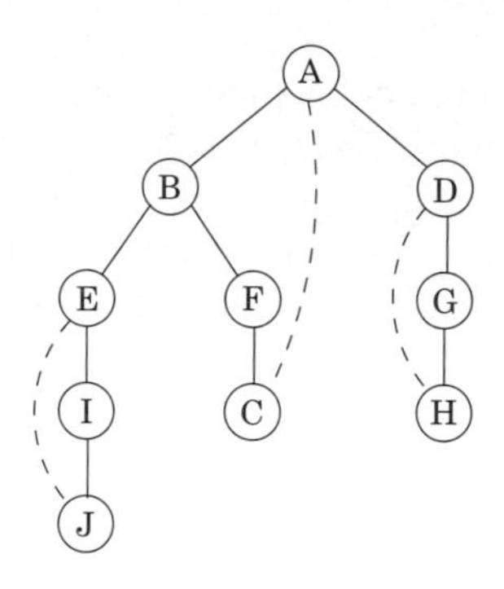

For instance, while B was being visited, the edge $B - E$ was noticed and, since E was as yet unknown, was traversed via a call to explore(E). These solid edges form a tree (a connected graph with no cycles) and are therefore called *tree edges*. The dotted edges were ignored because they led back to familiar terrain, to vertices previously visited. They are called *back edges*.

3.2.2 Depth-first search

The explore procedure visits only the portion of the graph reachable from its starting point. To examine the rest of the graph, we need to restart the procedure elsewhere, at some vertex that has not yet been visited. The algorithm of Figure 3.5, called *depth-first search* (DFS), does this repeatedly until the entire graph has been traversed.

Figure 3.5 Depth-first search.

```
procedure dfs(G)

for all v ∈ V:
   visited(v) = false

for all v ∈ V:
   if not visited(v): explore(v)
```

The first step in analyzing the running time of DFS is to observe that each vertex is explore'd just once, thanks to the visited array (the chalk marks). During the exploration of a vertex, there are the following steps:

1. Some fixed amount of work—marking the spot as visited, and the pre/postvisit.

2. A loop in which adjacent edges are scanned, to see if they lead somewhere new.

This loop takes a different amount of time for each vertex, so let's consider all vertices together. The total work done in step 1 is then $O(|V|)$. In step 2, over the course of the entire DFS, each edge $\{x, y\} \in E$ is examined exactly *twice*, once during `explore(x)` and once during `explore(y)`. The overall time for step 2 is therefore $O(|E|)$ and so the depth-first search has a running time of $O(|V| + |E|)$, linear in the size of its input. This is as efficient as we could possibly hope for, since it takes this long even just to read the adjacency list.

Figure 3.6 (a) A 12-node graph. (b) DFS search forest.

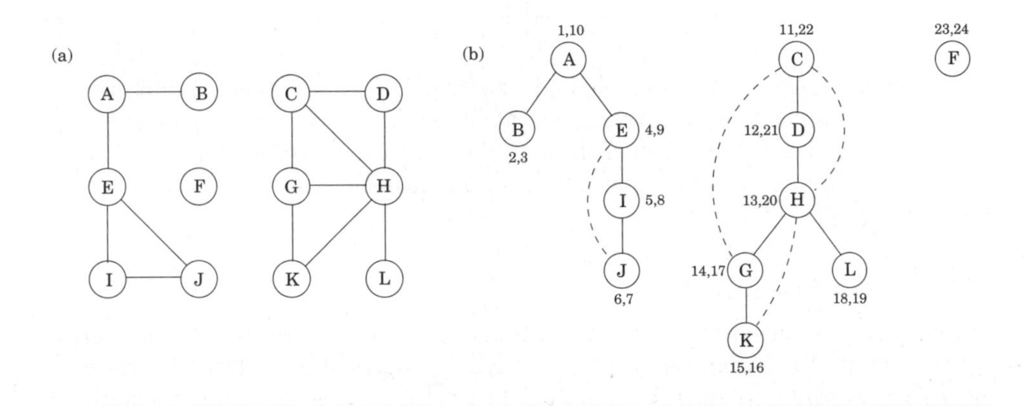

Figure 3.6 shows the outcome of depth-first search on a 12-node graph, once again breaking ties alphabetically (ignore the pairs of numbers for the time being). The outer loop of DFS calls `explore` three times, on A, C, and finally F. As a result, there are three trees, each rooted at one of these starting points. Together they constitute a *forest*.

3.2.3 Connectivity in undirected graphs

An undirected graph is *connected* if there is a path between any pair of vertices. The graph of Figure 3.6 is *not* connected because, for instance, there is no path from A to K. However, it does have three disjoint connected regions, corresponding to the following sets of vertices:

$$\{A, B, E, I, J\} \qquad \{C, D, G, H, K, L\} \qquad \{F\}.$$

These regions are called *connected components*: each of them is a subgraph that is internally connected but has no edges to the remaining vertices. When `explore` is started at a particular vertex, it identifies precisely the connected component containing that vertex. And each time the DFS outer loop calls `explore`, a new connected component is picked out.

Thus depth-first search is trivially adapted to check if a graph is connected and, more generally, to assign each node v an integer `ccnum[`v`]` identifying the connected component to which it belongs. All it takes is

```
procedure previsit(v)
ccnum[v] = cc
```

where `cc` needs to be initialized to zero and to be incremented each time the DFS procedure calls `explore`.

3.2.4 Previsit and postvisit orderings

We have seen how depth-first search—a few unassuming lines of code—is able to uncover the connectivity structure of an undirected graph in just linear time. But it is far more versatile than this. In order to stretch it further, we will collect a little more information during the exploration process: for each node, we will note down the times of two important events, the moment of first discovery (corresponding to `previsit`) and that of final departure (`postvisit`). Figure 3.6 shows these numbers for our earlier example, in which there are 24 events. The fifth event is the discovery of I. The 21st event consists of leaving D behind for good.

One way to generate arrays `pre` and `post` with these numbers is to define a simple counter `clock`, initially set to 1, which gets updated as follows.

```
procedure previsit(v)
pre[v] = clock
clock = clock + 1

procedure postvisit(v)
post[v] = clock
clock = clock + 1
```

These timings will soon take on larger significance. Meanwhile, you might have noticed from Figure 3.4 that:

Property *For any nodes u and v, the two intervals* [*pre*(u), *post*(u)] *and* [*pre*(v), *post*(v)] *are either disjoint or one is contained within the other.*

Why? Because [pre(u), post(u)] is essentially the time during which vertex u was on the stack. The last-in, first-out behavior of a stack explains the rest.

3.3 Depth-first search in directed graphs

3.3.1 Types of edges

Our depth-first search algorithm can be run verbatim on directed graphs, taking care to traverse edges only in their prescribed directions. Figure 3.7 shows an example and the search tree that results when vertices are considered in lexicographic order.

Figure 3.7 DFS on a directed graph.

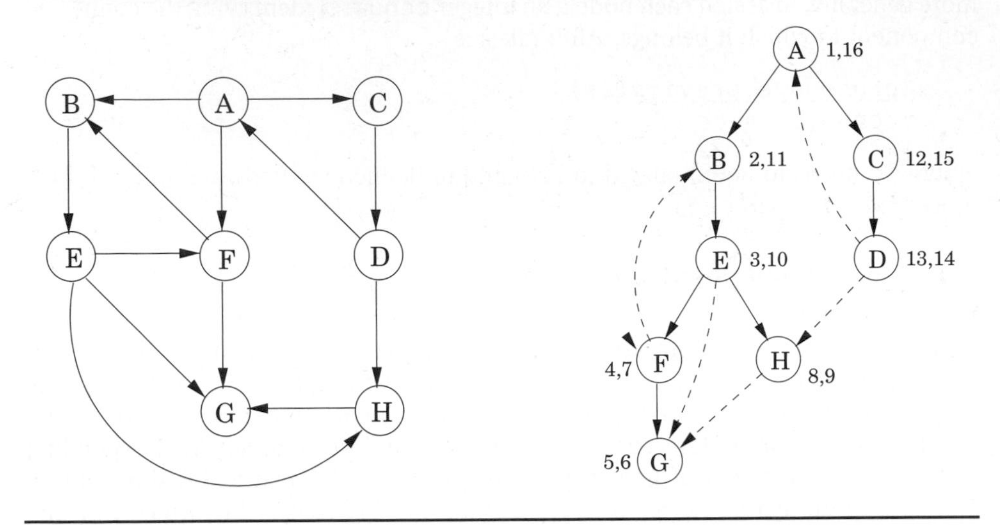

In further analyzing the directed case, it helps to have terminology for important relationships between nodes of a tree. A is the *root* of the search tree; everything else is its *descendant*. Similarly, E has descendants F, G, and H, and conversely, is an *ancestor* of these three nodes. The family analogy is carried further: C is the *parent* of D, which is its *child*.

For undirected graphs we distinguished between tree edges and nontree edges. In the directed case, there is a slightly more elaborate taxonomy:

DFS tree

A
B
C
D
Back
Forward
Tree
Cross

Tree edges are actually part of the DFS forest.

Forward edges lead from a node to a *nonchild* descendant in the DFS tree.

Back edges lead to an ancestor in the DFS tree.

Cross edges lead to neither descendant nor ancestor; they therefore lead to a node that has already been completely explored (that is, already postvisited).

Figure 3.7 has two forward edges, two back edges, and two cross edges. Can you spot them?

Ancestor and descendant relationships, as well as edge types, can be read off directly from `pre` and `post` numbers. Because of the depth-first exploration strategy, vertex u is an ancestor of vertex v exactly in those cases where u is discovered first and

v is discovered during explore(u). This is to say pre(u) < pre(v) < post(v) < post(u), which we can depict pictorially as two nested intervals:

$$\underset{u}{[}\quad\underset{v}{[}\quad\quad\underset{v}{]}\quad\underset{u}{]}$$

The case of descendants is symmetric, since u is a descendant of v if and only if v is an ancestor of u. And since edge categories are based entirely on ancestor-descendant relationships, it follows that they, too, can be read off from pre and post numbers. Here is a summary of the various possibilities for an edge (u, v):

pre/post *ordering for* (u, v)	*Edge type*
$\underset{u}{[}\ \underset{v}{[}\ \ \underset{v}{]}\ \underset{u}{]}$	Tree/forward
$\underset{v}{[}\ \underset{u}{[}\ \ \underset{u}{]}\ \underset{v}{]}$	Back
$\underset{v}{[}\ \ \underset{v}{]}\ \underset{u}{[}\ \ \underset{u}{]}$	Cross

You can confirm each of these characterizations by consulting the diagram of edge types. Do you see why no other orderings are possible?

3.3.2 Directed acyclic graphs

A *cycle* in a directed graph is a circular path $v_0 \to v_1 \to v_2 \to \cdots \to v_k \to v_0$. Figure 3.7 has quite a few of them, for example, $B \to E \to F \to B$. A graph without cycles is *acyclic*. It turns out we can test for acyclicity in linear time, with a single depth-first search.

Property *A directed graph has a cycle if and only if its depth-first search reveals a back edge.*

Proof. One direction is quite easy: if (u, v) is a back edge, then there is a cycle consisting of this edge together with the path from v to u in the search tree.

Conversely, if the graph has a cycle $v_0 \to v_1 \to \cdots \to v_k \to v_0$, look at the *first* node on this cycle to be discovered (the node with the lowest pre number). Suppose it is v_i. All the other v_j on the cycle are reachable from it and will therefore be its descendants in the search tree. In particular, the edge $v_{i-1} \to v_i$ (or $v_k \to v_0$ if $i = 0$) leads from a node to its ancestor and is thus by definition a back edge. ∎

Directed acyclic graphs, or *dags* for short, come up all the time. They are good for modeling relations like causalities, hierarchies, and temporal dependencies. For example, suppose that you need to perform many tasks, but some of them cannot begin until certain others are completed (you have to wake up before you can get

Figure 3.8 A directed acyclic graph with one source, two sinks, and four possible linearizations.

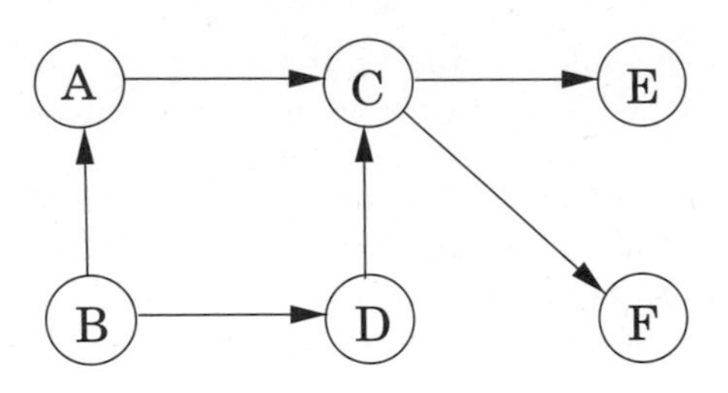

out of bed; you have to be out of bed, but not yet dressed, to take a shower; and so on). The question then is, what is a valid order in which to perform the tasks?

Such constraints are conveniently represented by a directed graph in which each task is a node, and there is an edge from u to v if u is a precondition for v. In other words, before performing a task, all the tasks pointing to it must be completed. If this graph has a cycle, there is no hope: no ordering can possibly work. If on the other hand the graph is a dag, we would like if possible to *linearize* (or *topologically sort*) it, to order the vertices one after the other in such a way that each edge goes from an earlier vertex to a later vertex, so that all precedence constraints are satisfied. In Figure 3.8, for instance, one valid ordering is B, A, D, C, E, F. (Can you spot the other three?)

What types of dags can be linearized? Simple: *All of them*. And once again depth-first search tells us exactly how to do it: simply perform tasks in *decreasing* order of their `post` numbers. After all, the only edges (u, v) in a graph for which $\texttt{post}(u) < \texttt{post}(v)$ are back edges (recall the table of edge types on page 88)—and we have seen that a dag cannot have back edges. Therefore:

Property *In a dag, every edge leads to a vertex with a lower* `post` *number.*

This gives us a linear-time algorithm for ordering the nodes of a dag. And, together with our earlier observations, it tells us that three rather different-sounding properties—acyclicity, linearizability, and the absence of back edges during a depth-first search—are in fact one and the same thing.

Since a dag is linearized by decreasing `post` numbers, the vertex with the smallest `post` number comes last in this linearization, and it must be a *sink*—no outgoing edges. Symmetrically, the one with the highest `post` is a *source*, a node with no incoming edges.

Property *Every dag has at least one source and at least one sink.*

The guaranteed existence of a source suggests an alternative approach to linearization:

Find a source, output it, and delete it from the graph.

Repeat until the graph is empty.

Can you see why this generates a valid linearization for any dag? What happens if the graph has cycles? And, how can this algorithm be implemented in linear time? (Exercise 3.14.)

3.4 Strongly connected components

3.4.1 Defining connectivity for directed graphs

Connectivity in undirected graphs is pretty straightforward: a graph that is not connected can be decomposed in a natural and obvious manner into several connected components (Figure 3.6 is a case in point). As we saw in Section 3.2.3, depth-first search does this handily, with each restart marking a new connected component.

In directed graphs, connectivity is more subtle. In some primitive sense, the directed graph of Figure 3.9(a) is "connected"—it can't be "pulled apart," so to speak, without breaking edges. But this notion is hardly interesting or informative. The graph cannot be considered connected, because for instance there is no path from G to B or from F to A. The right way to define connectivity for directed graphs is this:

Two nodes u and v of a directed graph are connected if there is a path from u to v and a path from v to u.

Figure 3.9 (a) A directed graph and its strongly connected components. (b) The meta-graph.

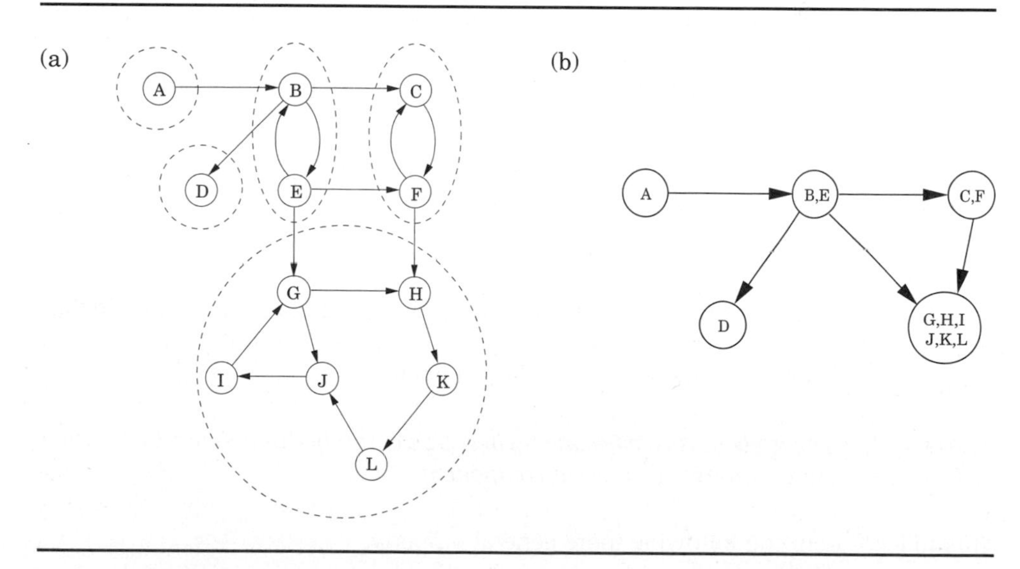

This relation partitions V into disjoint sets (Exercise 3.30) that we call *strongly connected components*. The graph of Figure 3.9(a) has five of them.

Now shrink each strongly connected component down to a single meta-node, and draw an edge from one meta-node to another if there is an edge (in the same direction) between their respective components (Figure 3.9(b)). The resulting *meta-graph* must be a dag. The reason is simple: a cycle containing several strongly connected components would merge them all into a single, strongly connected component. Restated,

Property *Every directed graph is a dag of its strongly connected components.*

This tells us something important: The connectivity structure of a directed graph is two-tiered. At the top level we have a dag, which is a rather simple structure—for instance, it can be linearized. If we want finer detail, we can look inside one of the nodes of this dag and examine the full-fledged strongly connected component within.

3.4.2 An efficient algorithm

The decomposition of a directed graph into its strongly connected components is very informative and useful. It turns out, fortunately, that it can be found in linear time by making further use of depth-first search. The algorithm is based on some properties we have already seen but which we will now pinpoint more closely.

Property 1 *If the* `explore` *subroutine is started at node u, then it will terminate precisely when all nodes reachable from u have been visited.*

Therefore, if we call `explore` on a node that lies somewhere in a *sink* strongly connected component (a strongly connected component that is a sink in the meta-graph), then we will retrieve exactly that component. Figure 3.9 has two sink strongly connected components. Starting `explore` at node K, for instance, will completely traverse the larger of them and then stop.

This suggests a way of finding one strongly connected component, but still leaves open two major problems: (A) how do we find a node that we know for sure lies in a sink strongly connected component and (B) how do we continue once this first component has been discovered?

Let's start with problem (A). There is not an easy, direct way to pick out a node that is guaranteed to lie in a sink strongly connected component. But there is a way to get a node in a *source* strongly connected component.

Property 2 *The node that receives the highest* `post` *number in a depth-first search must lie in a source strongly connected component.*

This follows from the following more general property.

Figure 3.10 The *reverse* of the graph from Figure 3.9.

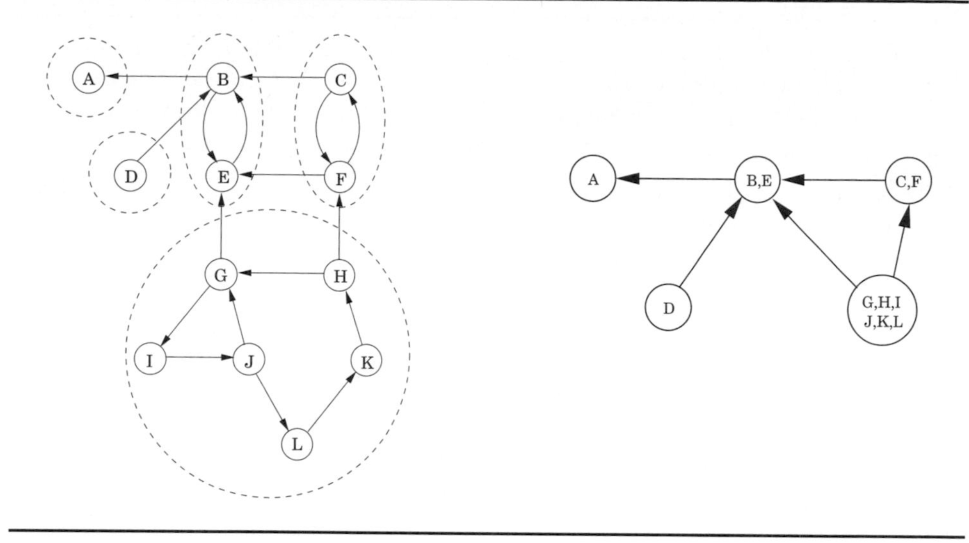

Property 3 *If C and C' are strongly connected components, and there is an edge from a node in C to a node in C', then the highest* `post` *number in C is bigger than the highest* `post` *number in C'.*

Proof. In proving Property 3, there are two cases to consider. If the depth-first search visits component C before component C', then clearly all of C and C' will be traversed before the procedure gets stuck (see Property 1). Therefore the first node visited in C will have a higher `post` number than any node of C'. On the other hand, if C' gets visited first, then the depth-first search will get stuck after seeing all of C' but before seeing any of C, in which case the property follows immediately. ∎

Property 3 can be restated as saying that *the strongly connected components can be linearized by arranging them in decreasing order of their highest* `post` *numbers.* This is a generalization of our earlier algorithm for linearizing dags; in a dag, each node is a singleton strongly connected component.

Property 2 helps us find a node in the source strongly connected component of G. However, what we need is a node in the *sink* component. Our means seem to be the opposite of our needs! But consider the *reverse* graph G^R, the same as G but with all edges reversed (Figure 3.10). G^R has exactly the same strongly connected components as G (why?). So, if we do a depth-first search of G^R, the node with the highest `post` number will come from a source strongly connected component in G^R, which is to say a sink strongly connected component in G. We have solved problem (A)!

Onward to problem (B). How do we continue after the first sink component is identified? The solution is also provided by Property 3. Once we have found the first strongly connected component and deleted it from the graph, the node with

Crawling fast

All this assumes that the graph is neatly given to us, with vertices numbered 1 to n and edges tucked in adjacency lists. The realities of the World Wide Web are very different. The nodes of the Web graph are not known in advance, and they have to be discovered one by one during the process of search. And, of course, recursion is out of the question.

Still, crawling the Web is done by algorithms very similar to depth-first search. An explicit stack is maintained, containing all nodes that have been discovered (as endpoints of hyperlinks) but not yet explored. In fact, this "stack" is not exactly a last-in, first-out list. It gives highest priority not to the nodes that were inserted most recently (nor the ones that were inserted earliest, that would be a *breadth-first search*, see Chapter 2), but to the ones that look most "interesting"—a heuristic criterion whose purpose is to keep the stack from overflowing and, in the worst case, to leave unexplored only nodes that are very unlikely to lead to vast new expanses.

In fact, crawling is typically done by many computers running `explore` simultaneously: each one takes the next node to be explored from the top of the stack, downloads the http file (the kind of Web files that point to each other), and scans it for hyperlinks. But when a new http document is found at the end of a hyperlink, no recursive calls are made: instead, the new vertex is inserted in the central stack.

But one question remains: When we see a "new" document, how do we know that it is indeed new, that we have not seen it before in our crawl? And how do we give it a *name*, so it can be inserted in the stack and recorded as "already seen"? The answer is *by hashing*.

Incidentally, researchers have run the strongly connected components algorithm on the Web and have discovered some very interesting structure.

the highest `post` number among those remaining will belong to a sink strongly connected component of whatever remains of G. Therefore we can keep using the `post` numbering from our initial depth-first search on G^R to successively output the second strongly connected component, the third strongly connected component, and so on. The resulting algorithm is this.

1. Run depth-first search on G^R.
2. Run the undirected connected components algorithm (from Section 3.2.3) on G, and during the depth-first search, process the vertices in decreasing order of their `post` numbers from step 1.

This algorithm is linear-time, only the constant in the linear term is about twice that of straight depth-first search. (Question: How does one construct an adjacency list representation of G^R in linear time? And how, in linear time, does one order the vertices of G by decreasing `post` values?)

Let's run this algorithm on the graph of Figure 3.9. If step 1 considers vertices in lexicographic order, then the ordering it sets up for the second step (namely, decreasing post numbers in the depth-first search of G^R) is: $G, I, J, L, K, H, D, C, F, B, E, A$. Then step 2 peels off components in the following sequence: $\{G, H, I, J, K, L\}, \{D\}, \{C, F\}, \{B, E\}, \{A\}$.

Exercises

3.1. Perform a depth-first search on the following graph; whenever there's a choice of vertices, pick the one that is alphabetically first. Classify each edge as a tree edge or back edge, and give the pre and post number of each vertex.

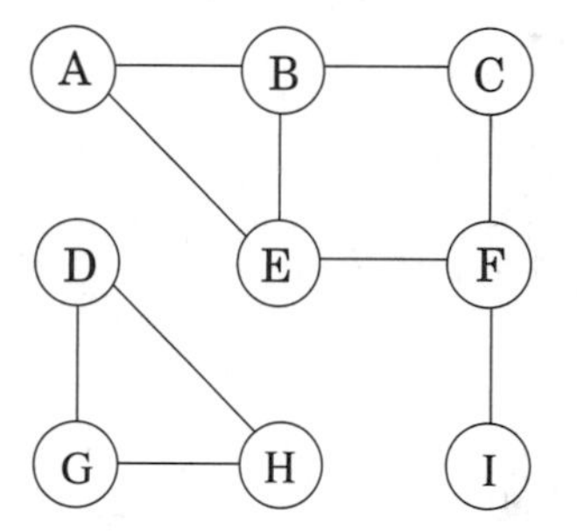

3.2. Perform depth-first search on each of the following graphs; whenever there's a choice of vertices, pick the one that is alphabetically first. Classify each edge as a tree edge, forward edge, back edge, or cross edge, and give the pre and post number of each vertex.

(a)

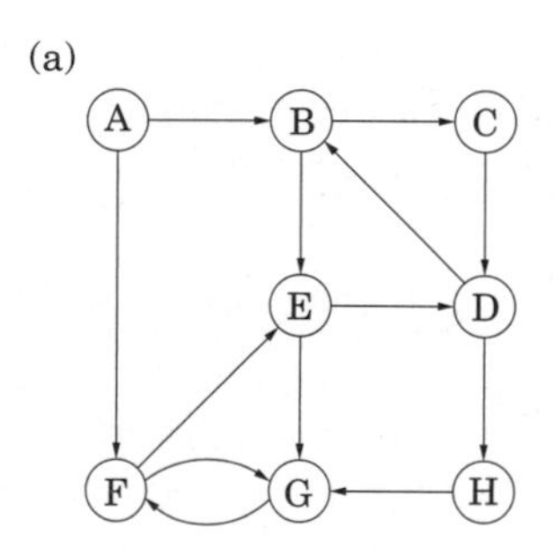

(b)

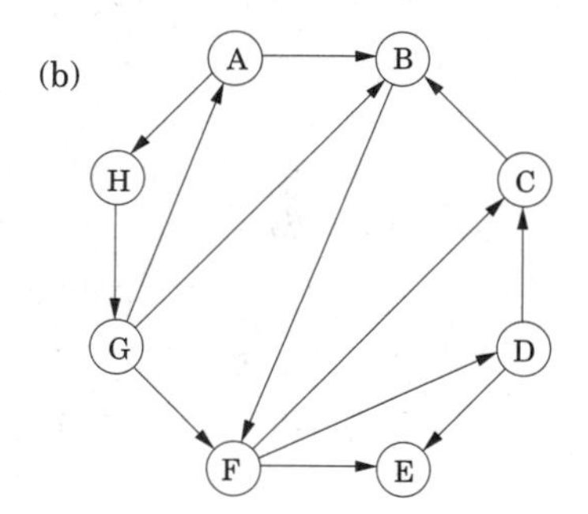

3.3. Run the DFS-based topological ordering algorithm on the following graph. Whenever you have a choice of vertices to explore, always pick the one that is alphabetically first.

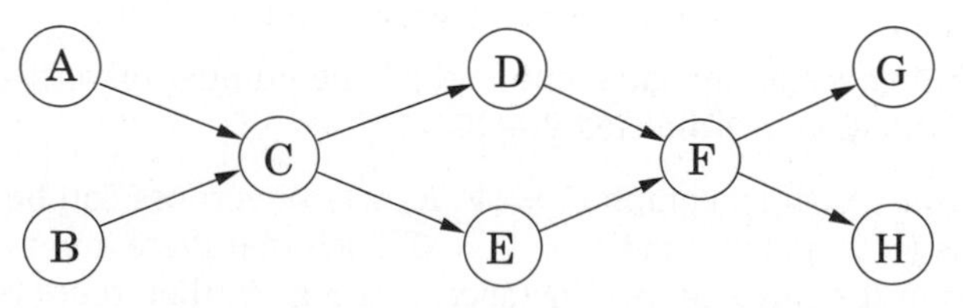

(a) Indicate the pre and post numbers of the nodes.

(b) What are the sources and sinks of the graph?

(c) What topological ordering is found by the algorithm?

(d) How many topological orderings does this graph have?

3.4. Run the strongly connected components algorithm on the following directed graphs G. When doing DFS on G^R: whenever there is a choice of vertices to explore, always pick the one that is alphabetically first.

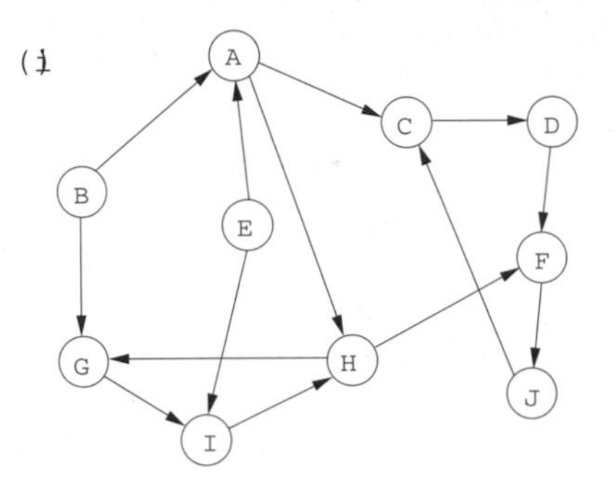

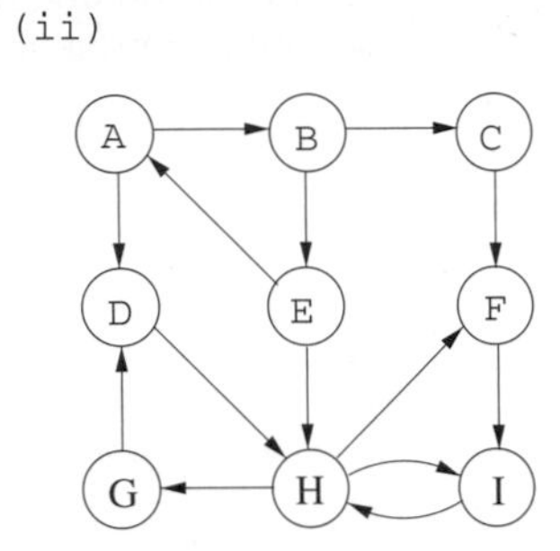

In each case answer the following questions.

(a) In what order are the strongly connected components (SCCs) found?

(b) Which are source SCCs and which are sink SCCs?

(c) Draw the "metagraph" (each meta-node is an SCC of G).

(d) What is the minimum number of edges you must add to this graph to make it strongly connected?

3.5. The *reverse* of a directed graph $G = (V, E)$ is another directed graph $G^R = (V, E^R)$ on the same vertex set, but with all edges reversed; that is, $E^R = \{(v, u) : (u, v) \in E\}$.

Give a linear-time algorithm for computing the reverse of a graph in adjacency list format.

3.6. In an undirected graph, the *degree* $d(u)$ of a vertex u is the number of neighbors u has, or equivalently, the number of edges incident upon it. In a directed graph, we distinguish between the *indegree* $d_{in}(u)$, which is the number of edges into u, and the *outdegree* $d_{out}(u)$, the number of edges leaving u.

(a) Show that in an undirected graph, $\sum_{u \in V} d(u) = 2|E|$.

(b) Use part (a) to show that in an undirected graph, there must be an even number of vertices whose degree is odd.

(c) Does a similar statement hold for the number of vertices with odd indegree in a directed graph?

3.7. A *bipartite graph* is a graph $G = (V, E)$ whose vertices can be partitioned into two sets ($V = V_1 \cup V_2$ and $V_1 \cap V_2 = \emptyset$) such that there are no edges between vertices in the same set (for instance, if $u, v \in V_1$, then there is no edge between u and v).

(a) Give a linear-time algorithm to determine whether an undirected graph is bipartite.

(b) There are many other ways to formulate this property. For instance, an undirected graph is bipartite if and only if it can be colored with just two colors.

Prove the following formulation: an undirected graph is bipartite if and only if it contains no cycles of odd length.

(c) At most how many colors are needed to color in an undirected graph with exactly *one* odd-length cycle?

3.8. *Pouring water.* We have three containers whose sizes are 10 pints, 7 pints, and 4 pints, respectively. The 7-pint and 4-pint containers start out full of water, but the 10-pint container is initially empty. We are allowed one type of operation: pouring the contents of one container into another, stopping only when the source container is empty or the destination container is full. We want to know if there is a sequence of pourings that leaves exactly 2 pints in the 7- or 4-pint container.

(a) Model this as a graph problem: give a precise definition of the graph involved and state the specific question about this graph that needs to be answered.

(b) What algorithm should be applied to solve the problem?

3.9. For each node u in an undirected graph, let `twodegree`$[u]$ be the sum of the degrees of u's neighbors. Show how to compute the entire array of `twodegree`$[\cdot]$ values in linear time, given a graph in adjacency list format.

3.10. Rewrite the `explore` procedure (Figure 3.3) so that it is non-recursive (that is, explicitly use a stack). The calls to `previsit` and `postvisit` should be positioned so that they have the same effect as in the recursive procedure.

3.11. Design a linear-time algorithm which, given an undirected graph G and a particular edge e in it, determines whether G has a cycle containing e.

3.12. Either prove or give a counterexample: if $\{u, v\}$ is an edge in an undirected graph, and during depth-first search `post`$(u) <$ `post`(v), then v is an ancestor of u in the DFS tree.

3.13. *Undirected vs. directed connectivity.*

(a) Prove that in any connected undirected graph $G = (V, E)$ there is a vertex $v \in V$ whose removal leaves G connected. (*Hint:* Consider the DFS search tree for G.)

(b) Give an example of a strongly connected directed graph $G = (V, E)$ such that, for every $v \in V$, removing v from G leaves a directed graph that is not strongly connected.

(c) In an undirected graph with 2 connected components it is always possible to make the graph connected by adding only one edge. Give an example of a directed graph with two strongly connected components such that no addition of one edge can make the graph strongly connected.

3.14. The chapter suggests an alternative algorithm for linearization (topological sorting), which repeatedly removes source nodes from the graph (page 90). Show that this algorithm can be implemented in linear time.

3.15. The police department in the city of Computopia has made all streets one-way. The mayor contends that there is still a way to drive legally from any intersection in the city to any other intersection, but the opposition is not convinced. A computer program is needed to determine whether the mayor is right. However, the city elections are coming up soon, and there is just enough time to run a *linear-time* algorithm.

(a) Formulate this problem graph-theoretically, and explain why it can indeed be solved in linear time.

(b) Suppose it now turns out that the mayor's original claim is false. She next claims something weaker: if you start driving from town hall, navigating one-way streets, then no matter where you reach, there is always a way to drive legally back to the town hall. Formulate this weaker property as a graph-theoretic problem, and carefully show how it too can be checked in linear time.

3.16. Suppose a CS curriculum consists of n courses, all of them mandatory. The prerequisite graph G has a node for each course, and an edge from course v to course w if and only if v is a prerequisite for w. Find an algorithm that works directly with this graph representation, and computes the minimum number of semesters necessary to complete the curriculum (assume that a student can take any number of courses in one semester). The running time of your algorithm should be linear.

3.17. *Infinite paths.* Let $G = (V, E)$ be a directed graph with a designated "start vertex" $s \in V$, a set $V_G \subseteq V$ of "good" vertices, and a set $V_B \subseteq V$ of "bad" vertices. An *infinite* trace p of G is an infinite sequence $v_0 v_1 v_2 \cdots$ of vertices $v_i \in V$ such that (1) $v_0 = s$, and (2) for all $i \geq 0$, $(v_i, v_{i+1}) \in E$. That is, p is an infinite path in G starting at vertex s. Since the set V of vertices is finite, every infinite trace of G must visit some vertices infinitely often.

(a) If p is an infinite trace, let $\mathit{Inf}(p) \subseteq V$ be the set of vertices that occur infinitely often in p. Show that $\mathit{Inf}(p)$ is a subset of a strongly connected component of G.

(b) Describe an algorithm that determines if G has an infinite trace.

(c) Describe an algorithm that determines if G has an infinite trace that visits some good vertex in V_G infinitely often.

(d) Describe an algorithm that determines if G has an infinite trace that visits some good vertex in V_G infinitely often, but visits no bad vertex in V_B infinitely often.

3.18. You are given a binary tree $T = (V, E)$ (in adjacency list format), along with a designated root node $r \in V$. Recall that u is said to be an *ancestor* of v in the rooted tree, if the path from r to v in T passes through u.

You wish to preprocess the tree so that queries of the form "is u an ancestor of v?" can be answered in constant time. The preprocessing itself should take linear time. How can this be done?

3.19. As in the previous problem, you are given a binary tree $T = (V, E)$ with designated root node. In addition, there is an array $x[\cdot]$ with a value for each node in V. Define a new array $z[\cdot]$ as follows: for each $u \in V$,

$$z[u] = \textit{the maximum of the x-values associated with u's descendants.}$$

Give a linear-time algorithm which calculates the entire z-array.

3.20. You are given a tree $T = (V, E)$ along with a designated root node $r \in V$. The *parent* of any node $v \neq r$, denoted $p(v)$, is defined to be the node adjacent to v in the path from r to v. By convention, $p(r) = r$. For $k > 1$, define $p^k(v) = p^{k-1}(p(v))$ and $p^1(v) = p(v)$ (so $p^k(v)$ is the kth ancestor of v).

Each vertex v of the tree has an associated non-negative integer label $l(v)$. Give a linear-time algorithm to update the labels of all the vertices in T according to the following rule: $l_{new}(v) = l(p^{l(v)}(v))$.

3.21. Give a linear-time algorithm to find an odd-length cycle in a *directed* graph. (*Hint:* First solve this problem under the assumption that the graph is strongly connected.)

3.22. Give an efficient algorithm which takes as input a directed graph $G = (V, E)$, and determines whether or not there is a vertex $s \in V$ from which all other vertices are reachable.

3.23. Give an efficient algorithm that takes as input a directed acyclic graph $G = (V, E)$, and two vertices $s, t \in V$, and outputs the number of different directed paths from s to t in G.

3.24. Give a linear-time algorithm for the following task.

Input: A directed acyclic graph G
Question: Does G contain a directed path that touches every vertex exactly once?

3.25. You are given a directed graph in which each node $u \in V$ has an associated *price* p_u which is a positive integer. Define the array `cost` as follows: for each $u \in V$,

`cost`$[u]$ = price of the cheapest node reachable from u (including u itself).

For instance, in the graph below (with prices shown for each vertex), the `cost` values of the nodes A, B, C, D, E, F are 2, 1, 4, 1, 4, 5, respectively.

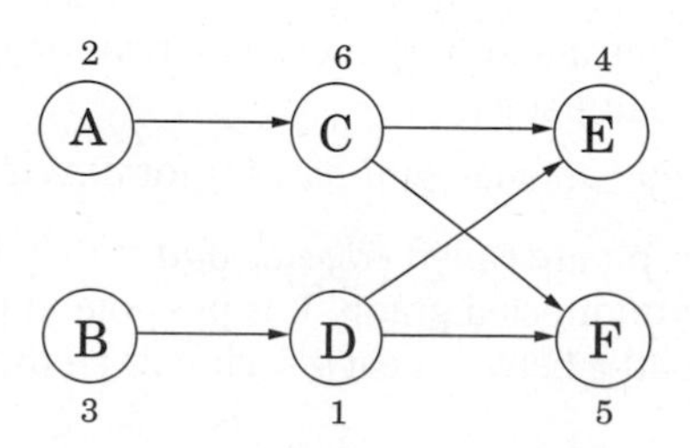

Your goal is to design an algorithm that fills in the *entire* `cost` array (i.e., for all vertices).

(a) Give a linear-time algorithm that works for directed *acyclic* graphs. (*Hint:* Handle the vertices in a particular *order*.)

(b) Extend this to a linear-time algorithm that works for all directed graphs. (*Hint:* Recall the "two-tiered" structure of directed graphs.)

3.26. An *Eulerian tour* in an undirected graph is a cycle that is allowed to pass through each vertex multiple times, but must use each edge exactly once.

This simple concept was used by Euler in 1736 to solve the famous Konigsberg bridge problem, which launched the field of graph theory. The city of Konigsberg (now called Kaliningrad, in western Russia) is the meeting point of two rivers with a small island in the middle. There are seven bridges across the rivers, and a popular recreational question of the time was to determine whether it is possible to perform a tour in which each bridge is crossed *exactly once*.

Euler formulated the relevant information as a graph with four nodes (denoting land masses) and seven edges (denoting bridges), as shown here.

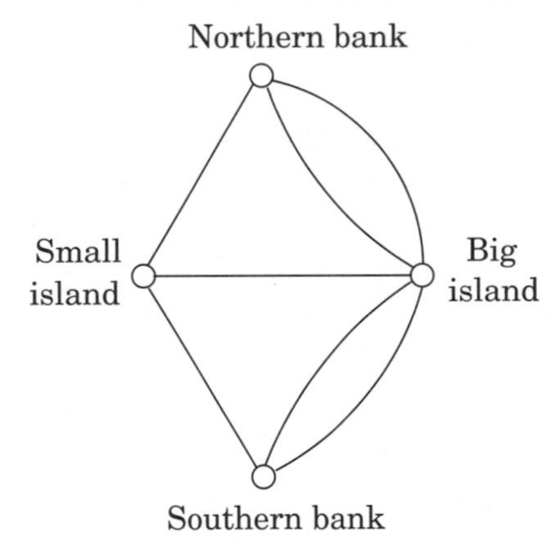

Notice an unusual feature of this problem: multiple edges between certain pairs of nodes.

(a) Show that an undirected graph has an Eulerian tour if and only if all its vertices have even degree. Conclude that there is no Eulerian tour of the Konigsberg bridges.

(b) An *Eulerian path* is a path which uses each edge exactly once. Can you give a similar if-and-only-if characterization of which undirected graphs have Eulerian paths?

(c) Can you give an analog of part (a) for *directed* graphs?

3.27. Two paths in a graph are called *edge-disjoint* if they have no edges in common. Show that in any undirected graph, it is possible to pair up the vertices of odd degree and find paths between each such pair so that all these paths are edge-disjoint.

3.28. In the 2SAT problem, you are given a set of *clauses*, where each clause is the disjunction (OR) of two literals (a literal is a Boolean variable or the negation of a Boolean variable). You are looking for a way to assign a value `true` or `false` to each of the variables so that *all* clauses are satisfied—that is, there is at least one true literal in each clause. For example, here's an instance of 2SAT:

$$(x_1 \vee \overline{x}_2) \wedge (\overline{x}_1 \vee \overline{x}_3) \wedge (x_1 \vee x_2) \wedge (\overline{x}_3 \vee x_4) \wedge (\overline{x}_1 \vee x_4).$$

This instance has a satisfying assignment: set x_1, x_2, x_3, and x_4 to `true`, `false`, `false`, and `true`, respectively.

(a) Are there other satisfying truth assignments of this 2SAT formula? If so, find them all.

(b) Give an instance of 2SAT with four variables, and with no satisfying assignment.

The purpose of this problem is to lead you to a way of solving 2SAT efficiently by reducing it to the problem of finding the strongly connected components of a directed graph. Given an instance I of 2SAT with n variables and m clauses, construct a directed graph $G_I = (V, E)$ as follows.

- G_I has $2n$ nodes, one for each variable and its negation.
- G_I has $2m$ edges: for each clause $(\alpha \vee \beta)$ of I (where α, β are literals), G_I has an edge from the negation of α to β, and one from the negation of β to α.

Note that the clause $(\alpha \vee \beta)$ is equivalent to either of the implications $\overline{\alpha} \Rightarrow \beta$ or $\overline{\beta} \Rightarrow \alpha$. In this sense, G_I records all implications in I.

(c) Carry out this construction for the instance of 2SAT given above, and for the instance you constructed in (b).

(d) Show that if G_I has a strongly connected component containing both x and $\overline{x}$ for some variable x, then I has no satisfying assignment.

(e) Now show the converse of (d): namely, that if none of G_I's strongly connected components contain both a literal and its negation, then the instance I must be satisfiable. (*Hint:* Assign values to the variables as follows: repeatedly pick a sink strongly connected component of G_I. Assign value `true` to all literals in the sink, assign `false` to their negations, and delete all of these. Show that this ends up discovering a satisfying assignment.)

(f) Conclude that there is a linear-time algorithm for solving 2SAT.

3.29. Let S be a finite set. A binary *relation* on S is simply a collection R of ordered pairs $(x, y) \in S \times S$. For instance, S might be a set of people, and each such pair $(x, y) \in R$ might mean "x knows y."

An *equivalence relation* is a binary relation which satisfies three properties:

- Reflexivity: $(x, x) \in R$ for all $x \in S$
- Symmetry: if $(x, y) \in R$ then $(y, x) \in R$
- Transitivity: if $(x, y) \in R$ and $(y, z) \in R$ then $(x, z) \in R$

For instance, the binary relation "has the same birthday as" is an equivalence relation, whereas "is the father of" is not, since it violates all three properties.

Show that an equivalence relation partitions set S into disjoint groups $S_1, S_2, \ldots, S_k$ (in other words, $S = S_1 \cup S_2 \cup \cdots \cup S_k$ and $S_i \cap S_j = \emptyset$ for all $i \neq j$) such that:

- Any two members of a group are related, that is, $(x, y) \in R$ for any $x, y \in S_i$, for any i.
- Members of different groups are not related, that is, for all $i \neq j$, for all $x \in S_i$ and $y \in S_j$, we have $(x, y) \notin R$.

(*Hint:* Represent an equivalence relation by an undirected graph.)

3.30. On page 91, we defined the binary relation "connected" on the set of vertices of a *directed* graph. Show that this is an equivalence relation (see Exercise 3.29), and conclude that it partitions the vertices into disjoint strongly connected components.

3.31. *Biconnected components*. Let $G = (V, E)$ be an undirected graph. For any two edges $e, e' \in E$, we'll say $e \sim e'$ if either $e = e'$ or there is a (simple) cycle containing both e and e'.

(a) Show that $\sim$ is an equivalence relation (recall Exercise 3.29) on the edges.

The equivalence classes into which this relation partitions the edges are called the *biconnected components* of G. A *bridge* is an edge which is in a biconnected component all by itself.

A *separating vertex* is a vertex whose removal disconnects the graph.

(b) Partition the edges of the graph below into biconnected components, and identify the bridges and separating vertices.

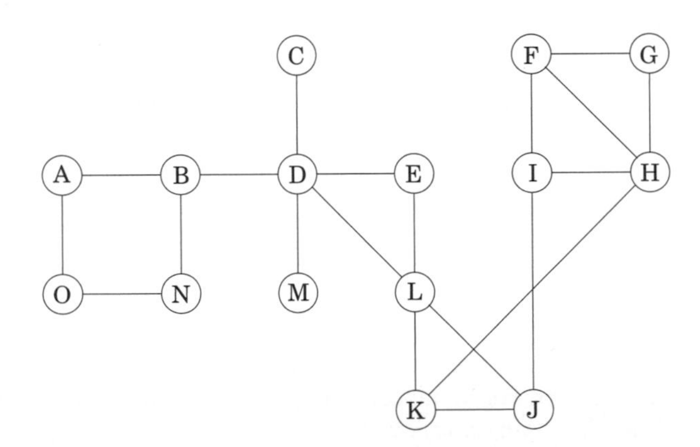

Not only do biconnected components partition the edges of the graph, they also *almost* partition the vertices in the following sense.

(c) Associate with each biconnected component all the vertices that are endpoints of its edges. Show that the vertices corresponding to two different biconnected components are either disjoint or intersect in a single separating vertex.

(d) Collapse each biconnected component into a single meta-node, and retain individual nodes for each separating vertex. (So there are edges between each component-node and its separating vertices.) Show that the resulting graph is a tree.

DFS can be used to identify the biconnected components, bridges, and separating vertices of a graph in linear time.

(e) Show that the root of the DFS tree is a separating vertex if and only if it has more than one child in the tree.

(f) Show that a non-root vertex v of the DFS tree is a separating vertex if and only if it has a child v' none of whose descendants (including itself) has a backedge to a proper ancestor of v.

(g) For each vertex u define:

$$\text{low}(u) = \min \begin{cases} \texttt{pre}(u) \\ \texttt{pre}(w) \text{ where } (v, w) \text{ is a backedge for} \\ \qquad\qquad \text{some descendant } v \text{ of } u \end{cases}$$

Show that the entire array of `low` values can be computed in linear time.

(h) Show how to compute all separating vertices, bridges, and biconnected components of a graph in linear time. (*Hint:* Use `low` to identify separating vertices, and run another DFS with an extra stack of edges to remove biconnected components one at a time.)

Chapter 4

Paths in graphs

4.1 Distances

Depth-first search readily identifies all the vertices of a graph that can be reached from a designated starting point. It also finds explicit paths to these vertices, summarized in its search tree (Figure 4.1). However, these paths might not be the most economical ones possible. In the figure, vertex C is reachable from S by traversing just one edge, while the DFS tree shows a path of length 3. This chapter is about algorithms for finding *shortest paths* in graphs.

Path lengths allow us to talk quantitatively about the extent to which different vertices of a graph are separated from each other:

> *The* distance *between two nodes is the length of the shortest path between them.*

To get a concrete feel for this notion, consider a physical realization of a graph that has a ball for each vertex and a piece of string for each edge. If you lift the ball for vertex s high enough, the other balls that get pulled up along with it are precisely the vertices reachable from s. And to find their distances from s, you need only measure how far below s they hang.

In Figure 4.2, for example, vertex B is at distance 2 from S, and there are two shortest paths to it. When S is held up, the strings along each of these paths become taut.

Figure 4.1 (a) A simple graph and (b) its depth-first search tree.

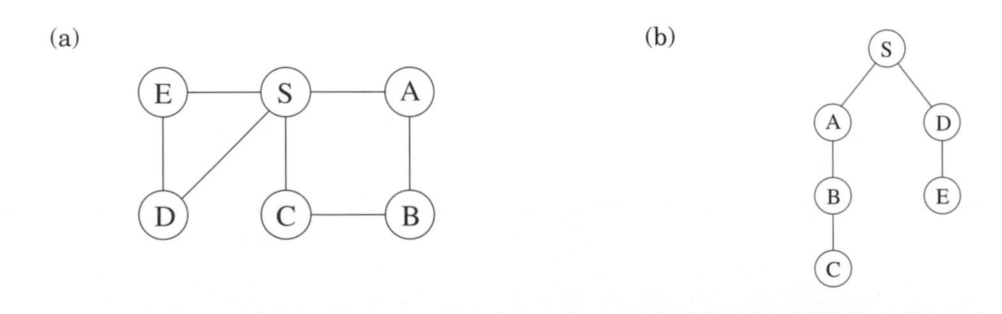

Figure 4.2 A physical model of a graph.

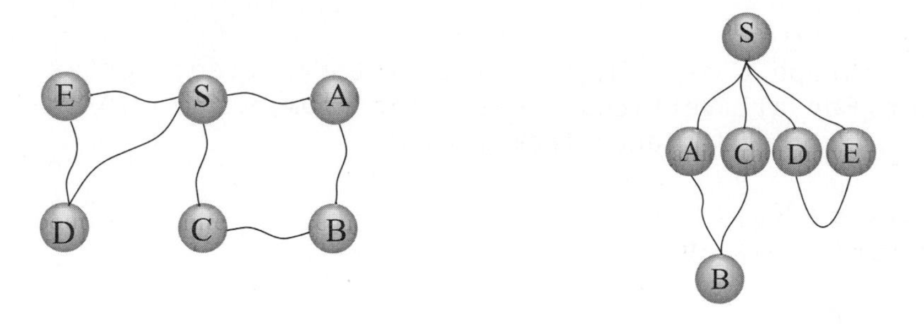

On the other hand, edge (D, E) plays no role in any shortest path and therefore remains slack.

4.2 Breadth-first search

In Figure 4.2, the lifting of s partitions the graph into layers: s itself, the nodes at distance 1 from it, the nodes at distance 2 from it, and so on. A convenient way to compute distances from s to the other vertices is to proceed layer by layer. Once we have picked out the nodes at distance $0, 1, 2, \ldots, d$, the ones at $d + 1$ are easily determined: they are precisely the as-yet-unseen nodes that are adjacent to the layer at distance d. This suggests an iterative algorithm in which two layers are active at any given time: some layer d, which has been fully identified, and $d + 1$, which is being discovered by scanning the neighbors of layer d.

Breadth-first search (BFS) directly implements this simple reasoning (Figure 4.3). Initially the queue Q consists only of s, the one node at distance 0. And for each subsequent distance $d = 1, 2, 3, \ldots,$ there is a point in time at which Q contains all the nodes at distance d and nothing else. As these nodes are processed (ejected off the front of the queue), their as-yet-unseen neighbors are injected into the end of the queue.

Let's try out this algorithm on our earlier example (Figure 4.1) to confirm that it does the right thing. If S is the starting point and the nodes are ordered alphabetically, they get visited in the sequence shown in Figure 4.4. The breadth-first search tree, on the right, contains the edges through which each node is initially discovered. Unlike the DFS tree we saw earlier, it has the property that all its paths from S are the shortest possible. It is therefore a *shortest-path tree*.

Correctness and efficiency

We have developed the basic intuition behind breadth-first search. In order to check that the algorithm works correctly, we need to make sure that it faithfully executes this intuition. What we expect, precisely, is that

> *For each $d = 0, 1, 2, \ldots,$ there is a moment at which (1) all nodes at distance $\leq d$ from s have their distances correctly set; (2) all other nodes have their distances set to ∞; and (3) the queue contains exactly the nodes at distance d.*

Figure 4.3 Breadth-first search.

```
procedure bfs(G, s)
Input:  Graph G = (V, E), directed or undirected; vertex s ∈ V
Output: For all vertices u reachable from s, dist(u) is set
        to the distance from s to u.

for all u ∈ V:
   dist(u) = ∞

dist(s) = 0
Q = [s] (queue containing just s)
while Q is not empty:
   u = eject(Q)
   for all edges (u, v) ∈ E:
      if dist(v) = ∞:
         inject(Q, v)
         dist(v) = dist(u) + 1
```

Figure 4.4 The result of breadth-first search on the graph of Figure 4.1.

Order of visitation	Queue contents after processing node
	$[S]$
S	$[A\ C\ D\ E]$
A	$[C\ D\ E\ B]$
C	$[D\ E\ B]$
D	$[E\ B]$
E	$[B]$
B	$[\,]$

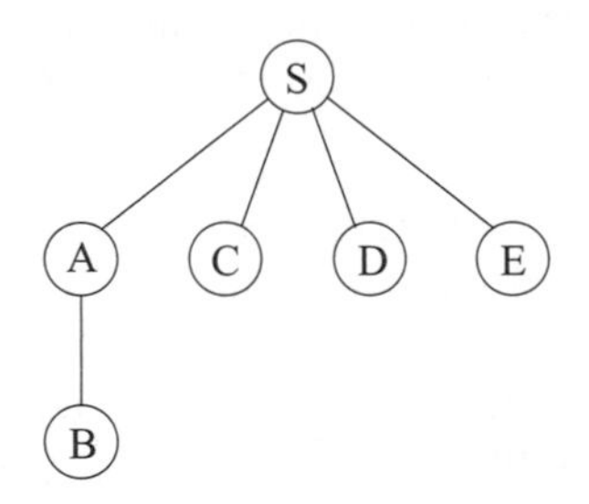

This has been phrased with an inductive argument in mind. We have already discussed both the base case and the inductive step. Can you fill in the details?

The overall running time of this algorithm is linear, $O(|V| + |E|)$, for exactly the same reasons as depth-first search. Each vertex is put on the queue exactly once, when it is first encountered, so there are $2\,|V|$ queue operations. The rest of the work is done in the algorithm's innermost loop. Over the course of execution, this loop looks at each edge once (in directed graphs) or twice (in undirected graphs), and therefore takes $O(|E|)$ time.

Now that we have both BFS and DFS before us: how do their exploration styles compare? Depth-first search makes deep incursions into a graph, retreating only when it runs out of new nodes to visit. This strategy gives it the wonderful, subtle, and extremely useful properties we saw in Chapter 3. But it also means that DFS can end up taking a long and convoluted route to a vertex that is actually very close by, as in Figure 4.1. Breadth-first search makes sure to visit vertices in increasing order of their distance from the starting point. This is a broader, shallower search, rather like the propagation of a wave upon water. And it is achieved using almost exactly the same code as DFS—but with a queue in place of a stack.

Also notice one stylistic difference from DFS: since we are only interested in distances from s, we do not restart the search in other connected components. Nodes not reachable from s are simply ignored.

4.3 Lengths on edges

Breadth-first search treats all edges as having the same length. This is rarely true in applications where shortest paths are to be found. For instance, suppose you are driving from San Francisco to Las Vegas, and want to find the quickest route. Figure 4.5 shows the major highways you might conceivably use. Picking the right combination of them is a shortest-path problem in which the length of each edge (each stretch of highway) is important. For the remainder of this chapter, we will deal with this more general scenario, annotating every edge $e \in E$ with a length l_e. If $e = (u, v)$, we will sometimes also write $l(u, v)$ or l_{uv}.

Figure 4.5 Edge lengths often matter.

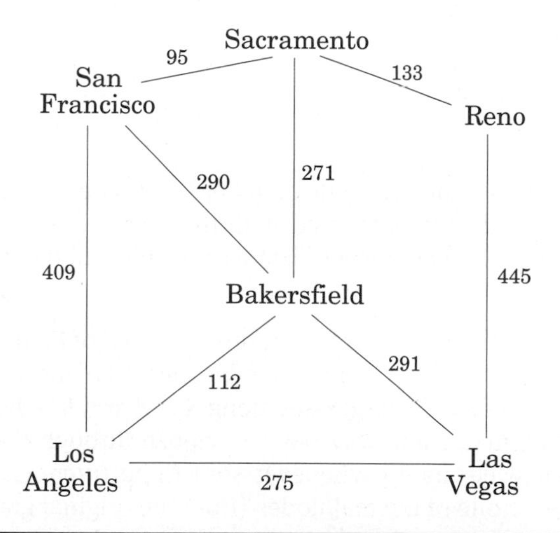

These l_e's do not have to correspond to physical lengths. They could denote time (driving time between cities) or money (cost of taking a bus), or any other quantity that we would like to conserve. In fact, there are cases in which we need to use negative lengths, but we will briefly overlook this particular complication.

4.4 Dijkstra's algorithm

4.4.1 An adaptation of breadth-first search

Breadth-first search finds shortest paths in any graph whose edges have unit length. Can we adapt it to a more general graph $G = (V, E)$ whose edge lengths l_e are *positive integers*?

A more convenient graph

Here is a simple trick for converting G into something BFS can handle: break G's long edges into unit-length pieces by introducing "dummy" nodes. Figure 4.6 shows an example of this transformation. To construct the new graph G',

> *For any edge $e = (u, v)$ of E, replace it by l_e edges of length 1, by adding $l_e - 1$ dummy nodes between u and v.*

Graph G' contains all the vertices V that interest us, and the distances between them are exactly the same as in G. Most importantly, the edges of G' all have unit length. Therefore, we can compute distances in G by running BFS on G'.

Figure 4.6 Breaking edges into unit-length pieces.

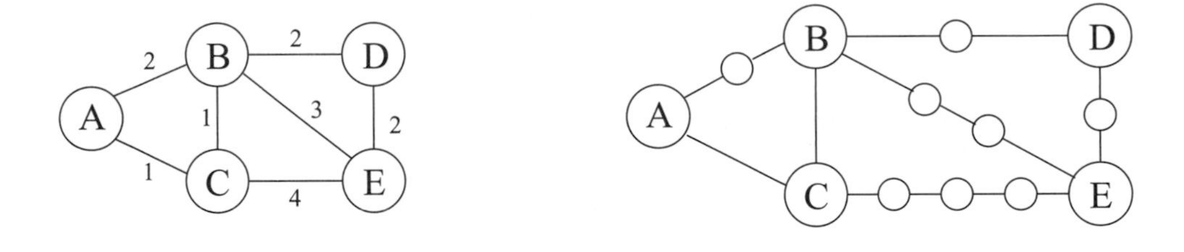

Alarm clocks

If efficiency were not an issue, we could stop here. But when G has very long edges, the G' it engenders is thickly populated with dummy nodes, and the BFS spends most of its time diligently computing distances to these nodes that we don't care about at all.

To see this more concretely, consider the graphs G and G' of Figure 4.7, and imagine that the BFS, started at node s of G', advances by one unit of distance per minute. For the first 99 minutes it tediously progresses along $S - A$ and $S - B$, an endless desert of dummy nodes. Is there some way we can snooze through these boring phases and have an alarm wake us up whenever something *interesting* is happening—specifically, whenever one of the real nodes (from the original graph G) is reached?

We do this by setting two alarms at the outset, one for node A, set to go off at time $T = 100$, and one for B, at time $T = 200$. These are *estimated times of arrival*, based upon the edges currently being traversed. We doze off and awake at $T = 100$ to find A has been discovered. At this point, the estimated time of arrival for B is adjusted to $T = 150$ and we change its alarm accordingly.

Figure 4.7 BFS on G' is mostly uneventful. The dotted lines show some early "wavefronts."

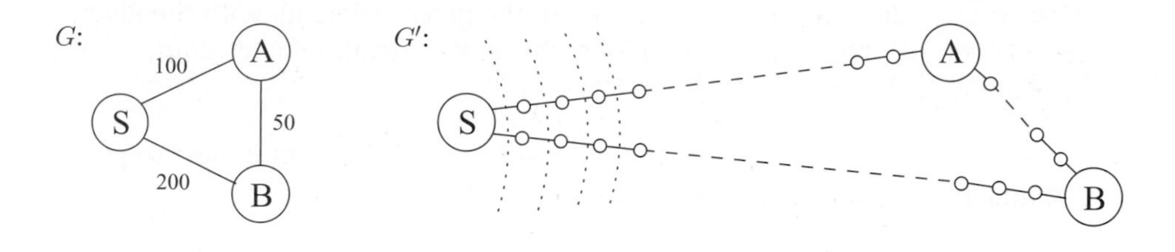

More generally, at any given moment the breadth-first search is advancing along certain edges of G, and there is an alarm for every endpoint node toward which it is moving, set to go off at the estimated time of arrival at that node. Some of these might be overestimates because BFS may later find shortcuts, as a result of future arrivals elsewhere. In the preceding example, a quicker route to B was revealed upon arrival at A. However, *nothing interesting can possibly happen before an alarm goes off*. The sounding of the next alarm must therefore signal the arrival of the wavefront to a real node $u \in V$ by BFS. At that point, BFS might also start advancing along some new edges out of u, and alarms need to be set for their endpoints.

The following "alarm clock algorithm" faithfully simulates the execution of BFS on G'.

- Set an alarm clock for node s at time 0.
- Repeat until there are no more alarms:

  Say the next alarm goes off at time T, for node u. Then:
  - The distance from s to u is T.
  - For each neighbor v of u in G:
    - If there is no alarm yet for v, set one for time $T + l(u, v)$.
    - If v's alarm is set for later than $T + l(u, v)$, then reset it to this earlier time.

Dijkstra's algorithm

The alarm clock algorithm computes distances in any graph with positive integral edge lengths. It is almost ready for use, except that we need to somehow implement the system of alarms. The right data structure for this job is a *priority queue* (usually implemented via a *heap*), which maintains a set of elements (nodes) with associated numeric key values (alarm times) and supports the following operations:

Insert. Add a new element to the set.

Decrease-key. Accommodate the decrease in key value of a particular element.[1]

[1]The name *decrease-key* is standard but is a little misleading: the priority queue typically does not itself change key values. What this procedure really does is to notify the queue that a certain key value has been decreased.

Delete-min. Return the element with the smallest key, and remove it from the set.

Make-queue. Build a priority queue out of the given elements with the given key values. (In many implementations, this is significantly faster than inserting the elements one by one.)

The first two let us set alarms, and the third tells us which alarm is next to go off. Putting this all together, we get Dijkstra's algorithm (Figure 4.8).

Figure 4.8 Dijkstra's shortest-path algorithm.

```
procedure dijkstra(G,l,s)
Input:  Graph G = (V,E), directed or undirected;
        positive edge lengths {l_e : e ∈ E}; vertex s ∈ V
Output: For all vertices u reachable from s, dist(u) is set
        to the distance from s to u.

for all u ∈ V:
   dist(u) = ∞
   prev(u) = nil
dist(s) = 0

H = makequeue(V)  (using dist-values as keys)
while H is not empty:
   u = deletemin(H)
   for all edges (u,v) ∈ E:
      if dist(v) > dist(u) + l(u,v):
         dist(v) = dist(u) + l(u,v)
         prev(v) = u
         decreasekey(H,v)
```

In the code, dist(u) refers to the current alarm clock setting for node u. A value of ∞ means the alarm hasn't so far been set. There is also a special array, prev, that holds one crucial piece of information for each node u: the identity of the node immediately before it on the shortest path from s to u. By following these back-pointers, we can easily reconstruct shortest paths, and so this array is a compact summary of all the paths found. A full example of the algorithm's operation, along with the final shortest-path tree, is shown in Figure 4.9.

In summary, we can think of Dijkstra's algorithm as just BFS, except it uses a priority queue instead of a regular queue, so as to prioritize nodes in a way that takes edge lengths into account. This viewpoint gives a concrete appreciation of how and why the algorithm works, but there is a more direct, more abstract derivation that

Figure 4.9 A complete run of Dijkstra's algorithm, with node A as the starting point. Also shown are the associated `dist` values and the final shortest-path tree.

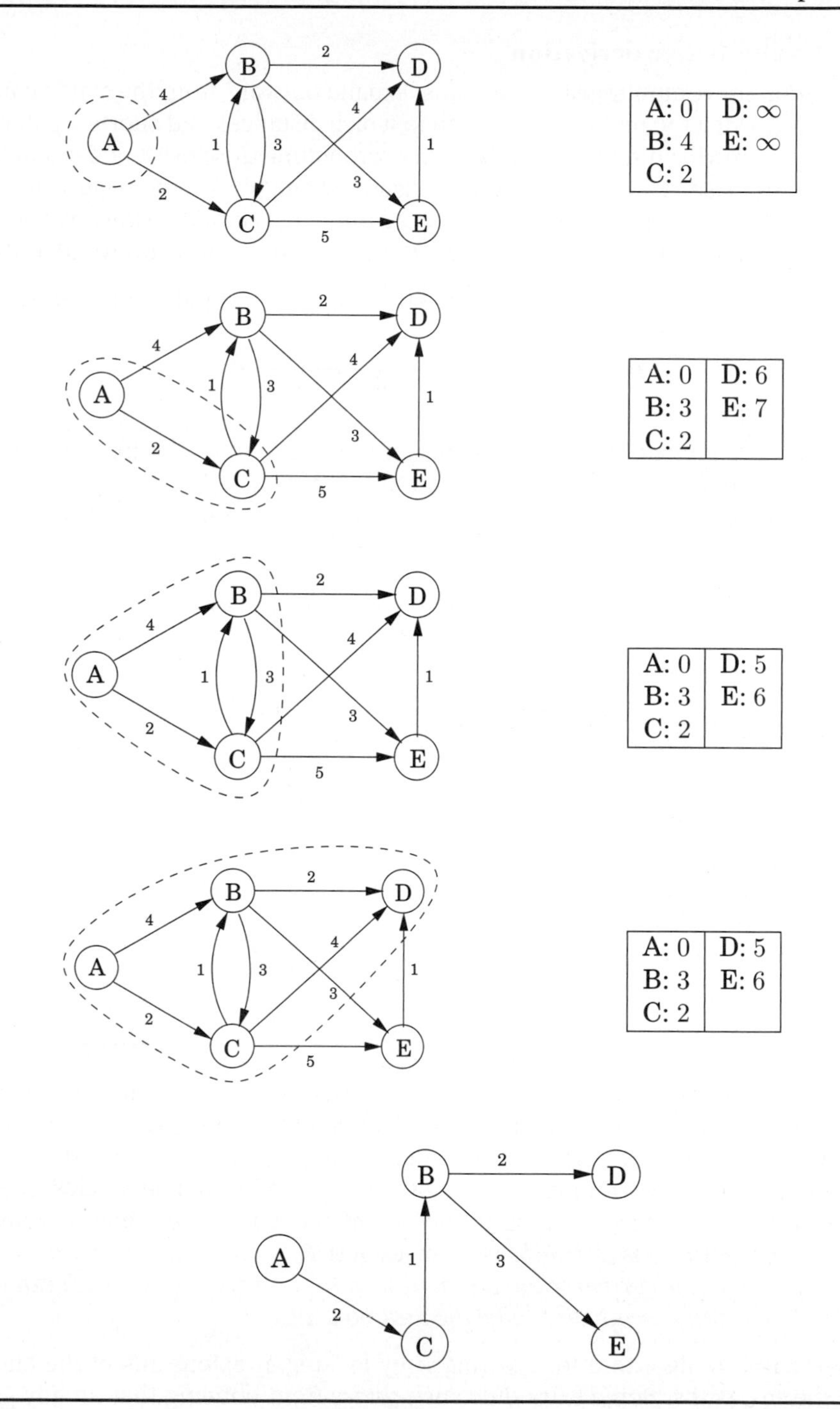

doesn't depend upon BFS at all. We now start from scratch with this complementary interpretation.

4.4.2 An alternative derivation

Here's a plan for computing shortest paths: expand outward from the starting point s, steadily growing the region of the graph to which distances and shortest paths are known. This growth should be orderly, first incorporating the closest nodes and then moving on to those further away. More precisely, when the "known region" is some subset of vertices R that includes s, the next addition to it should be *the node outside R that is closest to s*. Let us call this node v; the question is: how do we identify it?

To answer, consider u, the node just before v in the shortest path from s to v:

Since we are assuming that all edge lengths are positive, u must be closer to s than v is. This means that u is in R—otherwise it would contradict v's status as the closest node to s outside R. So, the shortest path from s to v is simply *a known shortest path extended by a single edge.*

Figure 4.10 Single-edge extensions of known shortest paths.

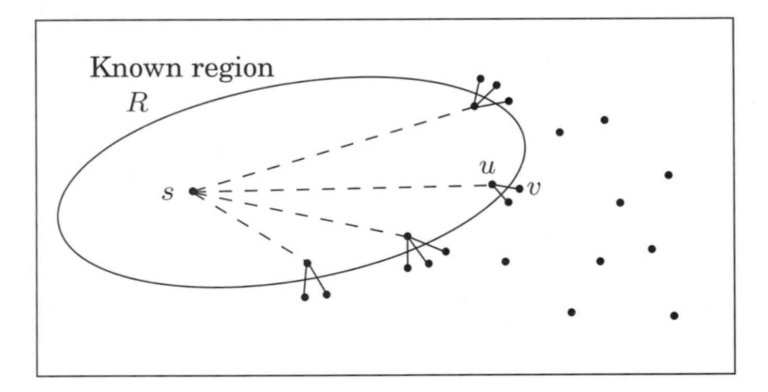

But there will typically be many single-edge extensions of the currently known shortest paths (Figure 4.10); which of these identifies v? The answer is, *the shortest of these extended paths*. Because, if an even shorter single-edge-extended path existed, this would once more contradict v's status as the node outside R closest to s. So, it's easy to find v: it is the node outside R for which the smallest value of $\text{distance}(s, u) + l(u, v)$ is attained, as u ranges over R. In other words, *try all single-edge extensions of the currently known shortest paths, find the shortest such extended path, and proclaim its endpoint to be the next node of R.*

We now have an algorithm for growing R by looking at extensions of the current set of shortest paths. Some extra efficiency comes from noticing that on any given iteration, the only *new* extensions are those involving the node most recently added to region R. All other extensions will have been assessed previously and do not need to be recomputed. In the following pseudocode, dist(v) is the length of the currently shortest single-edge-extended path leading to v; it is ∞ for nodes not adjacent to R.

```
Initialize dist(s) to 0, other dist(·) values to ∞
R = { } (the "known region∫)
while R ≠ V:
   Pick the node v ∉ R with smallest dist(·)
   Add v to R
   for all edges (v, z) ∈ E:
       if dist(z) > dist(v) + l(v, z):
          dist(z) = dist(v) + l(v, z)
```

Incorporating priority queue operations gives us back Dijkstra's algorithm (Figure 4.8).

To justify this algorithm formally, we would use a proof by induction, as with breadth-first search. Here's an appropriate inductive hypothesis.

> *At the end of each iteration of the while loop, the following conditions hold: (1) there is a value d such that all nodes in R are at distance $\leq d$ from s and all nodes outside R are at distance $\geq d$ from s, and (2) for every node u, the value* dist(u) *is the length of the shortest path from s to u whose intermediate nodes are constrained to be in R (if no such path exists, the value is ∞).*

The base case is straightforward (with $d = 0$), and the details of the inductive step can be filled in from the preceding discussion.

4.4.3 Running time

At the level of abstraction of Figure 4.8, Dijkstra's algorithm is structurally identical to breadth-first search. However, it is slower because the priority queue primitives are computationally more demanding than the constant-time `eject`'s and `inject`'s of BFS. Since `makequeue` takes at most as long as $|V|$ `insert` operations, we get a total of $|V|$ `deletemin` and $|V| + |E|$ `insert/decreasekey` operations. The time needed for these varies by implementation; for instance, a binary heap gives an overall running time of $O((|V| + |E|) \log |V|)$.

4.5 Priority queue implementations

4.5.1 Array

The simplest implementation of a priority queue is as an unordered array of key values for all potential elements (the vertices of the graph, in the case of Dijkstra's algorithm). Initially, these values are set to ∞.

Which heap is best?

The running time of Dijkstra's algorithm depends heavily on the priority queue implementation used. Here are the typical choices.

Implementation	`deletemin`	`insert`/ `decreasekey`	$\|V\| \times$ `deletemin` $+$ $(\|V\|+\|E\|) \times$ `insert`
Array	$O(\|V\|)$	$O(1)$	$O(\|V\|^2)$
Binary heap	$O(\log\|V\|)$	$O(\log\|V\|)$	$O((\|V\|+\|E\|)\log\|V\|)$
d-ary heap	$O\left(\frac{d\log\|V\|}{\log d}\right)$	$O\left(\frac{\log\|V\|}{\log d}\right)$	$O\left((\|V\|\cdot d+\|E\|)\frac{\log\|V\|}{\log d}\right)$
Fibonacci heap	$O(\log\|V\|)$	$O(1)$ (amortized)	$O(\|V\|\log\|V\|+\|E\|)$

So for instance, even a naive array implementation gives a respectable time complexity of $O(|V|^2)$, whereas with a binary heap we get $O((|V|+|E|)\log|V|)$. Which is preferable?

This depends on whether the graph is *sparse* (has few edges) or *dense* (has lots of them). For all graphs, $|E|$ is less than $|V|^2$. If it is $\Omega(|V|^2)$, then clearly the array implementation is the faster. On the other hand, the binary heap becomes preferable as soon as $|E|$ dips below $|V|^2/\log|V|$.

The d-ary heap is a generalization of the binary heap (which corresponds to $d=2$) and leads to a running time that is a function of d. The optimal choice is $d \approx |E|/|V|$; in other words, to optimize we must set the degree of the heap to be equal to the *average degree* of the graph. This works well for both sparse and dense graphs. For very sparse graphs, in which $|E| = O(|V|)$, the running time is $O(|V|\log|V|)$, as good as with a binary heap. For dense graphs, $|E| = \Omega(|V|^2)$ and the running time is $O(|V|^2)$, as good as with a linked list. Finally, for graphs with intermediate density $|E| = |V|^{1+\delta}$, the running time is $O(|E|)$, linear!

The last line in the table gives running times using a sophisticated data structure called a *Fibonacci heap*. Although its efficiency is impressive, this data structure requires considerably more work to implement than the others, and this tends to dampen its appeal in practice. We will say little about it except to mention a curious feature of its time bounds. Its *insert* operations take varying amounts of time but are guaranteed to *average* $O(1)$ over the course of the algorithm. In such situations (one of which we shall encounter in Chapter 5) we say that the *amortized* cost of heap *insert*'s is $O(1)$.

An `insert` or `decreasekey` is fast, because it just involves adjusting a key value, an $O(1)$ operation. To `deletemin`, on the other hand, requires a linear-time scan of the list.

4.5.2 Binary heap

Here elements are stored in a *complete* binary tree, namely, a binary tree in which each level is filled in from left to right, and must be full before the next level is started. In addition, a special ordering constraint is enforced: *the key value of any node of the tree is less than or equal to that of its children.* In particular, therefore, the root always contains the smallest element. See Figure 4.11(a) for an example.

To `insert`, place the new element at the bottom of the tree (in the first available position), and let it "bubble up." That is, if it is smaller than its parent, swap the two and repeat (Figure 4.11(b)–(d)). The number of swaps is at most the height of the tree, which is $\lfloor \log_2 n \rfloor$ when there are n elements. A `decreasekey` is similar, except that the element is already in the tree, so we let it bubble up from its current position.

To `deletemin`, return the root value. To then remove this element from the heap, take the last node in the tree (in the rightmost position in the bottom row) and place it at the root. Let it "sift down": if it is bigger than either child, swap it with the smaller child and repeat (Figure 4.11(e)–(g)). Again this takes $O(\log n)$ time.

The regularity of a complete binary tree makes it easy to represent using an array. The tree nodes have a natural ordering: row by row, starting at the root and moving left to right within each row. If there are n nodes, this ordering specifies their positions $1, 2, \ldots, n$ within the array. Moving up and down the tree is easily simulated on the array, using the fact that node number j has parent $\lfloor j/2 \rfloor$ and children $2j$ and $2j + 1$ (Exercise 4.16).

4.5.3 *d*-ary heap

A d-ary heap is identical to a binary heap, except that nodes have d children instead of just two. This reduces the height of a tree with n elements to $\Theta(\log_d n)$ $= \Theta((\log n)/(\log d))$. Inserts are therefore speeded up by a factor of $\Theta(\log d)$. Deletemin operations, however, take a little longer, namely $O(d \log_d n)$ (do you see why?).

The array representation of a binary heap is easily extended to the d-ary case. This time, node number j has parent $\lceil (j-1)/d \rceil$ and children $\{(j-1)d + 2, \ldots, \min\{n, (j-1)d + d + 1\}\}$ (Exercise 4.16).

4.6 Shortest paths in the presence of negative edges

4.6.1 Negative edges

Dijkstra's algorithm works in part because the shortest path from the starting point s to any node v must pass exclusively through nodes that are closer than v. This no longer holds when edge lengths can be negative. In Figure 4.12, the shortest path from S to A passes through B, a node that is further away!

What needs to be changed in order to accommodate this new complication? To answer this, let's take a particular high-level view of Dijkstra's algorithm. A crucial invariant is that the `dist` values it maintains are always either overestimates or exactly correct. They start off at ∞, and the only way they ever change is by updating

Figure 4.11 (a) A binary heap with 10 elements. Only the key values are shown. (b)–(d) The intermediate "bubble-up" steps in inserting an element with key 7. (e)–(g) The "sift-down" steps in a delete-min operation.

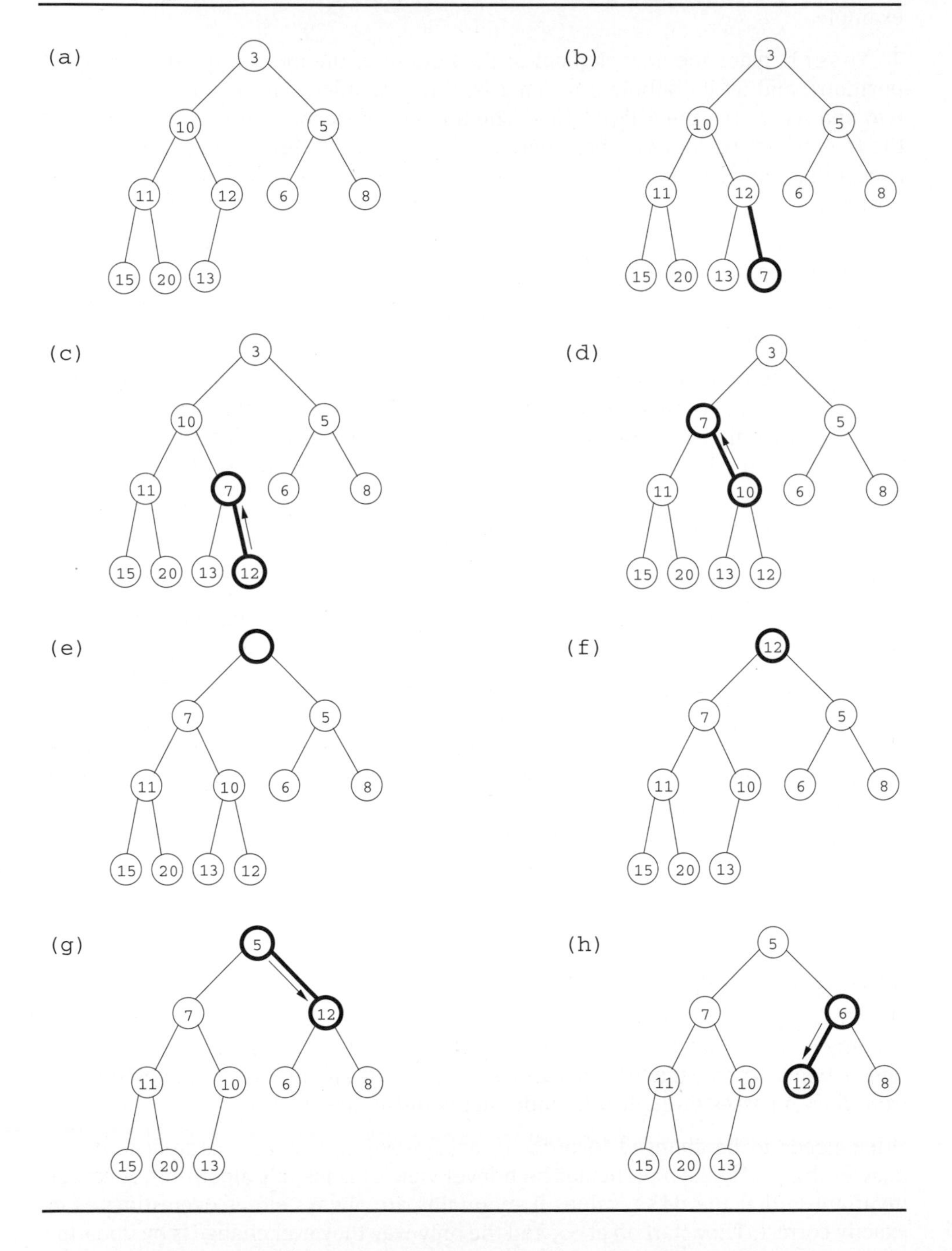

Figure 4.12 Dijkstra's algorithm will not work if there are negative edges.

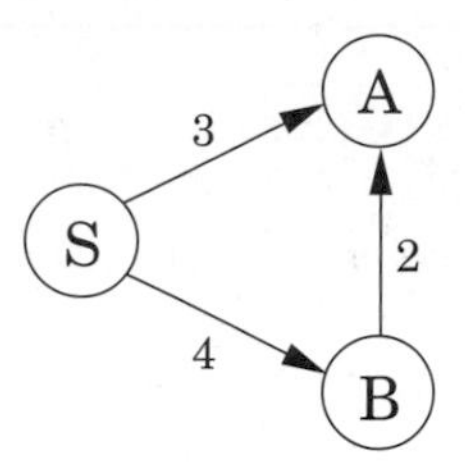

along an edge:

```
procedure update((u, v) ∈ E)
dist(v) = min{dist(v), dist(u) + l(u, v)}
```

This *update* operation is simply an expression of the fact that the distance to v cannot possibly be more than the distance to u, plus $l(u, v)$. It has the following properties.

1. It gives the correct distance to v in the particular case where u is the second-last node in the shortest path to v, and `dist`(u) is correctly set.
2. It will never make `dist`(v) too small, and in this sense it is *safe*. For instance, a slew of extraneous `update`'s can't hurt.

This operation is extremely useful: it is harmless, and if used carefully, will correctly set distances. In fact, Dijkstra's algorithm can be thought of simply as a sequence of `update`'s. We know this particular sequence doesn't work with negative edges, but is there some other sequence that does? To get a sense of the properties this sequence must possess, let's pick a node t and look at the shortest path to it from s.

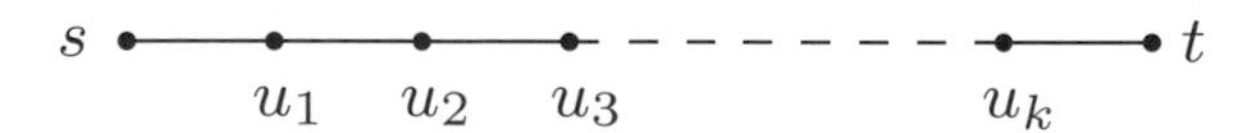

This path can have at most $|V| - 1$ edges (do you see why?). If the sequence of updates performed includes $(s, u_1), (u_1, u_2), (u_2, u_3), \ldots, (u_k, t)$, *in that order* (though not necessarily consecutively), then by the first property the distance to t will be correctly computed. It doesn't matter what other updates occur on these edges, or what happens in the rest of the graph, because updates are *safe*.

But still, if we don't know all the shortest paths beforehand, how can we be sure to update the right edges in the right order? Here is an easy solution: simply update *all* the edges, $|V| - 1$ times! The resulting $O(|V| \cdot |E|)$ procedure is called the Bellman-Ford algorithm and is shown in Figure 4.13, with an example run in Figure 4.14.

Figure 4.13 The Bellman-Ford algorithm for single-source shortest paths in general graphs.

```
procedure shortest-paths(G, l, s)
Input:  Directed graph G = (V, E);
        edge lengths {l_e : e ∈ E} with no negative cycles;
        vertex s ∈ V
Output: For all vertices u reachable from s, dist(u) is set
        to the distance from s to u.

for all u ∈ V:
   dist(u) = ∞
   prev(u) = nil

dist(s) = 0
repeat |V| − 1 times:
   for all e ∈ E:
      update(e)
```

Figure 4.14 The Bellman-Ford algorithm illustrated on a sample graph.

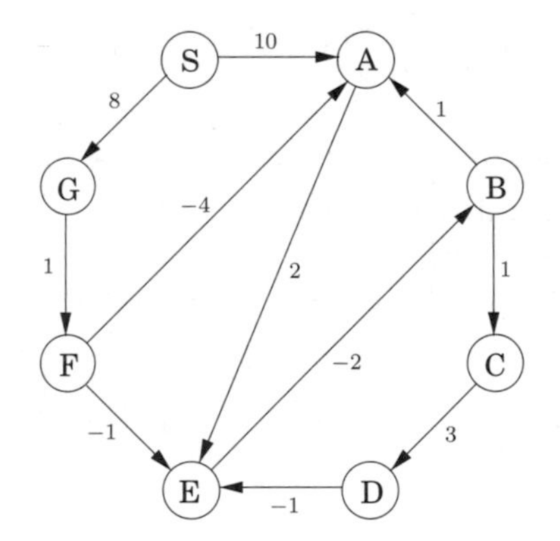

Node	Iteration 0	1	2	3	4	5	6	7
S	0	0	0	0	0	0	0	0
A	∞	10	10	5	5	5	5	5
B	∞	∞	∞	10	6	5	5	5
C	∞	∞	∞	∞	11	7	6	6
D	∞	∞	∞	∞	∞	14	10	9
E	∞	∞	12	8	7	7	7	7
F	∞	∞	9	9	9	9	9	9
G	∞	8	8	8	8	8	8	8

A note about implementation: for many graphs, the maximum number of edges in any shortest path is substantially less than $|V| - 1$, with the result that fewer rounds of updates are needed. Therefore, it makes sense to add an extra check to the shortest-path algorithm, to make it terminate immediately after any round in which no update occurred.

4.6.2 Negative cycles

If the length of edge (E, B) in Figure 4.14 were changed to -4, the graph would have a *negative cycle* $A \rightarrow E \rightarrow B \rightarrow A$. In such situations, it doesn't make sense to even ask about shortest paths. There is a path of length 2 from A to E. But going round the cycle, there's also a path of length 1, and going round multiple times, we find paths of lengths $0, -1, -2$, and so on.

The shortest-path problem is ill-posed in graphs with negative cycles. As might be expected, our algorithm from Section 4.6.1 works only in the absence of such cycles. But where did this assumption appear in the derivation of the algorithm? Well, it slipped in when we asserted the *existence* of a shortest path from s to t.

Fortunately, it is easy to automatically detect negative cycles and issue a warning. Such a cycle would allow us to endlessly apply rounds of `update` operations, reducing `dist` estimates every time. So instead of stopping after $|V| - 1$ iterations, perform one extra round. There is a negative cycle if and only if some `dist` value is reduced during this final round.

4.7 Shortest paths in dags

There are two subclasses of graphs that automatically exclude the possibility of negative cycles: graphs without negative edges, and graphs without cycles. We already know how to efficiently handle the former. We will now see how the single-source shortest-path problem can be solved in just linear time on directed acyclic graphs.

As before, we need to perform a sequence of updates that includes every shortest path as a subsequence. The key source of efficiency is that

> *In any path of a dag, the vertices appear in increasing linearized order.*

Figure 4.15 A single-source shortest-path algorithm for directed acyclic graphs.

```
procedure dag-shortest-paths(G, l, s)
Input:  Dag G = (V, E);
        edge lengths {l_e : e ∈ E}; vertex s ∈ V
Output: For all vertices u reachable from s, dist(u) is set
        to the distance from s to u.

for all u ∈ V:
   dist(u) = ∞
   prev(u) = nil

dist(s) = 0
Linearize G
for each u ∈ V, in linearized order:
   for all edges (u, v) ∈ E:
      update(u, v)
```

Therefore, it is enough to linearize (that is, topologically sort) the dag by depth-first search, and then visit the vertices in sorted order, updating the edges out of each. The algorithm is given in Figure 4.15.

Notice that our scheme doesn't require edges to be positive. In particular, we can find *longest paths* in a dag by the same algorithm: just negate all edge lengths.

Exercises

4.1. Suppose Dijkstra's algorithm is run on the following graph, starting at node A.

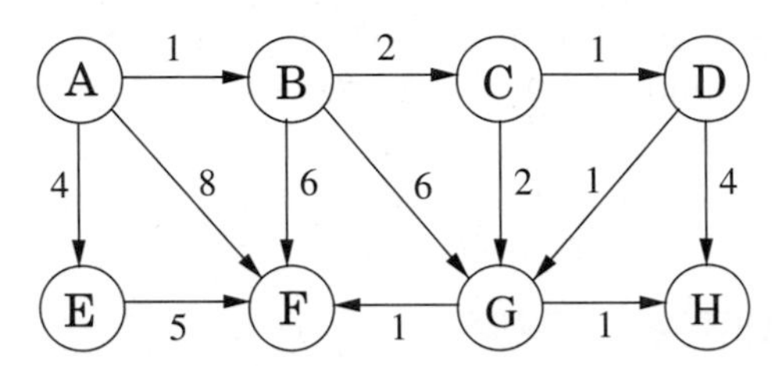

(a) Draw a table showing the intermediate distance values of all the nodes at each iteration of the algorithm.

(b) Show the final shortest-path tree.

4.2. Just like the previous problem, but this time with the Bellman-Ford algorithm.

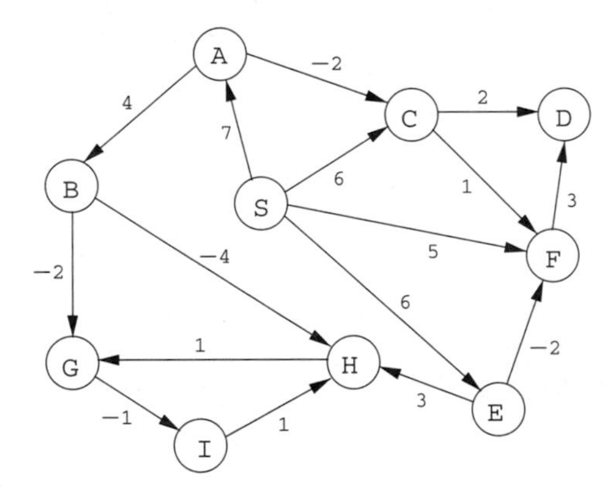

4.3. *Squares.* Design and analyze an algorithm that takes as input an undirected graph $G = (V, E)$ and determines whether G contains a simple cycle (that is, a cycle which doesn't intersect itself) of length four. Its running time should be at most $O(|V|^3)$.

You may assume that the input graph is represented either as an adjacency matrix or with adjacency lists, whichever makes your algorithm simpler.

4.4. Here's a proposal for how to find the length of the shortest cycle in an undirected graph with unit edge lengths.

When a back edge, say (v, w), is encountered during a depth-first search, it forms a cycle with the tree edges from w to v. The length of the cycle is level[v] − level[w] + 1, where the level of a vertex is its distance in the DFS tree from the root vertex. This suggests the following algorithm:

- *Do a depth-first search, keeping track of the level of each vertex.*
- *Each time a back edge is encountered, compute the cycle length and save it if it is smaller than the shortest one previously seen.*

Show that this strategy does not always work by providing a counterexample as well as a brief (one or two sentence) explanation.

4.5. Often there are multiple shortest paths between two nodes of a graph. Give a linear-time algorithm for the following task.

Input: Undirected graph $G = (V, E)$ with unit edge lengths; nodes $u, v \in V$.
Output: The number of distinct shortest paths from u to v.

4.6. Prove that for the array `prev` computed by Dijkstra's algorithm, the edges $\{u, \texttt{prev}[u]\}$ (for all $u \in V$) form a tree.

4.7. You are given a directed graph $G = (V, E)$ with (possibly negative) weighted edges, along with a specific node $s \in V$ and a tree $T = (V, E'), E' \subseteq E$. Give an algorithm that checks whether T is a shortest-path tree for G with starting point s. Your algorithm should run in linear time.

4.8. Professor F. Lake suggests the following algorithm for finding the shortest path from node s to node t in a directed graph with some negative edges: add a large constant to each edge weight so that all the weights become positive, then run Dijkstra's algorithm starting at node s, and return the shortest path found to node t.

Is this a valid method? Either prove that it works correctly, or give a counterexample.

4.9. Consider a directed graph in which the only negative edges are those that leave s; all other edges are positive. Can Dijkstra's algorithm, started at s, fail on such a graph? Prove your answer.

4.10. You are given a directed graph with (possibly negative) weighted edges, in which the shortest path between any two vertices is guaranteed to have at most k edges. Give an algorithm that finds the shortest path between two vertices u and v in $O(k|E|)$ time.

4.11. Give an algorithm that takes as input a directed graph with positive edge lengths, and returns the length of the shortest cycle in the graph (if the graph is acyclic, it should say so). Your algorithm should take time at most $O(|V|^3)$.

4.12. Give an $O(|V|^2)$ algorithm for the following task.

Input: An undirected graph $G = (V, E)$; edge lengths $l_e > 0$; an edge $e \in E$.
Output: The length of the shortest cycle containing edge e.

4.13. You are given a set of cities, along with the pattern of highways between them, in the form of an undirected graph $G = (V, E)$. Each stretch of highway $e \in E$ connects two of the cities, and you know its length in miles, l_e. You want to get from city s to city t. There's one problem: your car can only hold enough gas to cover L miles. There are gas stations in each city, but not between cities. Therefore, you can only take a route if every one of its edges has length $l_e \leq L$.

(a) Given the limitation on your car's fuel tank capacity, show how to determine in linear time whether there is a feasible route from s to t.

(b) You are now planning to buy a new car, and you want to know the minimum fuel tank capacity that is needed to travel from s to t. Give an $O((|V| + |E|) \log |V|)$ algorithm to determine this.

4.14. You are given a strongly connected directed graph $G = (V, E)$ with positive edge weights along with a particular node $v_0 \in V$. Give an efficient algorithm for finding shortest paths between *all pairs of nodes*, with the one restriction that these paths must all pass through v_0.

4.15. Shortest paths are not always unique: sometimes there are two or more different paths with the minimum possible length. Show how to solve the following problem in $O((|V| + |E|) \log |V|)$ time.

Input: An undirected graph $G = (V, E)$; edge lengths $l_e > 0$; starting vertex $s \in V$.
Output: A Boolean array `usp[·]`: for each node u, the entry `usp`$[u]$ should be `true` if and only if there is a *unique* shortest path from s to u. (Note: `usp`$[s]$ = `true`.)

4.16. Section 4.5.2 describes a way of storing a complete binary tree of n nodes in an array indexed by $1, 2, \ldots, n$.

(a) Consider the node at position j of the array. Show that its parent is at position $\lfloor j/2 \rfloor$ and its children are at $2j$ and $2j + 1$ (if these numbers are $\leq n$).

(b) What are the corresponding indices when a complete d-ary tree is stored in an array?

Figure 4.16 shows pseudocode for a binary heap, modeled on an exposition by R. E. Tarjan.[2] The heap is stored as an array h, which is assumed to support two constant-time operations:

- $|h|$, which returns the number of elements currently in the array;
- h^{-1}, which returns the position of an element within the array.

The latter can always be achieved by maintaining the values of h^{-1} as an auxiliary array.

(c) Show that the `makeheap` procedure takes $O(n)$ time when called on a set of n elements. What is the worst-case input? (*Hint:* Start by showing that the running time is at most $\sum_{i=1}^{n} \log(n/i)$.)

(d) What needs to be changed to adapt this pseudocode to d-ary heaps?

---

[2]See: R. E. Tarjan, *Data Structures and Network Algorithms*, Society for Industrial and Applied Mathematics, 1983.

Figure 4.16 Operations on a binary heap.

```
procedure insert(h, x)
bubbleup(h, x, |h| + 1)

procedure decreasekey(h, x)
bubbleup(h, x, h⁻¹(x))

function deletemin(h)
if |h| = 0:
   return null
else:
   x = h(1)
   siftdown(h, h(|h|), 1)
   return x

function makeheap(S)
h = empty array of size |S|
for x ∈ S:
   h(|h| + 1) = x
for i = |S| downto 1:
   siftdown(h, h(i), i)
return h

procedure bubbleup(h, x, i)
(place element x in position i of h, and let it bubble up)
p = ⌈i/2⌉
while i ≠ 1 and key(h(p)) > key(x):
   h(i) = h(p);  i = p;  p = ⌈i/2⌉
h(i) = x

procedure siftdown(h, x, i)
(place element x in position i of h, and let it sift down)
c = minchild(h, i)
while c ≠ 0 and key(h(c)) < key(x):
   h(i) = h(c);  i = c;  c = minchild(h, i)
h(i) = x

function minchild(h, i)
(return the index of the smallest child of h(i))
if 2i > |h|:
   return 0 (no children)
else:
   return arg min{key(h(j)) : 2i ≤ j ≤ min{|h|, 2i + 1}}
```

4.17. Suppose we want to run Dijkstra's algorithm on a graph whose edge weights are integers in the range $0, 1, \ldots, W$, where W is a relatively small number.

(a) Show how Dijkstra's algorithm can be made to run in time $O(W|V| + |E|)$.

(b) Show an alternative implementation that takes time just $O((|V| + |E|) \log W)$.

4.18. In cases where there are several different shortest paths between two nodes (and edges have varying lengths), the most convenient of these paths is often *the one with fewest edges*. For instance, if nodes represent cities and edge lengths represent costs of flying between cities, there might be many ways to get from city s to city t which all have the same cost. The most convenient of these alternatives is the one which involves the fewest stopovers. Accordingly, for a specific starting node s, define

$$\texttt{best}[u] = \text{minimum number of edges in a shortest path from } s \text{ to } u.$$

In the example below, the `best` values for nodes S, A, B, C, D, E, F are $0, 1, 1, 1, 2, 2, 3$, respectively.

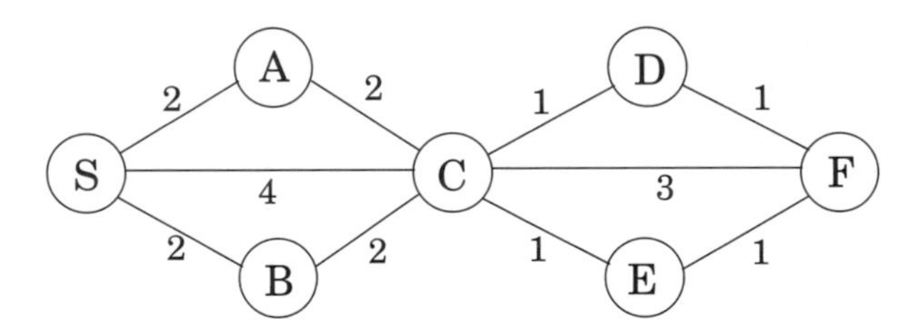

Give an efficient algorithm for the following problem.

Input: Graph $G = (V, E)$; positive edge lengths l_e; starting node $s \in V$.
Output: The values of $\texttt{best}[u]$ should be set for *all* nodes $u \in V$.

4.19. *Generalized shortest-paths problem.* In Internet routing, there are delays on lines but also, more significantly, delays at routers. This motivates a generalized shortest-paths problem.

Suppose that in addition to having edge lengths $\{l_e : e \in E\}$, a graph also has *vertex costs* $\{c_v : v \in V\}$. Now define the cost of a path to be the sum of its edge lengths, *plus* the costs of all vertices on the path (including the endpoints). Give an efficient algorithm for the following problem.

Input: A directed graph $G = (V, E)$; positive edge lengths l_e and positive vertex costs c_v; a starting vertex $s \in V$.
Output: An array $\texttt{cost}[\cdot]$ such that for every vertex u, $\texttt{cost}[u]$ is the least cost of any path from s to u (i.e., the cost of the cheapest path), under the definition above.

Notice that $\texttt{cost}[s] = c_s$.

4.20. There is a network of roads $G = (V, E)$ connecting a set of cities V. Each road in E has an associated length l_e. There is a proposal to add one new road to this network, and there is a list E' of pairs of cities between which the new road can be built. Each such potential road $e' \in E'$ has an associated length. As a designer for the public works department you are asked to determine the road $e' \in E'$ whose addition to the existing network G would result in the maximum decrease in the driving distance between two fixed cities s and t in the network. Give an efficient algorithm for solving this problem.

4.21. Shortest path algorithms can be applied in currency trading. Let $c_1, c_2, \ldots, c_n$ be various currencies; for instance, c_1 might be dollars, c_2 pounds, and c_3 lire. For any two currencies c_i and c_j, there is an exchange rate $r_{i,j}$; this means that you can purchase $r_{i,j}$ units of currency c_j in exchange for one unit of c_i. These exchange rates satisfy the condition that $r_{i,j} \cdot r_{j,i} < 1$, so that if you start with a unit of currency c_i, change it into currency c_j and then convert back to currency c_i, you end up with less than one unit of currency c_i (the difference is the cost of the transaction).

(a) Give an efficient algorithm for the following problem: Given a set of exchange rates $r_{i,j}$, and two currencies s and t, find the most advantageous sequence of currency exchanges for converting currency s into currency t. Toward this goal, you should represent the currencies and rates by a graph whose edge lengths are real numbers.

The exchange rates are updated frequently, reflecting the demand and supply of the various currencies. Occasionally the exchange rates satisfy the following property: there is a sequence of currencies $c_{i_1}, c_{i_2}, \ldots, c_{i_k}$ such that $r_{i_1,i_2} \cdot r_{i_2,i_3} \cdots r_{i_{k-1},i_k} \cdot r_{i_k,i_1} > 1$. This means that by starting with a unit of currency c_{i_1} and then successively converting it to currencies $c_{i_2}, c_{i_3}, \ldots, c_{i_k}$, and finally back to c_{i_1}, you would end up with more than one unit of currency c_{i_1}. Such anomalies last only a fraction of a minute on the currency exchange, but they provide an opportunity for risk-free profits.

(b) Give an efficient algorithm for detecting the presence of such an anomaly. Use the graph representation you found above.

4.22. *The tramp steamer problem.* You are the owner of a steamship that can ply between a group of port cities V. You make money at each port: a visit to city i earns you a profit of p_i dollars. Meanwhile, the transportation cost from port i to port j is $c_{ij} > 0$. You want to find a cyclic route in which the ratio of profit to cost is maximized.

To this end, consider a directed graph $G = (V, E)$ whose nodes are ports, and which has edges between each pair of ports. For any cycle C in this graph, the profit-to-cost ratio is

$$r(C) = \frac{\sum_{(i,j) \in C} p_j}{\sum_{(i,j) \in C} c_{ij}}.$$

Let r^* be the maximum ratio achievable by a simple cycle. One way to determine r^* is by binary search: by first guessing some ratio r, and then testing whether it is too large or too small.

Consider any positive $r > 0$. Give each edge (i, j) a weight of $w_{ij} = rc_{ij} - p_j$.

(a) Show that if there is a cycle of negative weight, then $r < r^*$.

(b) Show that if all cycles in the graph have strictly positive weight, then $r > r^*$.

(c) Give an efficient algorithm that takes as input a desired accuracy $\epsilon > 0$ and returns a simple cycle C for which $r(C) \geq r^* - \epsilon$. Justify the correctness of your algorithm and analyze its running time in terms of $|V|$, ϵ, and $R = \max_{(i,j)\in E}(p_j/c_{ij})$.

Chapter 5

Greedy algorithms

A game like chess can be won only by *thinking ahead*: a player who is focused entirely on immediate advantage is easy to defeat. But in many other games, such as Scrabble, it is possible to do quite well by simply making whichever move seems best at the moment and not worrying too much about future consequences.

This sort of myopic behavior is easy and convenient, making it an attractive algorithmic strategy. *Greedy* algorithms build up a solution piece by piece, always choosing the next piece that offers the most obvious and immediate benefit. Although such an approach can be disastrous for some computational tasks, there are many for which it is optimal. Our first example is that of minimum spanning trees.

5.1 Minimum spanning trees

Suppose you are asked to network a collection of computers by linking selected pairs of them. This translates into a graph problem in which nodes are computers, undirected edges are potential links, and the goal is to pick enough of these edges that the nodes are connected. But this is not all; each link also has a maintenance cost, reflected in that edge's weight. What is the cheapest possible network?

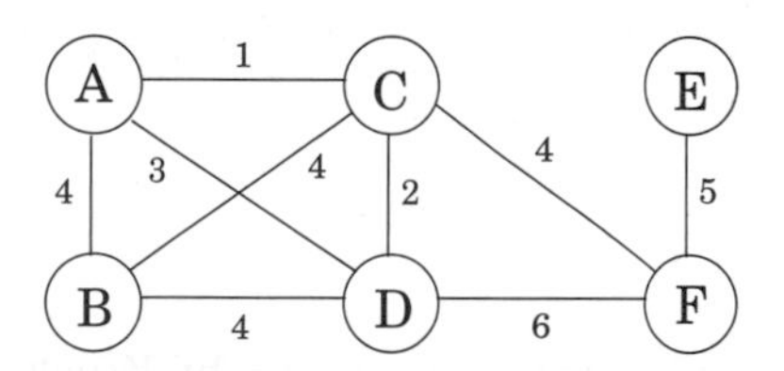

One immediate observation is that the optimal set of edges cannot contain a cycle, because removing an edge from this cycle would reduce the cost without compromising connectivity:

Property 1 *Removing a cycle edge cannot disconnect a graph.*

So the solution must be connected and acyclic: undirected graphs of this kind are called *trees*. The particular tree we want is the one with minimum total weight, known as the *minimum spanning tree*. Here is its formal definition.

Input: An undirected graph $G = (V, E)$; edge weights w_e.

Output: A tree $T = (V, E')$, with $E' \subseteq E$, that minimizes

$$\text{weight}(T) = \sum_{e \in E'} w_e.$$

In the preceding example, the minimum spanning tree has a cost of 16:

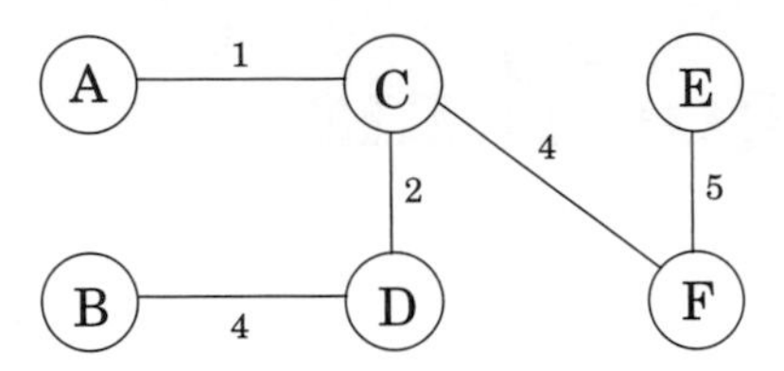

However, this is not the only optimal solution. Can you spot another?

5.1.1 A greedy approach

Kruskal's minimum spanning tree algorithm starts with the empty graph and then selects edges from E according to the following rule.

Repeatedly add the next lightest edge that doesn't produce a cycle.

In other words, it constructs the tree edge by edge and, apart from taking care to avoid cycles, simply picks whichever edge is cheapest at the moment. This is a *greedy* algorithm: every decision it makes is the one with the most obvious immediate advantage.

Figure 5.1 shows an example. We start with an empty graph and then attempt to add edges in increasing order of weight (ties are broken arbitrarily):

$$B-C,\ C-D,\ B-D,\ C-F,\ D-F,\ E-F,\ A-D,\ A-B,\ C-E,\ A-C.$$

The first two succeed, but the third, $B-D$, would produce a cycle if added. So we ignore it and move along. The final result is a tree with cost 14, the minimum possible.

Figure 5.1 The minimum spanning tree found by Kruskal's algorithm.

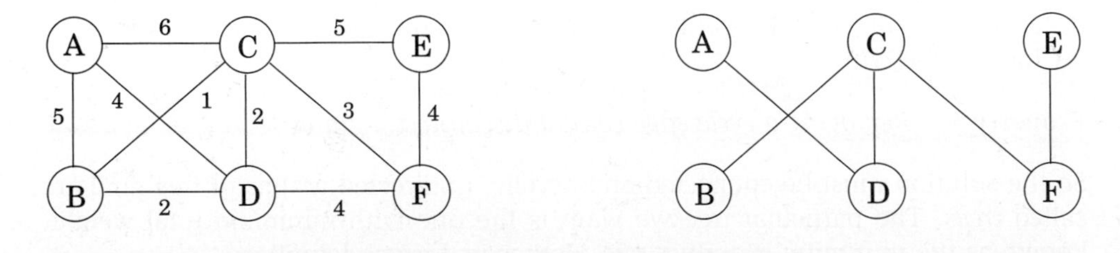

Trees

A *tree* is an undirected graph that is connected and acyclic. Much of what makes trees so useful is the simplicity of their structure. For instance,

Property 2 *A tree on n nodes has $n - 1$ edges.*

This can be seen by building the tree one edge at a time, starting from an empty graph. Initially each of the n nodes is disconnected from the others, in a connected component by itself. As edges are added, these components merge. Since each edge unites two different components, exactly $n - 1$ edges are added by the time the tree is fully formed.

In a little more detail: When a particular edge $\{u, v\}$ comes up, we can be sure that u and v lie in separate connected components, for otherwise there would already be a path between them and this edge would create a cycle. Adding the edge then merges these two components, thereby reducing the total number of connected components by one. Over the course of this incremental process, the number of components decreases from n to one, meaning that $n - 1$ edges must have been added along the way.

The converse is also true.

Property 3 *Any connected, undirected graph $G = (V, E)$ with $|E| = |V| - 1$ is a tree.*

We just need to show that G is acyclic. One way to do this is to run the following iterative procedure on it: while the graph contains a cycle, remove one edge from this cycle. The process terminates with some graph $G' = (V, E')$, $E' \subseteq E$, which is acyclic and, by Property 1 (from page 127), is also connected. Therefore G' is a tree, whereupon $|E'| = |V| - 1$ by Property 2. So $E' = E$, no edges were removed, and G was acyclic to start with.

In other words, we can tell whether a connected graph is a tree just by counting how many edges it has. Here's another characterization.

Property 4 *An undirected graph is a tree if and only if there is a unique path between any pair of nodes.*

In a tree, any two nodes can only have one path between them; for if there were two paths, the union of these paths would contain a cycle.

On the other hand, if a graph has a path between any two nodes, then it is connected. If these paths are unique, then the graph is also acyclic (since a cycle has two paths between any pair of nodes).

Figure 5.2 $T \cup \{e\}$. The addition of e (dotted) to T (solid lines) produces a cycle. This cycle must contain at least one other edge, shown here as e', across the cut $(S, V - S)$.

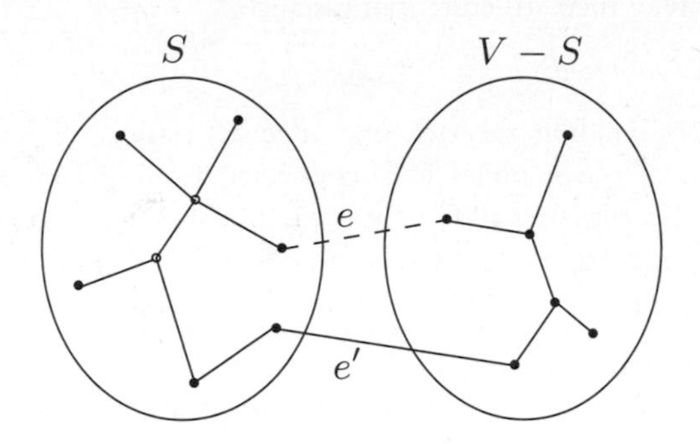

The correctness of Kruskal's method follows from a certain *cut property*, which is general enough to also justify a whole slew of other minimum spanning tree algorithms.

5.1.2 The cut property

Say that in the process of building a minimum spanning tree (MST), we have already chosen some edges and are so far on the right track. Which edge should we add next? The following lemma gives us a lot of flexibility in our choice.

Cut property *Suppose edges X are part of a minimum spanning tree of $G = (V, E)$. Pick any subset of nodes S for which X does not cross between S and $V - S$, and let e be the lightest edge across this partition. Then $X \cup \{e\}$ is part of some MST.*

A *cut* is any partition of the vertices into two groups, S and $V - S$. What this property says is that it is always safe to add the lightest edge across any cut (that is, between a vertex in S and one in $V - S$), provided X has no edges across the cut.

Let's see why this holds. Edges X are part of some MST T; if the new edge e also happens to be part of T, then there is nothing to prove. So assume e is not in T. We will construct a different MST T' containing $X \cup \{e\}$ by altering T slightly, changing just one of its edges.

Add edge e to T. Since T is connected, it already has a path between the endpoints of e, so adding e creates a cycle. This cycle must also have some other edge e' across the cut $(S, V - S)$ (Figure 5.2). If we now remove this edge, we are left with $T' = T \cup \{e\} - \{e'\}$, which we will show to be a tree. T' is connected by Property 1, since e' is a cycle edge. And it has the same number of edges as T; so by Properties 2 and 3, it is also a tree.

Moreover, T' is a minimum spanning tree. Compare its weight to that of T:

$$\text{weight}(T') = \text{weight}(T) + w(e) - w(e').$$

Figure 5.3 The cut property at work. (a) An undirected graph. (b) Set X has three edges, and is part of the MST T on the right. (c) If $S = \{A, B, C, D\}$, then one of the minimum-weight edges across the cut $(S, V - S)$ is $e = \{D, E\}$. $X \cup \{e\}$ is part of MST T', shown on the right.

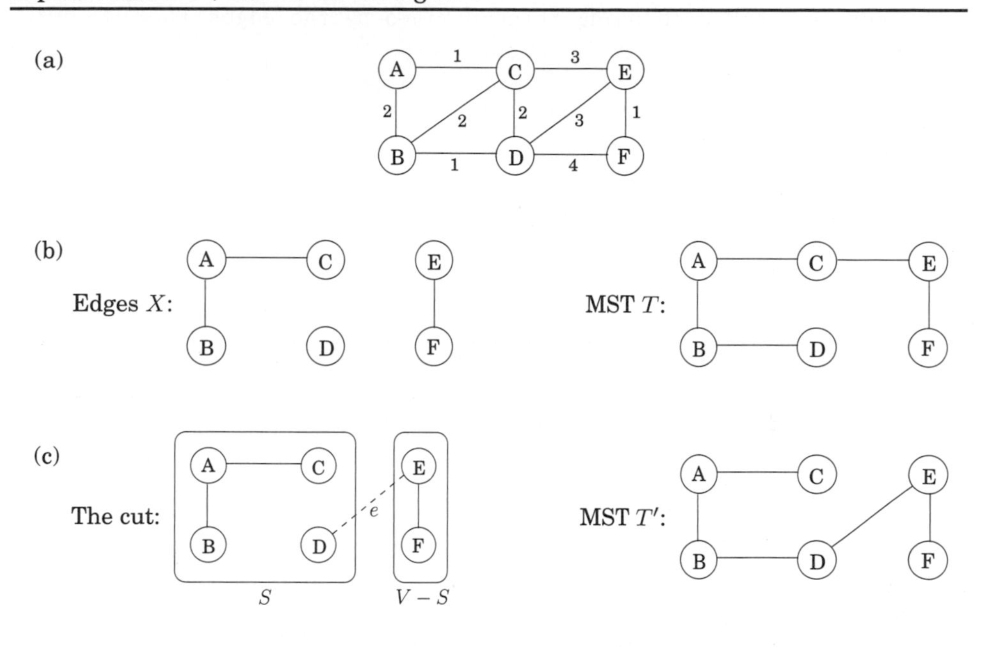

Both e and e' cross between S and $V - S$, and e is specifically the lightest edge of this type. Therefore $w(e) \leq w(e')$, and weight$(T') \leq$ weight(T). Since T is an MST, it must be the case that weight$(T') =$ weight(T) and that T' is also an MST.

Figure 5.3 shows an example of the cut property. Which edge is e'?

5.1.3 Kruskal's algorithm

We are ready to justify Kruskal's algorithm. At any given moment, the edges it has already chosen form a partial solution, a collection of connected components each of which has a tree structure. The next edge e to be added connects two of these components; call them T_1 and T_2. Since e is the lightest edge that doesn't produce a cycle, it is certain to be the lightest edge between T_1 and $V - T_1$ and therefore satisfies the cut property.

Now we fill in some implementation details. At each stage, the algorithm chooses an edge to add to its current partial solution. To do so, it needs to test each candidate edge $u - v$ to see whether the endpoints u and v lie in different components; otherwise the edge produces a cycle. And once an edge is chosen, the corresponding components need to be merged. What kind of data structure supports such operations?

Figure 5.4 Kruskal's minimum spanning tree algorithm.

```
procedure kruskal (G, w)
Input:  A connected undirected graph G = (V, E) with edge weights w_e
output: A minimum spanning three defined by the edges X

for all u ∈ V:
   makeset (u)

X = {}
sort the edges E by weight
for all edges {u, v} ∈ E, in increasing order of weight:
   if find(u) ≠ find(v):
      add edge {u, v} to X
      union(u, v)
```

We will model the algorithm's state as a collection of *disjoint sets*, each of which contains the nodes of a particular component. Initially each node is in a component by itself:

> `makeset`(x): create a singleton set containing just x.

We repeatedly test pairs of nodes to see if they belong to the same set.

> `find`(x): to which set does x belong?

And whenever we add an edge, we are merging two components.

> `union`(x, y): merge the sets containing x and y.

The final algorithm is shown in Figure 5.4. It uses $|V|$ `makeset`, $2|E|$ `find`, and $|V| - 1$ `union` operations.

5.1.4 A data structure for disjoint sets

Union by rank:
One way to store a set is as a directed tree (Figure 5.5). Nodes of the tree are elements of the set, arranged in no particular order, and each has parent pointers that eventually lead up to the root of the tree. This root element is a convenient *representative*, or *name*, for the set. It is distinguished from the other elements by the fact that its parent pointer is a self-loop.

In addition to a parent pointer π, each node also has a *rank* that, for the time being, should be interpreted as the height of the subtree hanging from that node.

```
procedure makeset(x)
π(x) = x
rank(x) = 0

function find(x)
while x ≠ π(x):  x = π(x)
return x
```

Figure 5.5 A directed-tree representation of two sets $\{B, E\}$ and $\{A, C, D, F, G, H\}$.

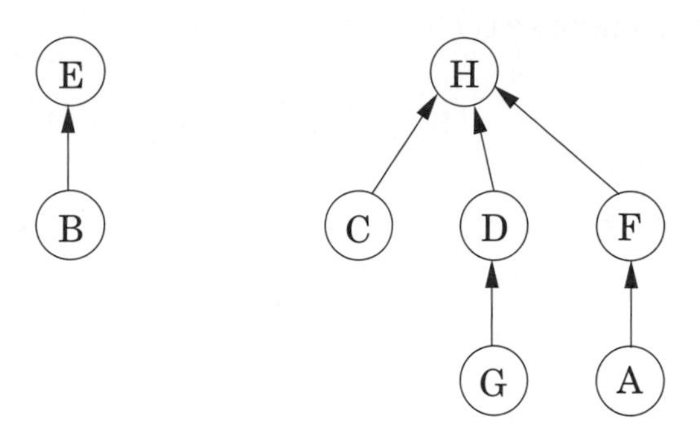

As can be expected, `makeset` is a constant-time operation. On the other hand, `find` follows parent pointers to the root of the tree and therefore takes time proportional to the height of the tree. The tree actually gets built via the third operation, `union`, and so we must make sure that this procedure keeps trees shallow.

Merging two sets is easy: make the root of one point to the root of the other. But we have a choice here. If the representatives (roots) of the sets are r_x and r_y, do we make r_x point to r_y or the other way around? Since tree height is the main impediment to computational efficiency, a good strategy is to *make the root of the shorter tree point to the root of the taller tree*. This way, the overall height increases only if the two trees being merged are equally tall. Instead of explicitly computing heights of trees, we will use the *rank* numbers of their root nodes—which is why this scheme is called *union by rank*.

```
procedure union(x, y)
r_x = find(x)
r_y = find(y)
if r_x = r_y: return
if rank(r_x) > rank(r_y):
    π(r_y) = r_x
else:
    π(r_x) = r_y
    if rank(r_x) = rank(r_y):  rank(r_y) = rank(r_y) + 1
```

See Figure 5.6 for an example.

By design, the *rank* of a node is exactly the height of the subtree rooted at that node. This means, for instance, that as you move up a path toward a root node, the *rank* values along the way are strictly increasing.

Property 1 *For any* x, $rank(x) < rank(\pi(x))$.

Figure 5.6 A sequence of disjoint-set operations. Superscripts denote rank.

After `makeset`(A),`makeset`(B),...,`makeset`(G):

After `union`(A,D),`union`(B,E),`union`(C,F):

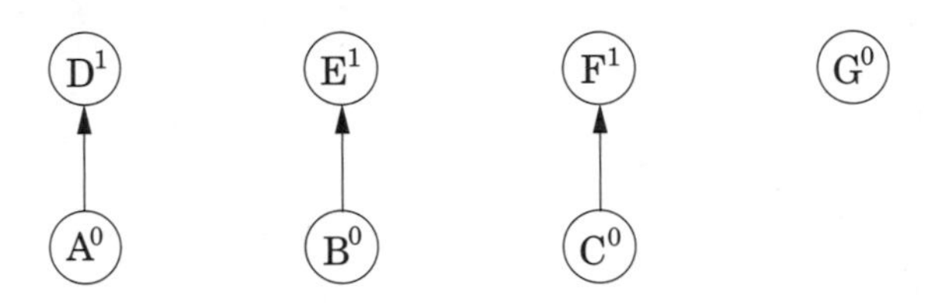

After `union`(C,G),`union`(E,A):

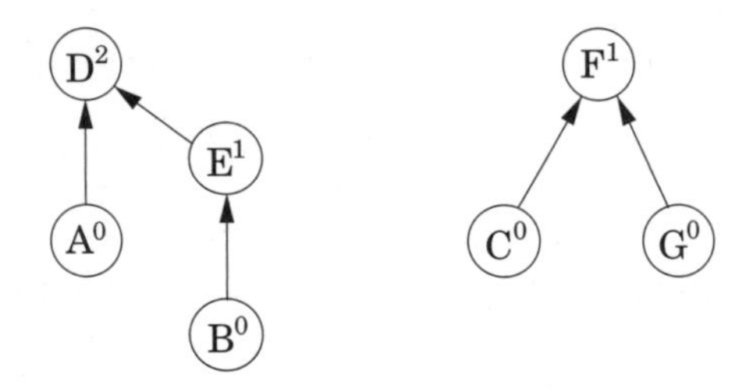

After `union`(B,G):

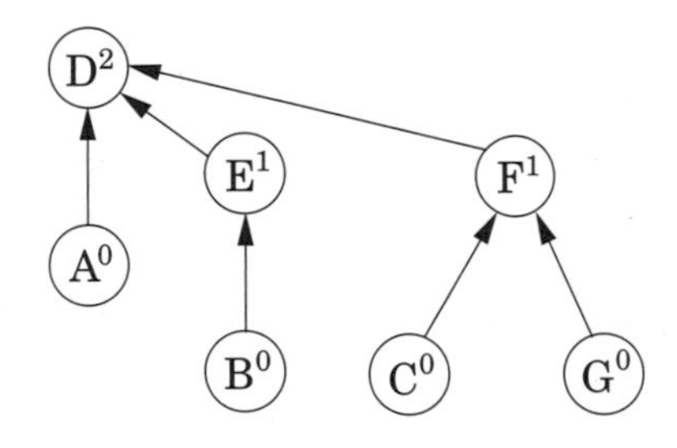

A root node with rank k is created by the merger of two trees with roots of rank $k-1$. It follows by induction (try it!) that

Property 2 *Any root node of rank k has at least* 2^k *nodes in its tree.*

This extends to internal (nonroot) nodes as well: a node of rank k has at least 2^k descendants. After all, any internal node was once a root, and neither its rank nor its set of descendants has changed since then. Moreover, different rank-k nodes cannot have common descendants, since by Property 1 any element has at most one ancestor of rank k. Which means

Property 3 *If there are n elements overall, there can be at most $n/2^k$ nodes of rank k.*

This last observation implies, crucially, that the maximum rank is $\log n$. Therefore, all the trees have height $\leq \log n$, and this is an upper bound on the running time of `find` and `union`.

Path compression:
With the data structure as presented so far, the total time for Kruskal's algorithm becomes $O(|E|\log|V|)$ for sorting the edges (remember, $\log|E| \approx \log|V|$) plus another $O(|E|\log|V|)$ for the `union` and `find` operations that dominate the rest of the algorithm. So there seems to be little incentive to make our data structure any more efficient.

But what if the edges are given to us sorted? Or if the weights are small (say, $O(|E|)$) so that sorting can be done in linear time? Then the data structure part becomes the bottleneck, and it is useful to think about improving its performance beyond $\log n$ per operation. As it turns out, the improved data structure is useful in many other applications.

But how can we perform `union`'s and `find`'s faster than $\log n$? The answer is, by being a little more careful to maintain our data structure in good shape. As any housekeeper knows, a little extra effort put into routine maintenance can pay off handsomely in the long run, by forestalling major calamities. We have in mind a particular maintenance operation for our union-find data structure, intended to keep the trees short—during each `find`, when a series of parent pointers is followed up to the root of a tree, we will change all these pointers so that they point directly to the root (Figure 5.7). This *path compression* heuristic only slightly increases the time needed for a `find` and is easy to code.

```
function find(x)
if x ≠ π(x): π(x) = find(π(x))
return π(x)
```

The benefit of this simple alteration is long-term rather than instantaneous and thus necessitates a particular kind of analysis: we need to look at *sequences* of `find` and `union` operations, starting from an empty data structure, and determine the average time per operation. This *amortized cost* turns out to be just barely more than $O(1)$, down from the earlier $O(\log n)$.

Think of the data structure as having a "top level" consisting of the root nodes, and below it, the insides of the trees. There is a division of labor: `find` operations (with or without path compression) only touch the insides of trees, whereas `union`'s only look at the top level. Thus path compression has no effect on `union` operations and leaves the top level unchanged.

We now know that the ranks of root nodes are unaltered, but what about *nonroot* nodes? The key point here is that once a node ceases to be a root, it never resurfaces,

Figure 5.7 The effect of path compression: `find`(I) followed by `find`(K).

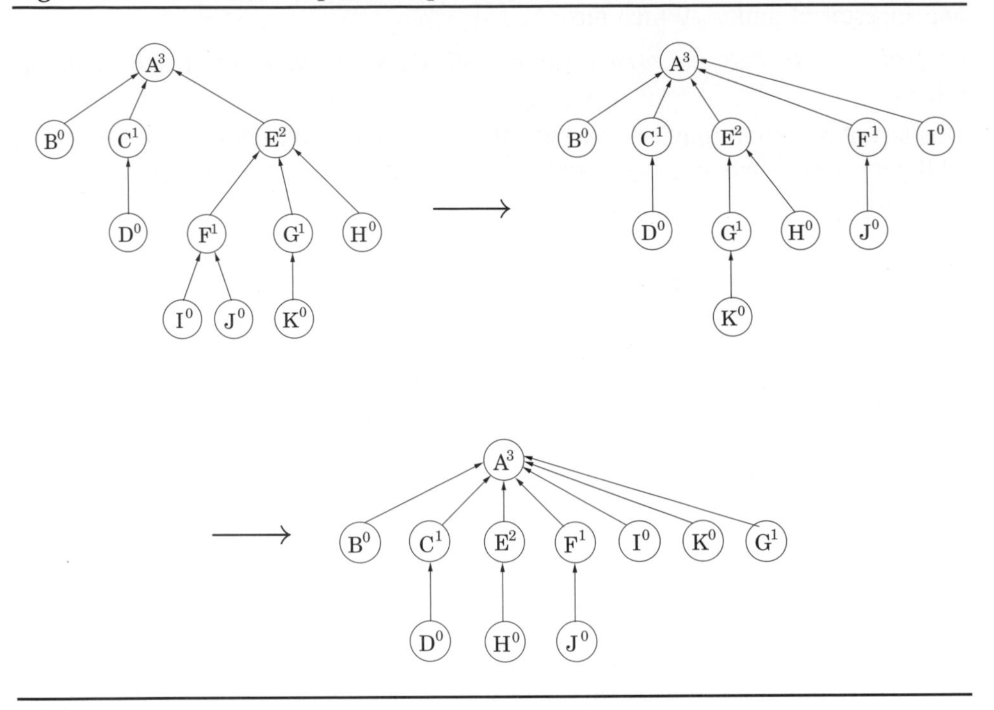

and its rank is forever fixed. Therefore the ranks of all nodes are unchanged by path compression, even though these numbers can no longer be interpreted as tree heights. In particular, properties 1–3 (from page 133) still hold.

If there are n elements, their rank values can range from 0 to $\log n$ by Property 3. Let's divide the nonzero part of this range into certain carefully chosen intervals, for reasons that will soon become clear:

$$\{1\}, \{2\}, \{3, 4\}, \{5, 6, \ldots, 16\}, \{17, 18, \ldots, 2^{16} = 65536\}, \{65537, 65538, \ldots, 2^{65536}\}, \ldots$$

Each group is of the form $\{k+1, k+2, \ldots, 2^k\}$, where k is a power of 2. The number of groups is $\log^* n$, which is defined to be the number of successive log operations that need to be applied to n to bring it down to 1 (or below 1). For instance, $\log^* 1000 = 4$ since $\log\log\log\log 1000 \leq 1$. In practice there will just be the first five of the intervals shown; more are needed only if $n \geq 2^{65536}$, in other words never.

In a sequence of `find` operations, some may take longer than others. We'll bound the overall running time using some creative accounting. Specifically, we will give each node a certain amount of pocket money, such that the total money doled out is at most $n \log^* n$ dollars. We will then show that each `find` takes $O(\log^* n)$ steps, plus some additional amount of time that can be "paid for" using the pocket money of the nodes involved—one dollar per unit of time. Thus the overall time for m `find`'s is $O(m \log^* n)$ plus at most $O(n \log^* n)$.

In more detail, a node receives its allowance as soon as it ceases to be a root, at which point its rank is fixed. If this rank lies in the interval $\{k+1, \ldots, 2^k\}$, the node receives 2^k dollars. By Property 3, the number of nodes with rank $> k$ is bounded by

$$\frac{n}{2^{k+1}} + \frac{n}{2^{k+2}} + \cdots \leq \frac{n}{2^k}.$$

Therefore the total money given to nodes in this particular interval is at most n dollars, and since there are $\log^* n$ intervals, the total money disbursed to all nodes is $\leq n \log^* n$.

Now, the time taken by a specific `find` is simply the number of pointers followed. Consider the ascending rank values along this chain of nodes up to the root. Nodes x on the chain fall into two categories: either the rank of $\pi(x)$ is in a higher interval than the rank of x, or else it lies in the same interval. There are at most $\log^* n$ nodes of the first type (do you see why?), so the work done on them takes $O(\log^* n)$ time. The remaining nodes—whose parents' ranks are in the same interval as theirs—have to pay a dollar out of their pocket money for their processing time.

This only works if the initial allowance of each node x is enough to cover all of its payments in the sequence of `find` operations. Here's the crucial observation: each time x pays a dollar, its parent changes to one of higher rank. Therefore, if x's rank lies in the interval $\{k+1, \ldots, 2^k\}$, it has to pay at most 2^k dollars before its parent's rank is in a higher interval; whereupon it never has to pay again.

5.1.5 Prim's algorithm

Let's return to our discussion of minimum spanning tree algorithms. What the cut property tells us in most general terms is that any algorithm conforming to the following greedy schema is guaranteed to work.

```
X = { } (edges picked so far)
repeat until |X| = |V| − 1:
  pick a set S ⊂ V for which X has no edges between S and V − S
  let e ∈ E be the minimum-weight edge between S and V − S
  X = X ∪ {e}
```

A popular alternative to Kruskal's algorithm is Prim's, in which the intermediate set of edges X always forms a subtree, and S is chosen to be the set of this tree's vertices.

On each iteration, the subtree defined by X *grows* by one edge, namely, the lightest edge between a vertex in S and a vertex outside S (Figure 5.8). We can equivalently think of S as growing to include the vertex $v \notin S$ of smallest `cost`:

$$\texttt{cost}(v) = \min_{u \in S} w(u, v).$$

Figure 5.8 Prim's algorithm: the edges X form a tree, and S consists of its vertices.

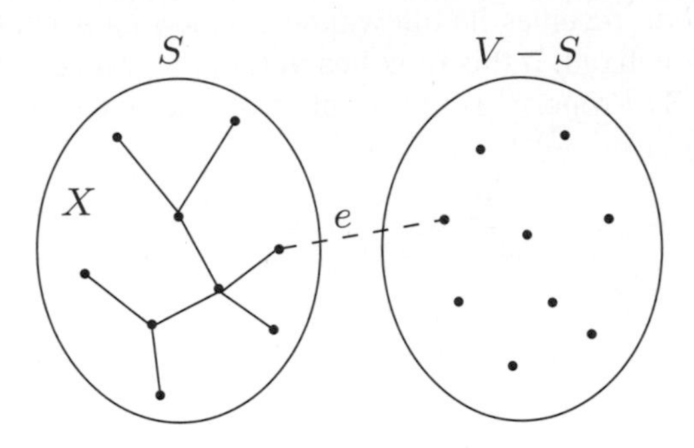

This is strongly reminiscent of Dijkstra's algorithm, and in fact the pseudocode (Figure 5.9) is almost identical. The only difference is in the key values by which the priority queue is ordered. In Prim's algorithm, the value of a node is the weight of the lightest incoming edge from set S, whereas in Dijkstra's it is the length of an entire path to that node from the starting point. Nonetheless, the two algorithms are similar enough that they have the same running time, which depends on the particular priority queue implementation.

Figure 5.9 shows Prim's algorithm at work, on a small six-node graph. Notice how the final MST is completely specified by the `prev` array.

5.2 Huffman encoding

In the MP3 audio compression scheme, a sound signal is encoded in three steps.

1. It is digitized by sampling at regular intervals, yielding a sequence of real numbers $s_1, s_2, \ldots, s_T$. For instance, at a rate of 44,100 samples per second, a 50-minute symphony would correspond to $T = 50 \times 60 \times 44{,}100 \approx 130$ million measurements.[1]
2. Each real-valued sample s_t is *quantized*: approximated by a nearby number from a finite set Γ. This set is carefully chosen to exploit human perceptual limitations, with the intention that the approximating sequence is indistinguishable from $s_1, s_2, \ldots, s_T$ by the human ear.
3. The resulting string of length T over alphabet Γ is encoded in binary.

It is in the last step that Huffman encoding is used. To understand its role, let's look at a toy example in which T is 130 million and the alphabet Γ consists of just four

[1]For stereo sound, two channels would be needed, doubling the number of samples.

Figure 5.9 *Top:* Prim's minimum spanning tree algorithm. *Below:* An illustration of Prim's algorithm, starting at node *A*. Also shown are a table of `cost/prev` values, and the final MST.

```
procedure prim (G, w)
Input:  A connected undirected graph G = (V, E) with edge
        weights w_e
output: A minimum spanning tree defined by the array prev

for all u ∈ V:
   cost(u) = ∞
   prev(u) = nil
pick any initial node u_0
cost(u_0) = 0

H = makequeue (V)  (priority queue, using cost-values as keys)
while H is not empty:
    v = deletemin (H)
    for each {v, z} ∈ E:
       if cost(z) > w(v, z):
          cost(z) > w(v, z)
          prev(z) = v
```

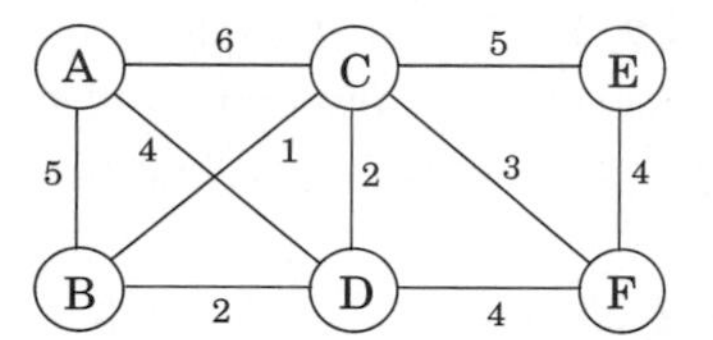

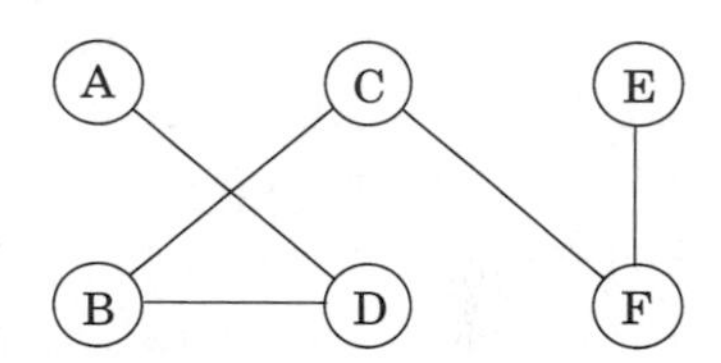

Set S	A	B	C	D	E	F
$\{\}$	0/nil	∞/nil	∞/nil	∞/nil	∞/nil	∞/nil
A		$5/A$	$6/A$	$4/A$	∞/nil	∞/nil
A, D		$2/D$	$2/D$		∞/nil	$4/D$
A, D, B			$1/B$		∞/nil	$4/D$
A, D, B, C					$5/C$	$3/C$
A, D, B, C, F					$4/F$	

values, denoted by the symbols A, B, C, D. What is the most economical way to write this long string in binary? The obvious choice is to use 2 bits per symbol—say codeword 00 for A, 01 for B, 10 for C, and 11 for D—in which case 260 megabits are needed in total. Can there possibly be a better encoding than this?

A randomized algorithm for minimum cut

We have already seen that spanning trees and cuts are intimately related. Here is another connection. Let's remove the last edge that Kruskal's algorithm adds to the spanning tree; this breaks the tree into two components, thus defining a cut $(S, \overline{S})$ in the graph. What can we say about this cut? Suppose the graph we were working with was unweighted, and that its edges were ordered uniformly at random for Kruskal's algorithm to process them. Here is a remarkable fact: with probability at least $1/n^2$, $(S, \overline{S})$ is the minimum cut in the graph, where the size of a cut $(S, \overline{S})$ is the number of edges crossing between S and $\overline{S}$. This means that repeating the process $O(n^2)$ times and outputting the smallest cut found yields the minimum cut in G with high probability: an $O(mn^2 \log n)$ algorithm for unweighted minimum cuts. Some further tuning gives the $O(n^2 \log n)$ minimum cut algorithm, invented by David Karger, which is the fastest known algorithm for this important problem.

So let us see why the cut found in each iteration is the minimum cut with probability at least $1/n^2$. At any stage of Kruskal's algorithm, the vertex set V is partitioned into connected components. The only edges eligible to be added to the tree have their two endpoints in distinct components. The number of edges incident to each component must be at least C, the size of the minimum cut in G (since we could consider a cut that separated this component from the rest of the graph). So if there are k components in the graph, the number of eligible edges is at least $kC/2$ (each of the k components has at least C edges leading out of it, and we need to compensate for the double-counting of each edge). Since the edges were randomly ordered, the chance that the next eligible edge in the list is from the minimum cut is at most $C/(kC/2) = 2/k$. Thus, with probability at least $1 - 2/k = (k-2)/k$, the choice leaves the minimum cut intact. But now the chance that Kruskal's algorithm leaves the minimum cut intact all the way up to the choice of the last spanning tree edge is at least

$$\frac{n-2}{n} \cdot \frac{n-3}{n-1} \cdot \frac{n-4}{n-2} \cdots \frac{2}{4} \cdot \frac{1}{3} = \frac{1}{n(n-1)}.$$

In search of inspiration, we take a closer look at our particular sequence and find that the four symbols are not equally abundant.

Symbol	Frequency
A	70 million
B	3 million
C	20 million
D	37 million

Is there some sort of *variable-length encoding*, in which just *one* bit is used for the frequently occurring symbol A, possibly at the expense of needing three or more bits for less common symbols?

A danger with having codewords of different lengths is that the resulting encoding may not be uniquely decipherable. For instance, if the codewords are $\{0, 01, 11, 001\}$,

Figure 5.10 A prefix-free encoding. Frequencies are shown in square brackets.

Symbol	Codeword
A	0
B	100
C	101
D	11

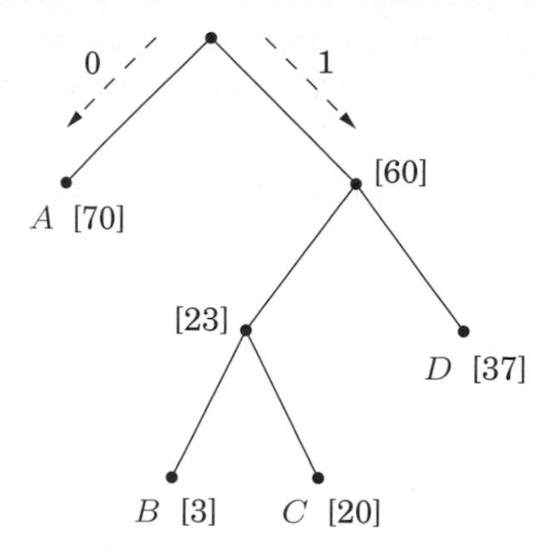

the decoding of strings like 001 is ambiguous. We will avoid this problem by insisting on the *prefix-free* property: no codeword can be a prefix of another codeword.

Any prefix-free encoding can be represented by a *full* binary tree—that is, a binary tree in which every node has either zero or two children—where the symbols are at the leaves, and where each codeword is generated by a path from root to leaf, interpreting left as 0 and right as 1 (Exercise 5.29). Figure 5.10 shows an example of such an encoding for the four symbols A, B, C, D. Decoding is unique: a string of bits is decrypted by starting at the root, reading the string from left to right to move downward, and, whenever a leaf is reached, outputting the corresponding symbol and returning to the root. It is a simple scheme and pays off nicely for our toy example, where (under the codes of Figure 5.10) the total size of the binary string drops to 213 megabits, a 17% improvement.

In general, how do we find the optimal coding tree, given the frequencies $f_1, f_2, \ldots, f_n$ of n symbols? To make the problem precise, we want a tree whose leaves each correspond to a symbol and which minimizes the overall length of the encoding,

$$\text{cost of tree} = \sum_{i=1}^{n} f_i \cdot (\text{depth of } i\text{th symbol in tree})$$

(the number of bits required for a symbol is exactly its depth in the tree).

There is another way to write this cost function that is very helpful. Although we are only given frequencies for the leaves, we can define the frequency of any *internal* node to be the sum of the frequencies of its descendant leaves; this is, after all, the number of times the internal node is visited during encoding or decoding. During the encoding process, each time we move down the tree, one bit gets output for every nonroot node through which we pass. So the total cost—the total number of bits which are output—can also be expressed thus:

> *The cost of a tree is the sum of the frequencies of all leaves and internal nodes, except the root.*

The first formulation of the cost function tells us that *the two symbols with the smallest frequencies must be at the bottom of the optimal tree,* as children of the lowest internal node (this internal node has two children since the tree is *full*). Otherwise, swapping these two symbols with whatever is lowest in the tree would improve the encoding.

This suggests that we start constructing the tree *greedily*: find the two symbols with the smallest frequencies, say i and j, and make them children of a new node, which then has frequency $f_i + f_j$. To keep the notation simple, let's just assume these are f_1 and f_2. By the second formulation of the cost function, any tree in which f_1 and f_2 are sibling-leaves has cost $f_1 + f_2$ plus the cost for a tree with $n-1$ leaves of frequencies $(f_1 + f_2), f_3, f_4, \ldots, f_n$:

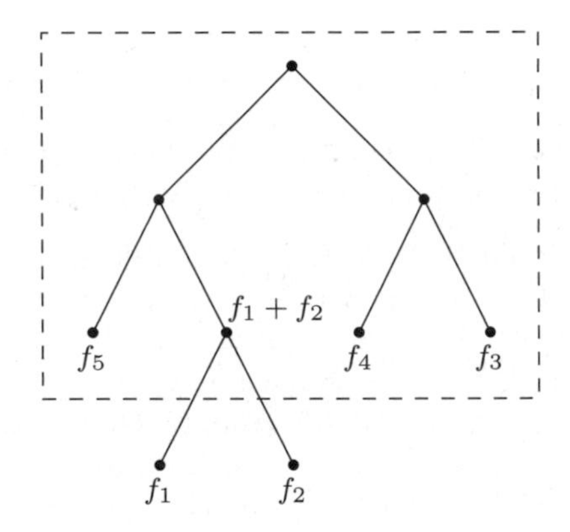

The latter problem is just a smaller version of the one we started with. So we pull f_1 and f_2 off the list of frequencies, insert $(f_1 + f_2)$, and loop. The resulting algorithm can be described in terms of priority queue operations (as defined on page 109) and takes $O(n \log n)$ time if a binary heap (Section 4.5.2) is used.

```
procedure Huffman(f)
Input:  An array f[1···n] of frequencies
Output: An encoding tree with n leaves

let H be a priority queue of integers, ordered by f
for i = 1 to n: insert(H, i)
for k = n+1 to 2n−1:
   i = deletemin(H), j = deletemin(H)
   create a node numbered k with children i, j
   f[k] = f[i] + f[j]
   insert(H, k)
```

Returning to our toy example: can you tell if the tree of Figure 5.10 is optimal?

Entropy

The annual county horse race is bringing in three thoroughbreds who have never competed against one another. Excited, you study their past 200 races and summarize these as probability distributions over four outcomes: `first` ("first place"), `second`, `third`, and `other`.

Outcome	Aurora	Whirlwind	Phantasm
`first`	0.15	0.30	0.20
`second`	0.10	0.05	0.30
`third`	0.70	0.25	0.30
`other`	0.05	0.40	0.20

Which horse is the most predictable? One quantitative approach to this question is to look at *compressibility*. Write down the history of each horse as a string of 200 values (`first`, `second`, `third`, `other`). The total number of bits needed to encode these track-record strings can then be computed using Huffman's algorithm. This works out to 290 bits for Aurora, 380 for Whirlwind, and 420 for Phantasm (check it!). Aurora has the shortest encoding and is therefore in a strong sense the most predictable.

The inherent unpredictability, or *randomness*, of a probability distribution can be measured by the extent to which it is possible to compress data drawn from that distribution.

$$\text{more compressible} \equiv \text{less random} \equiv \text{more predictable}$$

Suppose there are n possible outcomes, with probabilities $p_1, p_2, \ldots, p_n$. If a sequence of m values is drawn from the distribution, then the ith outcome will pop up roughly mp_i times (if m is large). For simplicity, assume these are exactly the observed frequencies, and moreover that the p_i's are all powers of 2 (that is, of the form $1/2^k$). It can be seen by induction (Exercise 5.19) that the number of bits needed to encode the sequence is $\sum_{i=1}^n mp_i \log(1/p_i)$. Thus the average number of bits needed to encode a single draw from the distribution is

$$\sum_{i=1}^{n} p_i \log \frac{1}{p_i}.$$

This is the *entropy* of the distribution, a measure of how much randomness it contains.

For example, a fair coin has two outcomes, each with probability $1/2$. So its entropy is

$$\tfrac{1}{2} \log 2 + \tfrac{1}{2} \log 2 = 1.$$

This is natural enough: the coin flip contains one bit of randomness. But what if the coin is not fair, if it has a $3/4$ chance of turning up heads? Then the entropy is

$$\tfrac{3}{4} \log \tfrac{4}{3} + \tfrac{1}{4} \log 4 = 0.81.$$

A biased coin is more predictable than a fair coin, and thus has lower entropy. As the bias becomes more pronounced, the entropy drops toward zero.

We explore these notions further in Exercises 5.18 and 5.19.

5.3 Horn formulas

In order to display human-level intelligence, a computer must be able to perform at least some modicum of logical reasoning. Horn formulas are a particular framework for doing this, for expressing logical facts and deriving conclusions.

The most primitive object in a Horn formula is a *Boolean variable,* taking value either `true` or `false`. For instance, variables x, y, and z might denote the following possibilities.

$$x \equiv \text{the murder took place in the kitchen}$$
$$y \equiv \text{the butler is innocent}$$
$$z \equiv \text{the colonel was asleep at 8 pm}$$

A *literal* is either a variable x or its negation $\overline{x}$ ("NOT x"). In Horn formulas, knowledge about variables is represented by two kinds of *clauses*:

1. *Implications,* whose left-hand side is an AND of any number of positive literals and whose right-hand side is a single positive literal. These express statements of the form "if the conditions on the left hold, then the one on the right must also be true." For instance,

   $$(z \wedge w) \Rightarrow u$$

   might mean "if the colonel was asleep at 8 pm and the murder took place at 8 pm then the colonel is innocent." A degenerate type of implication is the *singleton* "$\Rightarrow x$," meaning simply that x is `true`: "the murder definitely occurred in the kitchen."
2. Pure *negative clauses*, consisting of an OR of any number of negative literals, as in

   $$(\overline{u} \vee \overline{v} \vee \overline{y})$$

   ("they can't all be innocent").

Given a set of clauses of these two types, the goal is to determine whether there is a consistent explanation: an assignment of `true/false` values to the variables that satisfies all the clauses. This is also called a *satisfying assignment.*

The two kinds of clauses pull us in different directions. The implications tell us to set some of the variables to `true`, while the negative clauses encourage us to make them `false`. Our strategy for solving a Horn formula is this: We start with all variables `false`. We then proceed to set some of them to `true`, one by one, but very reluctantly, and only if we absolutely have to because an implication would otherwise be violated. Once we are done with this phase and all implications are satisfied, only then do we turn to the negative clauses and make sure they are all satisfied.

In other words, our algorithm for Horn clauses is the following greedy scheme (*stingy* is perhaps more descriptive):

```
Input:  a Horn formula
Output: a satisfying assignment, if one exists

set all variables to false

while there is an implication that is not satisfied:
  set the right-hand variable of the implication to true

if all pure negative clauses are satisfied:
  return the assignment
else: return "formula is not satisfiable"
```

For instance, suppose the formula is

$$(w \wedge y \wedge z) \Rightarrow x, \quad (x \wedge z) \Rightarrow w, \quad x \Rightarrow y, \quad \Rightarrow x, \quad (x \wedge y) \Rightarrow w, \quad (\overline{w} \vee \overline{x} \vee \overline{y}), \quad (\overline{z}).$$

We start with everything `false` and then notice that x must be `true` on account of the singleton implication $\Rightarrow x$. Then we see that y must also be `true`, because of $x \Rightarrow y$. And so on.

To see why the algorithm is correct, notice that if it returns an assignment, this assignment satisfies both the implications and the negative clauses, and so it is indeed a satisfying truth assignment of the input Horn formula. So we only have to convince ourselves that if the algorithm finds no satisfying assignment, then there really is none. This is so because our "stingy" rule maintains the following invariant:

> *If a certain set of variables is set to* `true`*, then they must be* `true` *in* any *satisfying assignment.*

Hence, if the truth assignment found after the *while* loop does not satisfy the negative clauses, there can be no satisfying truth assignment.

Horn formulas lie at the heart of Prolog ("programming by logic"), a language in which you program by specifying desired properties of the output, using simple logical expressions. The workhorse of Prolog interpreters is our greedy satisfiability algorithm. Conveniently, it can be implemented in time linear in the length of the formula; do you see how (Exercise 5.33)?

5.4 Set cover

The dots in Figure 5.11 represent a collection of towns. This county is in its early stages of planning and is deciding where to put schools. There are only two constraints: each school should be in a town, and no one should have to travel more than 30 miles to reach one of them. What is the minimum number of schools needed?

This is a typical *set cover* problem. For each town x, let S_x be the set of towns within 30 miles of it. A school at x will essentially "cover" these other towns. The question

Figure 5.11 (a) Eleven towns. (b) Towns that are within 30 miles of each other.

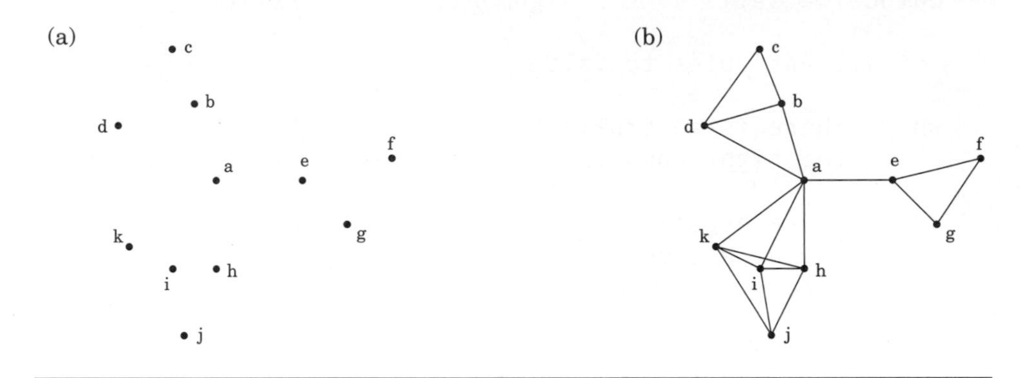

is then, how many sets S_x must be picked in order to cover all the towns in the county?

Set Cover

Input: A set of elements B; sets $S_1, \ldots, S_m \subseteq B$

Output: A selection of the S_i whose union is B.

Cost: Number of sets picked.

(In our example, the elements of B are the towns.) This problem lends itself immediately to a greedy solution:

Repeat until all elements of B are covered:

Pick the set S_i with the largest number of uncovered elements.

This is extremely natural and intuitive. Let's see what it would do on our earlier example: It would first place a school at town a, since this covers the largest number of other towns. Thereafter, it would choose three more schools—c, j, and either f or g—for a total of four. However, there exists a solution with just three schools, at b, e, and i. The greedy scheme is not optimal!

But luckily, it isn't too far from optimal.

Claim *Suppose B contains n elements and that the optimal cover consists of k sets. Then the greedy algorithm will use at most $k \ln n$ sets.*[2]

Let n_t be the number of elements still not covered after t iterations of the greedy algorithm (so $n_0 = n$). Since these remaining elements are covered by the optimal

[2] ln means "natural logarithm," that is, to the base e.

k sets, there must be some set with at least n_t/k of them. Therefore, the greedy strategy will ensure that

$$n_{t+1} \le n_t - \frac{n_t}{k} = n_t\left(1 - \frac{1}{k}\right),$$

which by repeated application implies $n_t \le n_0(1 - 1/k)^t$. A more convenient bound can be obtained from the useful inequality

$$1 - x \le e^{-x} \text{ for all } x, \text{ with equality if and only if } x = 0,$$

which is most easily proved by a picture:

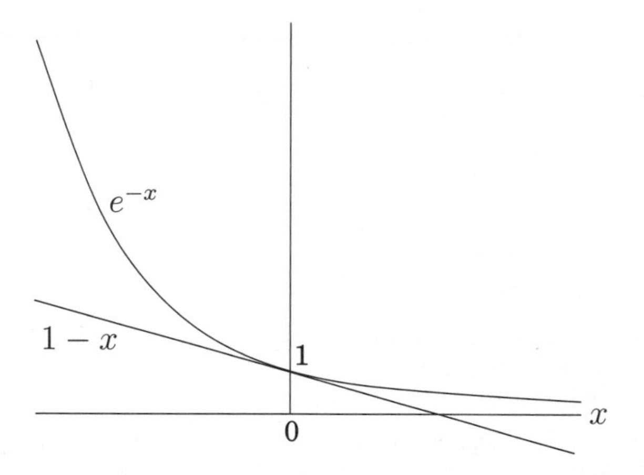

Thus

$$n_t \le n_0\left(1 - \frac{1}{k}\right)^t < n_0(e^{-1/k})^t = ne^{-t/k}$$

At $t = k \ln n$, therefore, n_t is strictly less than $ne^{-\ln n} = 1$, which means no elements remain to be covered.

The ratio between the greedy algorithm's solution and the optimal solution varies from input to input but is always less than $\ln n$. And there are certain inputs for which the ratio is very close to $\ln n$ (Exercise 5.34). We call this maximum ratio the *approximation factor* of the greedy algorithm. There seems to be a lot of room for improvement, but in fact such hopes are unjustified: it turns out that under certain widely-held complexity assumptions (which will be clearer when we reach Chapter 8), there is provably no polynomial-time algorithm with a smaller approximation factor.

Exercises

5.1. Consider the following graph.

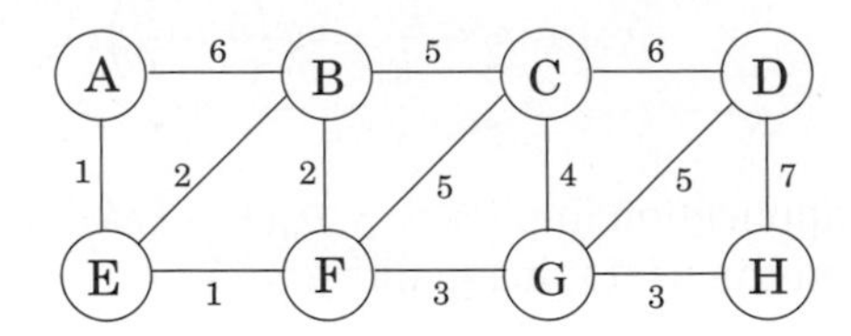

(a) What is the cost of its minimum spanning tree?

(b) How many minimum spanning trees does it have?

(c) Suppose Kruskal's algorithm is run on this graph. In what order are the edges added to the MST? For each edge in this sequence, give a cut that justifies its addition.

5.2. Suppose we want to find the minimum spanning tree of the following graph.

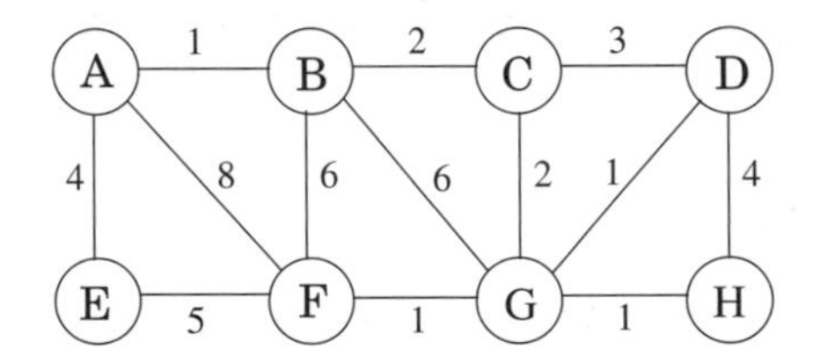

(a) Run Prim's algorithm; whenever there is a choice of nodes, always use alphabetic ordering (e.g., start from node A). Draw a table showing the intermediate values of the `cost` array.

(b) Run Kruskal's algorithm on the same graph. Show how the disjoint-sets data structure looks at every intermediate stage (including the structure of the directed trees), assuming path compression is used.

5.3. Design a linear-time algorithm for the following task.

Input: A connected, undirected graph G.

Question: Is there an edge you can remove from G while still leaving G connected?

Can you reduce the running time of your algorithm to $O(|V|)$?

5.4. Show that if an undirected graph with n vertices has k connected components, then it has at least $n - k$ edges.

5.5. Consider an undirected graph $G = (V, E)$ with nonnegative edge weights $w_e \geq 0$. Suppose that you have computed a minimum spanning tree of G, and that you have also computed shortest paths to all nodes from a particular node $s \in V$. Now suppose each edge weight is increased by 1: the new weights are $w'_e = w_e + 1$.

(a) Does the minimum spanning tree change? Give an example where it changes or prove it cannot change.

(b) Do the shortest paths change? Give an example where they change or prove they cannot change.

5.6. Let $G = (V, E)$ be an undirected graph. Prove that if all its edge weights are distinct, then it has a unique minimum spanning tree.

5.7. Show how to find the *maximum* spanning tree of a graph, that is, the spanning tree of largest total weight.

5.8. Suppose you are given a connected weighted graph $G = (V, E)$ with a distinguished vertex s and where all edge weights are positive and distinct. Is it possible for a tree of shortest paths from s and a minimum spanning tree in G to not share any edges? If so, give an example. If not, give a reason.

5.9. The following statements may or may not be correct. In each case, either prove it (if it is correct) or give a counterexample (if it isn't correct). Always assume that the graph $G = (V, E)$ is undirected and connected. Do not assume that edge weights are distinct unless this is specifically stated.

(a) If graph G has more than $|V| - 1$ edges, and there is a unique heaviest edge, then this edge cannot be part of a minimum spanning tree.

(b) If G has a cycle with a unique heaviest edge e, then e cannot be part of any MST.

(c) Let e be any edge of minimum weight in G. Then e must be part of some MST.

(d) If the lightest edge in a graph is unique, then it must be part of every MST.

(e) If e is part of some MST of G, then it must be a lightest edge across some cut of G.

(f) If G has a cycle with a unique lightest edge e, then e must be part of every MST.

(g) The shortest-path tree computed by Dijkstra's algorithm is necessarily an MST.

(h) The shortest path between two nodes is necessarily part of some MST.

(i) Prim's algorithm works correctly when there are negative edges.

(j) (For any $r > 0$, define an *r-path* to be a path whose edges all have weight $< r$.) If G contains an r-path from node s to t, then every MST of G must also contain an r-path from node s to node t.

5.10. Let T be an MST of graph G. Given a connected subgraph H of G, show that $T \cap H$ is contained in some MST of H.

5.11. Give the state of the disjoint-sets data structure after the following sequence of operations, starting from singleton sets $\{1\}, \ldots, \{8\}$. Use path compression. In case of ties, always make the lower numbered root point to the higher numbered one.

```
union(1, 2), union(3, 4), union(5, 6), union(7, 8), union(1, 4),
    union(6, 7), union(4, 5), find(1)
```

5.12. Suppose you implement the disjoint-sets data structure using union-by-rank but not path compression. Give a sequence of m `union` and `find` operations on n elements that take $\Omega(m \log n)$ time.

5.13. A long string consists of the four characters A, C, G, T; they appear with frequency 31%, 20%, 9%, and 40%, respectively. What is the Huffman encoding of these four characters?

5.14. Suppose the symbols a, b, c, d, e occur with frequencies 1/2, 1/4, 1/8, 1/16, 1/16, respectively.

(a) What is the Huffman encoding of the alphabet?

(b) If this encoding is applied to a file consisting of 1,000,000 characters with the given frequencies, what is the length of the encoded file in bits?

5.15. We use Huffman's algorithm to obtain an encoding of alphabet $\{a, b, c\}$ with frequencies f_a, f_b, f_c. In each of the following cases, either give an example of frequencies (f_a, f_b, f_c) that would yield the specified code, or explain why the code cannot possibly be obtained (no matter what the frequencies are).

(a) Code: $\{0, 10, 11\}$

(b) Code: $\{0, 1, 00\}$

(c) Code: $\{10, 01, 00\}$

5.16. Prove the following two properties of the Huffman encoding scheme.

(a) If some character occurs with frequency more than 2/5, then there is guaranteed to be a codeword of length 1.

(b) If all characters occur with frequency less than 1/3, then there is guaranteed to be no codeword of length 1.

5.17. Under a Huffman encoding of n symbols with frequencies $f_1, f_2, \ldots, f_n$, what is the longest a codeword could possibly be? Give an example set of frequencies that would produce this case.

5.18. The following table gives the frequencies of the letters of the English language (including the blank for separating words) in a particular corpus.

blank	18.3%	r	4.8%	y	1.6%
e	10.2%	d	3.5%	p	1.6%
t	7.7%	l	3.4%	b	1.3%
a	6.8%	c	2.6%	v	0.9%
o	5.9%	u	2.4%	k	0.6%
i	5.8%	m	2.1%	j	0.2%
n	5.5%	w	1.9%	x	0.2%
s	5.1%	f	1.8%	q	0.1%
h	4.9%	g	1.7%	z	0.1%

(a) What is the optimum Huffman encoding of this alphabet?

(b) What is the expected number of bits per letter?

(c) Suppose now that we calculate the entropy of these frequencies

$$H = \sum_{i=0}^{26} p_i \log \frac{1}{p_i}$$

(see the box in page 143). Would you expect it to be larger or smaller than your answer above? Explain.

(d) Do you think that this is the limit of how much English text can be compressed? What features of the English language, besides letters and their frequencies, should a better compression scheme take into account?

5.19. *Entropy.* Consider a distribution over n possible outcomes, with probabilities $p_1, p_2, \ldots, p_n$.

(a) Just for this part of the problem, assume that each p_i is a power of 2 (that is, of the form $1/2^k$). Suppose a long sequence of m samples is drawn from the distribution and that for all $1 \leq i \leq n$, the ith outcome occurs exactly mp_i times in the sequence. Show that if Huffman encoding is applied to this sequence, the resulting encoding will have length

$$\sum_{i=1}^{n} mp_i \log \frac{1}{p_i}.$$

(b) Now consider arbitrary distributions—that is, the probabilities p_i are not restricted to powers of 2. The most commonly used measure of the *amount of randomness* in the distribution is the *entropy*

$$\sum_{i=1}^{n} p_i \log \frac{1}{p_i}.$$

For what distribution (over n outcomes) is the entropy the largest possible? The smallest possible?

5.20. Give a linear-time algorithm that takes as input a tree and determines whether it has a *perfect matching*: a set of edges that touches each node exactly once.

5.21. A *feedback edge set* of an undirected graph $G = (V, E)$ is a subset of edges $E' \subseteq E$ that intersects every cycle of the graph. Thus, removing the edges E' will render the graph acyclic.

Give an efficient algorithm for the following problem:

Input: Undirected graph $G = (V, E)$ with positive edge weights w_e
Output: A feedback edge set $E' \subseteq E$ of minimum total weight $\sum_{e \in E'} w_e$

5.22. In this problem, we will develop a new algorithm for finding minimum spanning trees. It is based upon the following property:

Pick any cycle in the graph, and let e be the heaviest edge in that cycle. Then there is a minimum spanning tree that does not contain e.

(a) Prove this property carefully.

(b) Here is the new MST algorithm. The input is some undirected graph $G = (V, E)$ (in adjacency list format) with edge weights $\{w_e\}$.

```
sort the edges according to their weights
for each edge e ∈ E, in decreasing order of w_e:
   if e is part of a cycle of G:
      G = G − e (that is, remove e from G)
return G
```

Prove that this algorithm is correct.

(c) On each iteration, the algorithm must check whether there is a cycle containing a specific edge e. Give a linear-time algorithm for this task, and justify its correctness.

(d) What is the overall time taken by this algorithm, in terms of $|E|$?

5.23. You are given a graph $G = (V, E)$ with positive edge weights, and a minimum spanning tree $T = (V, E')$ with respect to these weights; you may assume G and T are given as adjacency lists. Now suppose the weight of a particular edge $e \in E$ is modified from $w(e)$ to a new value $\hat{w}(e)$. You wish to quickly update the minimum spanning tree T to reflect this change, without recomputing the entire tree from scratch. There are four cases. In each case give a linear-time algorithm for updating the tree.

(a) $e \notin E'$ and $\hat{w}(e) > w(e)$.

(b) $e \notin E'$ and $\hat{w}(e) < w(e)$.

(c) $e \in E'$ and $\hat{w}(e) < w(e)$.

(d) $e \in E'$ and $\hat{w}(e) > w(e)$.

5.24. Sometimes we want light spanning trees with certain special properties. Here's an example.

Input: Undirected graph $G = (V, E)$; edge weights w_e; subset of vertices $U \subset V$
Output: The lightest spanning tree in which the nodes of U are leaves (there might be other leaves in this tree as well).

(The answer isn't necessarily a minimum spanning tree.)

Give an algorithm for this problem which runs in $O(|E| \log |V|)$ time. (*Hint:* When you remove nodes U from the optimal solution, what is left?)

5.25. A binary counter of unspecified length supports two operations: `increment` (which increases its value by one) and `reset` (which sets its value back to zero). Show that, starting from an initially zero counter, any sequence of n `increment` and `reset` operations takes time $O(n)$; that is, the amortized time per operation is $O(1)$.

5.26. Here's a problem that occurs in automatic program analysis. For a set of variables $x_1, \ldots, x_n$, you are given some *equality* constraints, of the form

"$x_i = x_j$" and some *disequality* constraints, of the form "$x_i \neq x_j$." Is it possible to satisfy all of them?

For instance, the constraints

$$x_1 = x_2, x_2 = x_3, x_3 = x_4, x_1 \neq x_4$$

cannot be satisfied. Give an efficient algorithm that takes as input m constraints over n variables and decides whether the constraints can be satisfied.

5.27. *Graphs with prescribed degree sequences.* Given a list of n positive integers $d_1, d_2, \ldots, d_n$, we want to efficiently determine whether there exists an undirected graph $G = (V, E)$ whose nodes have degrees precisely $d_1, d_2, \ldots, d_n$. That is, if $V = \{v_1, \ldots, v_n\}$, then the degree of v_i should be exactly d_i. We call $(d_1, \ldots, d_n)$ the *degree sequence of* G. This graph G should not contain self-loops (edges with both endpoints equal to the same node) or multiple edges between the same pair of nodes.

(a) Give an example of d_1, d_2, d_3, d_4 where all the $d_i \leq 3$ and $d_1 + d_2 + d_3 + d_4$ is even, but for which no graph with degree sequence (d_1, d_2, d_3, d_4) exists.

(b) Suppose that $d_1 \geq d_2 \geq \cdots \geq d_n$ and that there exists a graph $G = (V, E)$ with degree sequence $(d_1, \ldots, d_n)$. We want to show that there must exist a graph that has this degree sequence and where in addition the neighbors of v_1 are $v_2, v_3, \ldots, v_{d_1+1}$. The idea is to gradually transform G into a graph with the desired additional property.

i. Suppose the neighbors of v_1 in G are not $v_2, v_3, \ldots, v_{d_1+1}$. Show that there exists $i < j \leq n$ and $u \in V$ such that $\{v_1, v_i\}, \{u, v_j\} \notin E$ and $\{v_1, v_j\}, \{u, v_i\} \in E$.

ii. Specify the changes you would make to G to obtain a new graph $G' = (V, E')$ with the same degree sequence as G and where $(v_1, v_i) \in E'$.

iii. Now show that there must be a graph with the given degree sequence but in which v_1 has neighbors $v_2, v_3, \ldots, v_{d_1+1}$.

(c) Using the result from part (b), describe an algorithm that on input $d_1, \ldots, d_n$ (not necessarily sorted) decides whether there exists a graph with this degree sequence. Your algorithm should run in time polynomial in n.

5.28. Alice wants to throw a party and is deciding whom to call. She has n people to choose from, and she has made up a list of which pairs of these people know each other. She wants to pick as many people as possible, subject to two constraints: at the party, each person should have at least five other people whom they know *and* five other people whom they don't know.

Give an efficient algorithm that takes as input the list of n people and the list of pairs who know each other and outputs the best choice of party invitees. Give the running time in terms of n.

5.29. A *prefix-free encoding* of a finite alphabet Γ assigns each symbol in Γ a binary codeword, such that no codeword is a prefix of another codeword. A prefix-free encoding is *minimal* if it is not possible to arrive at another prefix-free encoding (of the same symbols) by contracting some of the keywords. For instance, the encoding $\{0, 101\}$ is not minimal since the codeword 101 can be contracted to 1 while still maintaining the prefix-free property.

Show that a minimal prefix-free encoding can be represented by a full binary tree in which each leaf corresponds to a unique element of Γ, whose codeword is generated by the path from the root to that leaf (interpreting a left branch as 0 and a right branch as 1).

5.30. *Ternary Huffman.* Trimedia Disks Inc. has developed "ternary" hard disks. Each cell on a disk can now store values 0, 1, or 2 (instead of just 0 or 1). To take advantage of this new technology, provide a modified Huffman algorithm for compressing sequences of characters from an alphabet of size n, where the characters occur with known frequencies $f_1, f_2, \ldots, f_n$. Your algorithm should encode each character with a variable-length codeword over the values 0, 1, 2 such that no codeword is a prefix of another codeword and so as to obtain the maximum possible compression. Prove that your algorithm is correct.

5.31. The basic intuition behind Huffman's algorithm, that frequent blocks should have short encodings and infrequent blocks should have long encodings, is also at work in English, where typical words like `I`, `you`, `is`, `and`, `to`, `from`, and so on are short, and rarely used words like `velociraptor` are longer.

However, words like `fire!`, `help!`, and `run!` are short not because they are frequent, but perhaps because time is precious in situations where they are used.

To make things theoretical, suppose we have a file composed of m different words, with frequencies $f_1, \ldots, f_m$. Suppose also that for the ith word, the cost per bit of encoding is c_i. Thus, if we find a prefix-free code where the ith word has a codeword of length l_i, then the total cost of the encoding will be $\sum_i f_i \cdot c_i \cdot l_i$.

Show how to find the prefix-free encoding of minimal total cost.

5.32. A server has n customers waiting to be served. The service time required by each customer is known in advance: it is t_i minutes for customer i. So if, for example, the customers are served in order of increasing i, then the ith customer has to wait $\sum_{j=1}^{i} t_j$ minutes.

We wish to minimize the total waiting time

$$T = \sum_{i=1}^{n} (\text{time spent waiting by customer } i).$$

Give an efficient algorithm for computing the optimal order in which to process the customers.

5.33. Show how to implement the stingy algorithm for Horn formula satisfiability (Section 5.3) in time that is linear in the length of the formula (the number of occurrences of literals in it).

5.34. Show that for any integer n that is a power of 2, there is an instance of the set cover problem (Section 5.4) with the following properties:

i. There are n elements in the base set.

ii. The optimal cover uses just two sets.

iii. The greedy algorithm picks at least $\log n$ sets.

Thus the approximation ratio we derived in the chapter is tight.

5.35. Show that an unweighted graph with n nodes has at most $n(n-1)$ distinct minimum cuts.

Chapter 6
Dynamic programming

In the preceding chapters we have seen some elegant design principles—such as divide-and-conquer, graph exploration, and greedy choice—that yield definitive algorithms for a variety of important computational tasks. The drawback of these tools is that they can only be used on very specific types of problems. We now turn to the two *sledgehammers* of the algorithms craft, *dynamic programming* and *linear programming*, techniques of very broad applicability that can be invoked when more specialized methods fail. Predictably, this generality often comes with a cost in efficiency.

6.1 Shortest paths in dags, revisited

At the conclusion of our study of shortest paths (Chapter 4), we observed that the problem is especially easy in directed acyclic graphs (dags). Let's recapitulate this case, because it lies at the heart of dynamic programming.

The special distinguishing feature of a dag is that its nodes can be *linearized*; that is, they can be arranged on a line so that all edges go from left to right (Figure 6.1). To see why this helps with shortest paths, suppose we want to figure out distances from node S to the other nodes. For concreteness, let's focus on node D. The only way to get to it is through its predecessors, B or C; so to find the shortest path to D, we need only compare these two routes:

$$\texttt{dist}(D) = \min\{\texttt{dist}(B) + 1, \texttt{dist}(C) + 3\}.$$

A similar relation can be written for every node. If we compute these dist values in the left-to-right order of Figure 6.1, we can always be sure that by the time we get to a node v, we already have all the information we need to compute $\texttt{dist}(v)$. We are therefore able to compute all distances in a single pass:

```
initialize all dist(·) values to ∞
dist(s) = 0
for each v ∈ V\{s}, in linearized order:
    dist(v) = min_{(u,v)∈E}{dist(u) + l(u, v)}
```

Notice that this algorithm is solving a collection of *subproblems*, $\{\texttt{dist}(u) : u \in V\}$. We start with the smallest of them, $\texttt{dist}(s)$, since we immediately know its answer

Figure 6.1 A dag and its linearization (topological ordering).

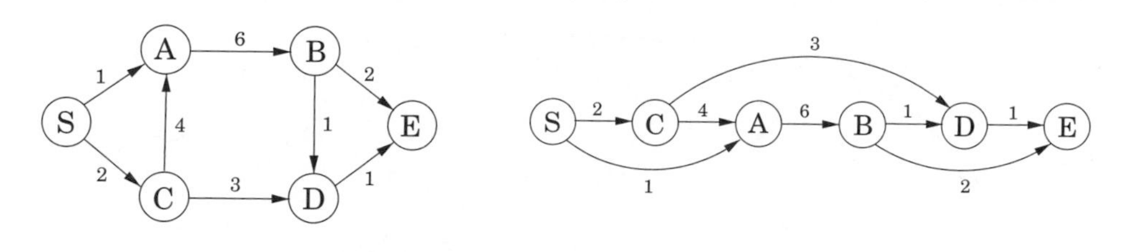

to be 0. We then proceed with progressively "larger" subproblems—distances to vertices that are further and further along in the linearization—where we are thinking of a subproblem as large if we need to have solved a lot of other subproblems before we can get to it.

This is a very general technique. At each node, we compute some function of the values of the node's predecessors. It so happens that our particular function is a minimum of sums, but we could just as well make it a *maximum*, in which case we would get *longest* paths in the dag. Or we could use a product instead of a sum inside the brackets, in which case we would end up computing the path with the smallest product of edge lengths.

Dynamic programming is a very powerful algorithmic paradigm in which a problem is solved by identifying a collection of subproblems and tackling them one by one, smallest first, using the answers to small problems to help figure out larger ones, until the whole lot of them is solved. In dynamic programming we are not given a dag; the dag is *implicit*. Its nodes are the subproblems we define, and its edges are the dependencies between the subproblems: if to solve subproblem B we need the answer to subproblem A, then there is a (conceptual) edge from A to B. In this case, A is thought of as a smaller subproblem than B—and it will always be smaller, in an obvious sense.

But it's time we saw an example.

6.2 Longest increasing subsequences

In the *longest increasing subsequence* problem, the input is a sequence of numbers $a_1, \ldots, a_n$. A *subsequence* is any subset of these numbers taken in order, of the form $a_{i_1}, a_{i_2}, \ldots, a_{i_k}$ where $1 \leq i_1 < i_2 < \cdots < i_k \leq n$, and an *increasing* subsequence is one in which the numbers are getting strictly larger. The task is to find the increasing subsequence of greatest length. For instance, the longest increasing subsequence of 5, 2, 8, 6, 3, 6, 9, 7 is 2, 3, 6, 9:

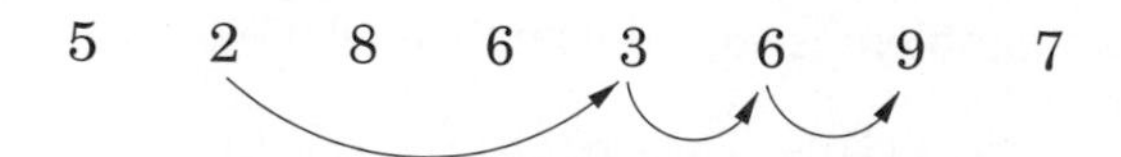

Figure 6.2 The dag of increasing subsequences.

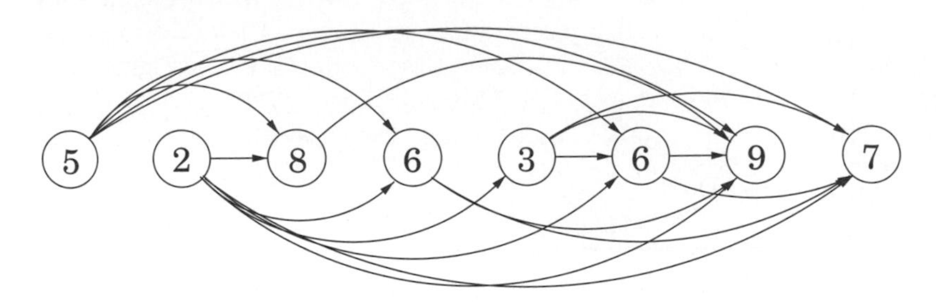

In this example, the arrows denote transitions between consecutive elements of the optimal solution. More generally, to better understand the solution space, let's create a graph of *all* permissible transitions: establish a node i for each element a_i, and add directed edges (i, j) whenever it is possible for a_i and a_j to be consecutive elements in an increasing subsequence, that is, whenever $i < j$ and $a_i < a_j$ (Figure 6.2).

Notice that (1) this graph $G = (V, E)$ is a dag, since all edges (i, j) have $i < j$, and (2) there is a one-to-one correspondence between increasing subsequences and paths in this dag. Therefore, our goal is simply to find the longest path in the dag!

Here is the algorithm:

```
for j = 1, 2, ..., n:
    L(j) = 1 + max{L(i) : (i, j) ∈ E}
return max_j L(j)
```

$L(j)$ is the length of the longest path—the longest increasing subsequence—ending at j (plus 1, since strictly speaking we need to count nodes on the path, not edges). By reasoning in the same way as we did for shortest paths, we see that any path to node j must pass through one of its predecessors, and therefore $L(j)$ is 1 plus the maximum $L(\cdot)$ value of these predecessors. If there are no edges into j, we take the maximum over the empty set, zero. And the final answer is the *largest* $L(j)$, since any ending position is allowed.

This is dynamic programming. In order to solve our original problem, we have defined a collection of subproblems $\{L(j) : 1 \le j \le n\}$ with the following key property that allows them to be solved in a single pass:

> (*) *There is an ordering on the subproblems, and a relation that shows how to solve a subproblem given the answers to "smaller" subproblems, that is, subproblems that appear earlier in the ordering.*

In our case, each subproblem is solved using the relation

$$L(j) = 1 + \max\{L(i) : (i, j) \in E\},$$

an expression which involves only smaller subproblems. How long does this step take? It requires the predecessors of j to be known; for this the adjacency list of the reverse graph G^R, constructible in linear time (recall Exercise 3.5), is handy. The computation of $L(j)$ then takes time proportional to the indegree of j, giving an overall running time linear in $|E|$. This is at most $O(n^2)$, the maximum being when the input array is sorted in increasing order. Thus the dynamic programming solution is both simple and efficient.

There is one last issue to be cleared up: the L-values only tell us the *length* of the optimal subsequence, so how do we recover the subsequence itself? This is easily managed with the same bookkeeping device we used for shortest paths in Chapter 4. While computing $L(j)$, we should also note down prev(j), the next-to-last node on the longest path to j. The optimal subsequence can then be reconstructed by following these backpointers.

6.3 Edit distance

When a spell checker encounters a possible misspelling, it looks in its dictionary for other words that are close by. What is the appropriate notion of closeness in this case?

A natural measure of the distance between two strings is the extent to which they can be *aligned*, or matched up. Technically, an alignment is simply a way of writing the strings one above the other. For instance, here are two possible alignments of SNOWY and SUNNY:

```
S − N O W Y          − S N O W − Y
S U N N − Y          S U N − − N Y
  Cost: 3               Cost: 5
```

The "−" indicates a "gap"; any number of these can be placed in either string. The *cost* of an alignment is the number of columns in which the letters differ. And the *edit distance* between two strings is the cost of their best possible alignment. Do you see that there is no better alignment of SNOWY and SUNNY than the one shown here with a cost of 3?

Edit distance is so named because it can also be thought of as the minimum number of *edits*—insertions, deletions, and substitutions of characters—needed to transform the first string into the second. For instance, the alignment shown on the left corresponds to three edits: insert U, substitute O $\rightarrow$ N, and delete W.

In general, there are so many possible alignments between two strings that it would be terribly inefficient to search through all of them for the best one. Instead we turn to dynamic programming.

A dynamic programming solution

When solving a problem by dynamic programming, the most crucial question is, *What are the subproblems?* As long as they are chosen so as to have the property

Recursion? No, thanks.

Returning to our discussion of longest increasing subsequences: the formula for $L(j)$ also suggests an alternative, recursive algorithm. Wouldn't that be even simpler?

Actually, recursion is a very bad idea: the resulting procedure would require exponential time! To see why, suppose that the dag contains edges (i, j) for *all* $i < j$—that is, the given sequence of numbers $a_1, a_2, \ldots, a_n$ is sorted. In that case, the formula for subproblem $L(j)$ becomes

$$L(j) = 1 + \max\{L(1), L(2), \ldots, L(j-1)\}.$$

The following figure unravels the recursion for $L(5)$. Notice that the same subproblems get solved over and over again!

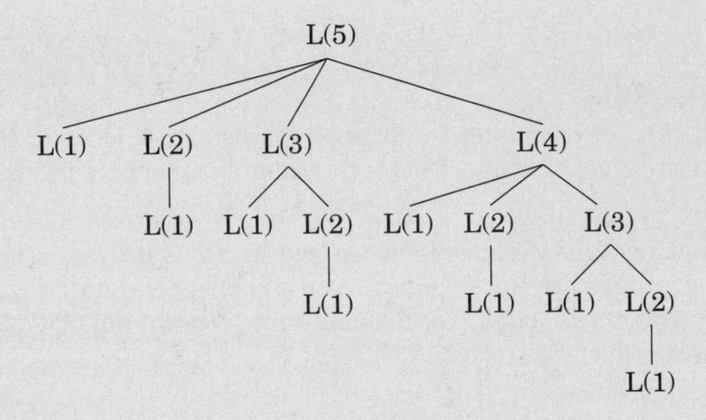

For $L(n)$ this tree has exponentially many nodes (can you bound it?), and so a recursive solution is disastrous.

Then why did recursion work so well with divide-and-conquer? The key point is that in divide-and-conquer, a problem is expressed in terms of subproblems that are *substantially smaller*, say half the size. For instance, `mergesort` sorts an array of size n by recursively sorting two subarrays of size $n/2$. Because of this sharp drop in problem size, the full recursion tree has only logarithmic depth and a polynomial number of nodes.

In contrast, in a typical dynamic programming formulation, a problem is reduced to subproblems that are only slightly smaller—for instance, $L(j)$ relies on $L(j-1)$. Thus the full recursion tree generally has polynomial depth and an exponential number of nodes. However, it turns out that most of these nodes are repeats, that there are not too many *distinct* subproblems among them. Efficiency is therefore obtained by explicitly enumerating the distinct subproblems and solving them in the right order.

Programming?

The origin of the term *dynamic programming* has very little to do with writing code. It was first coined by Richard Bellman in the 1950s, a time when computer programming was an esoteric activity practiced by so few people as to not even merit a name. Back then programming meant "planning," and "dynamic programming" was conceived to optimally plan multistage processes. The dag of Figure 6.2 can be thought of as describing the possible ways in which such a process can evolve: each node denotes a state, the leftmost node is the starting point, and the edges leaving a state represent possible actions, leading to different states in the next unit of time.

The etymology of *linear programming*, the subject of Chapter 7, is similar.

(*) from page 158. it is an easy matter to write down the algorithm: iteratively solve one subproblem after the other, in order of increasing size.

Our goal is to find the edit distance between two strings $x[1 \cdots m]$ and $y[1 \cdots n]$. What is a good subproblem? Well, it should go part of the way toward solving the whole problem; so how about looking at the edit distance between some *prefix* of the first string, $x[1 \cdots i]$, and some *prefix* of the second, $y[1 \cdots j]$? Call this subproblem $E(i, j)$ (see Figure 6.3). Our final objective, then, is to compute $E(m, n)$.

For this to work, we need to somehow express $E(i, j)$ in terms of smaller subproblems. Let's see—what do we know about the best alignment between $x[1 \cdots i]$ and $y[1 \cdots j]$? Well, its rightmost column can only be one of three things:

$$\begin{matrix} x[i] \\ - \end{matrix} \quad \text{or} \quad \begin{matrix} - \\ y[j] \end{matrix} \quad \text{or} \quad \begin{matrix} x[i] \\ y[j] \end{matrix}$$

The first case incurs a cost of 1 for this particular column, and it remains to align $x[1 \cdots i-1]$ with $y[1 \cdots j]$. But this is exactly the subproblem $E(i-1, j)$! We seem to be getting somewhere. In the second case, also with cost 1, we still need to align $x[1 \cdots i]$ with $y[1 \cdots j-1]$. This is again another subproblem, $E(i, j-1)$. And in the final case, which either costs 1 (if $x[i] \neq y[j]$) or 0 (if $x[i] = y[j]$), what's left is the subproblem $E(i-1, j-1)$. In short, we have expressed $E(i, j)$ in terms of

Figure 6.3 The subproblem $E(7, 5)$.

E X P O N E N	T I A L
P O L Y N	O M I A L

three *smaller* subproblems $E(i-1, j), E(i, j-1), E(i-1, j-1)$. We have no idea which of them is the right one, so we need to try them all and pick the best:

$$E(i, j) = \min\{1 + E(i-1, j),\ 1 + E(i, j-1),\ \text{diff}(i, j) + E(i-1, j-1)\}$$

where for convenience $\text{diff}(i, j)$ is defined to be 0 if $x[i] = y[j]$ and 1 otherwise.

For instance, in computing the edit distance between `EXPONENTIAL` and `POLYNOMIAL`, subproblem $E(4, 3)$ corresponds to the prefixes `EXPO` and `POL`. The rightmost column of their best alignment must be one of the following:

```
O         -         O
    or        or
-         L         L
```

Thus, $E(4, 3) = \min\{1 + E(3, 3),\ 1 + E(4, 2),\ 1 + E(3, 2)\}$.

The answers to all the subproblems $E(i, j)$ form a two-dimensional table, as in Figure 6.4. In what order should these subproblems be solved? Any order is fine, as long as $E(i-1, j)$, $E(i, j-1)$, and $E(i-1, j-1)$ are handled *before* $E(i, j)$. For instance, we could fill in the table one row at a time, from top row to bottom row, and moving left to right across each row. Or alternatively, we could fill it in column by column. Both methods would ensure that by the time we get around to computing a particular table entry, all the other entries we need are already filled in.

With both the subproblems and the ordering specified, we are almost done. There just remain the "base cases" of the dynamic programming, the very smallest subproblems. In the present situation, these are $E(0, \cdot)$ and $E(\cdot, 0)$, both of which are easily solved. $E(0, j)$ is the edit distance between the 0-length prefix of x,

Figure 6.4 (a) The table of subproblems. Entries $E(i-1, j-1)$, $E(i-1, j)$, and $E(i, j-1)$ are needed to fill in $E(i, j)$. (b) The final table of values found by dynamic programming.

(a)

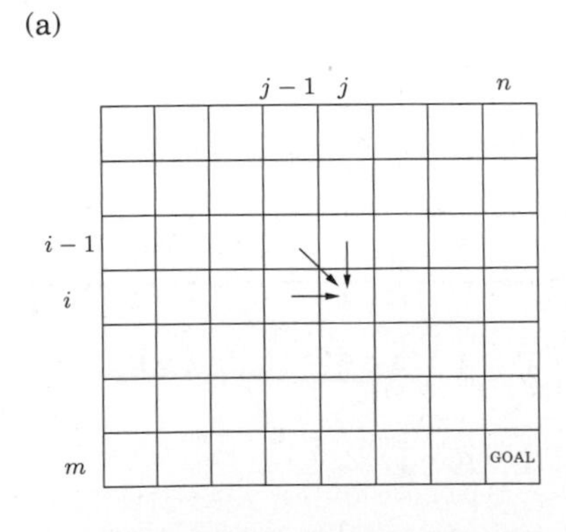

(b)

		P	O	L	Y	N	O	M	I	A	L
	0	1	2	3	4	5	6	7	8	9	10
E	1	1	2	3	4	5	6	7	8	9	10
X	2	2	2	3	4	5	6	7	8	9	10
P	3	2	3	3	4	5	6	7	8	9	10
O	4	3	2	3	4	5	5	6	7	8	9
N	5	4	3	3	4	4	5	6	7	8	9
E	6	5	4	4	4	5	5	6	7	8	9
N	7	6	5	5	5	4	5	6	7	8	9
T	8	7	6	6	6	5	5	6	7	8	9
I	9	8	7	7	7	6	6	6	6	7	8
A	10	9	8	8	8	7	7	7	7	6	7
L	11	10	9	8	9	8	8	8	8	7	6

namely the empty string, and the first j letters of y: clearly, j. And similarly, $E(i, 0) = i$.

At this point, the algorithm for edit distance basically writes itself.

```
for i = 0, 1, 2, ..., m:
    E(i, 0) = i
for j = 1, 2, ..., n:
    E(0, j) = j
for i = 1, 2, ..., m:
    for j = 1, 2, ..., n:
        E(i, j) = min{E(i − 1, j) + 1, E(i, j − 1) + 1, E(i − 1, j − 1) + diff(i, j)}
return E(m, n)
```

This procedure fills in the table row by row, and left to right within each row. Each entry takes constant time to fill in, so the overall running time is just the size of the table, $O(mn)$.

And in our example, the edit distance turns out to be 6:

```
E X P O N E N − T I A L
− − P O L Y N O M I A L
```

The underlying dag

Every dynamic program has an underlying dag structure: think of each node as representing a subproblem, and each edge as a precedence constraint on the order in which the subproblems can be tackled. Having nodes $u_1, \ldots, u_k$ point to v means "subproblem v can only be solved once the answers to $u_1, \ldots, u_k$ are known."

In our present edit distance application, the nodes of the underlying dag correspond to subproblems, or equivalently, to positions (i, j) in the table. Its edges are the precedence constraints, of the form $(i-1, j) \to (i, j)$, $(i, j-1) \to (i, j)$, and $(i-1, j-1) \to (i, j)$ (Figure 6.5). In fact, we can take things a little further and put weights on the edges so that the edit distances are given by shortest paths in the dag! To see this, set all edge lengths to 1, except for $\{(i-1, j-1) \to (i, j) : x[i] = y[j]\}$ (shown dotted in the figure), whose length is 0. The final answer is then simply the distance between nodes $s = (0, 0)$ and $t = (m, n)$. One possible shortest path is shown, the one that yields the alignment we found earlier. On this path, each move down is a deletion, each move right is an insertion, and each diagonal move is either a match or a substitution.

By altering the weights on this dag, we can allow generalized forms of edit distance, in which insertions, deletions, and substitutions have different associated costs.

Figure 6.5 The underlying dag, and a path of length 6.

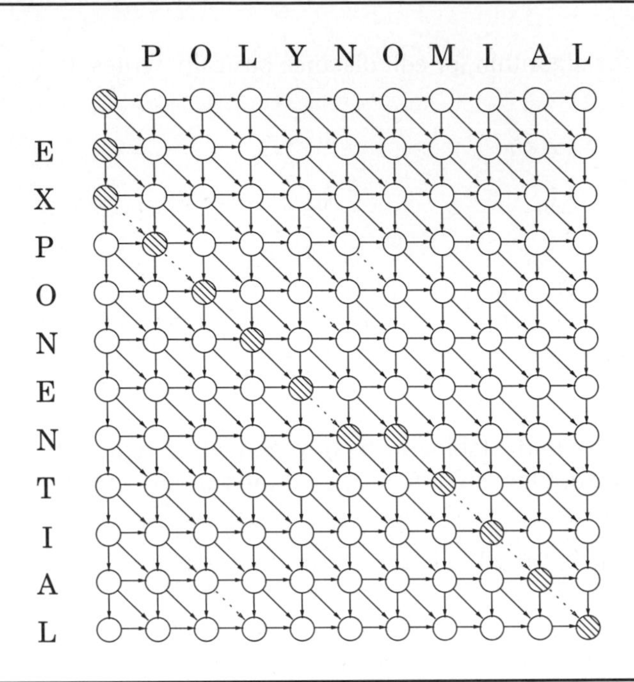

6.4 Knapsack

During a robbery, a burglar finds much more loot than he had expected and has to decide what to take. His bag (or "knapsack") will hold a total weight of at most W pounds. There are n items to pick from, of weight $w_1, \ldots, w_n$ and dollar value $v_1, \ldots, v_n$. What's the most valuable combination of items he can fit into his bag?[1]

For instance, take $W = 10$ and

Item	Weight	Value
1	6	\$30
2	3	\$14
3	4	\$16
4	2	\$9

[1] If this application seems frivolous, replace "weight" with "CPU time" and "only W pounds can be taken" with "only W units of CPU time are available." Or use "bandwidth" in place of "CPU time," etc. The knapsack problem generalizes a wide variety of resource-constrained selection tasks.

Common subproblems

Finding the right subproblem takes creativity and experimentation. But there are a few standard choices that seem to arise repeatedly in dynamic programming.

i. The input is $x_1, x_2, \ldots, x_n$ and a subproblem is $x_1, x_2, \ldots, x_i$.

$$\boxed{x_1 \quad x_2 \quad x_3 \quad x_4 \quad x_5 \quad x_6} \quad x_7 \quad x_8 \quad x_9 \quad x_{10}$$

The number of subproblems is therefore linear.

ii. The input is $x_1, \ldots, x_n$, and $y_1, \ldots, y_m$. A subproblem is $x_1, \ldots, x_i$ and $y_1, \ldots, y_j$.

$$\boxed{x_1 \quad x_2 \quad x_3 \quad x_4 \quad x_5 \quad x_6} \quad x_7 \quad x_8 \quad x_9 \quad x_{10}$$

$$\boxed{y_1 \quad y_2 \quad y_3 \quad y_4 \quad y_5} \quad y_6 \quad y_7 \quad y_8$$

The number of subproblems is $O(mn)$.

iii. The input is $x_1, \ldots, x_n$ and a subproblem is $x_i, x_{i+1}, \ldots, x_j$.

$$x_1 \quad x_2 \quad \boxed{x_3 \quad x_4 \quad x_5 \quad x_6} \quad x_7 \quad x_8 \quad x_9 \quad x_{10}$$

The number of subproblems is $O(n^2)$.

iv. The input is a rooted tree. A subproblem is a rooted subtree.

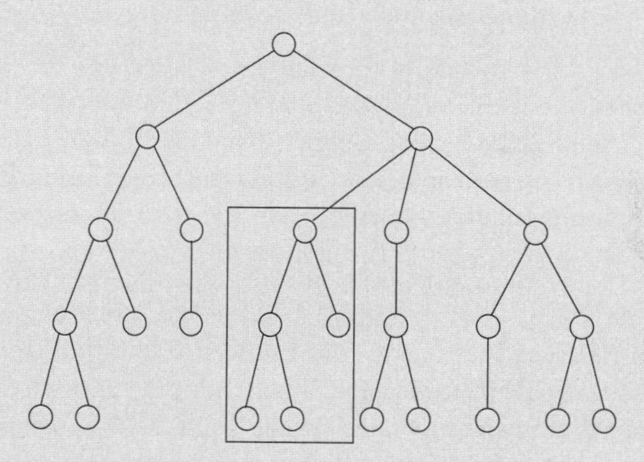

If the tree has n nodes, how many subproblems are there?

We've already encountered the first two cases, and the others are coming up shortly.

Of mice and men

Our bodies are extraordinary machines: flexible in function, adaptive to new environments, and able to interact and reproduce. All these capabilities are specified by a program unique to each of us, a string that is 3 billion characters long over the alphabet $\{A, C, G, T\}$—our DNA.

The DNA sequences of any two people differ by only about 0.1%. However, this still leaves 3 million positions on which they vary, more than enough to explain the vast range of human diversity. These differences are of great scientific and medical interest—for instance, they might help predict which people are prone to certain diseases.

DNA is a vast and seemingly inscrutable program, but it can be broken down into smaller units that are more specific in their role, rather like subroutines. These are called *genes*. Computers have become a crucial tool in understanding the genes of humans and other organisms, to the extent that *computational genomics* is now a field in its own right. Here are examples of typical questions that arise.

1. When a new gene is discovered, one way to gain insight into its function is to find known genes that match it closely. This is particularly helpful in transferring knowledge from well-studied species, such as mice, to human beings.

   A basic primitive in this search problem is to define an efficiently computable notion of when two strings approximately match. The biology suggests a generalization of edit distance, and dynamic programming can be used to compute it.

   Then there's the problem of searching through the vast thicket of known genes: the database GenBank already has a total length of over 10^{10}, and this number is growing rapidly. The current method of choice is BLAST, a clever combination of algorithmic tricks and biological intuitions that has made it the most widely used software in computational biology.

2. Methods for *sequencing* DNA (that is, determining the string of characters that constitute it) typically only find fragments of 500–700 characters. Billions of these randomly scattered fragments can be generated, but how can they be assembled into a coherent DNA sequence? For one thing, the position of any one fragment in the final sequence is unknown and must be inferred by piecing together overlapping fragments.

   A showpiece of these efforts is the draft of human DNA completed in 2001 by two groups simultaneously: the publicly funded Human Genome Consortium and the private Celera Genomics.

3. When a particular gene has been sequenced in each of several species, can this information be used to reconstruct the evolutionary history of these species?

We will explore these problems in the exercises at the end of this chapter. Dynamic programming has turned out to be an invaluable tool for some of them and for computational biology in general.

There are two versions of this problem. If there are unlimited quantities of each item available, the optimal choice is to pick item 1 and two of item 4 (total: \$48). On the other hand, if there is one of each item (the burglar has broken into an art gallery, say), then the optimal knapsack contains items 1 and 3 (total: \$46).

As we shall see in Chapter 8, neither version of this problem is likely to have a polynomial-time algorithm. However, using dynamic programming they can both be solved in $O(nW)$ time, which is reasonable when W is small, but is not polynomial since the input size is proportional to $\log W$ rather than W.

Knapsack with repetition

Let's start with the version that allows repetition. As always, the main question in dynamic programming is, what are the subproblems? In this case we can shrink the original problem in two ways: we can either look at smaller knapsack capacities $w \leq W$, or we can look at fewer items (for instance, items $1, 2, \ldots, j$, for $j \leq n$). It usually takes a little experimentation to figure out exactly what works.

The first restriction calls for smaller capacities. Accordingly, define

$$K(w) = \text{maximum value achievable with a knapsack of capacity } w.$$

Can we express this in terms of smaller subproblems? Well, if the optimal solution to $K(w)$ includes item i, then removing this item from the knapsack leaves an optimal solution to $K(w - w_i)$. In other words, $K(w)$ is simply $K(w - w_i) + v_i$, for some i. We don't know which i, so we need to try all possibilities.

$$K(w) = \max_{i:w_i \leq w} \{K(w - w_i) + v_i\},$$

where as usual our convention is that the maximum over an empty set is 0. We're done! The algorithm now writes itself, and it is characteristically simple and elegant.

```
K(0) = 0
for w = 1 to W:
    K(w) = max{K(w - w_i) + v_i : w_i ≤ w}
return K(W)
```

This algorithm fills in a one-dimensional table of length $W + 1$, in left-to-right order. Each entry can take up to $O(n)$ time to compute, so the overall running time is $O(nW)$.

As always, there is an underlying dag. Try constructing it, and you will be rewarded with a startling insight: this particular variant of knapsack boils down to finding the longest path in a dag!

Knapsack without repetition

On to the second variant: what if repetitions are not allowed? Our earlier subproblems now become completely useless. For instance, knowing that the value $K(w - w_n)$ is very high doesn't help us, because we don't know whether or not item n already got used up in this partial solution. We must therefore refine our concept of a subproblem to carry additional information about the items being used. We add a second parameter, $0 \le j \le n$:

$K(w, j)$ = maximum value achievable using a knapsack of capacity w and items $1, \ldots, j$.

The answer we seek is $K(W, n)$.

How can we express a subproblem $K(w, j)$ in terms of smaller subproblems? Quite simple: either item j is needed to achieve the optimal value, or it isn't needed:

$$K(w, j) = \max\{K(w - w_j, j - 1) + v_j, K(w, j - 1)\}.$$

(The first case is invoked only if $w_j \le w$.) In other words, we can express $K(w, j)$ in terms of subproblems $K(\cdot, j - 1)$.

The algorithm then consists of filling out a two-dimensional table, with $W + 1$ rows and $n + 1$ columns. Each table entry takes just constant time, so even though the table is much larger than in the previous case, the running time remains the same, $O(nW)$. Here's the code.

```
Initialize all K(0, j) = 0 and all K(w, 0) = 0
for j = 1 to n:
   for w = 1 to W:
      if w_j > w:  K(w, j) = K(w, j − 1)
      else:  K(w, j) = max{K(w, j − 1), K(w − w_j, j − 1) + v_j}
return K(W, n)
```

6.5 Chain matrix multiplication

Suppose that we want to multiply four matrices, $A \times B \times C \times D$, of dimensions 50×20, 20×1, 1×10, and 10×100, respectively (Figure 6.6). This will involve iteratively multiplying two matrices at a time. Matrix multiplication is not

Figure 6.6 $A \times B \times C \times D = (A \times (B \times C)) \times D$.

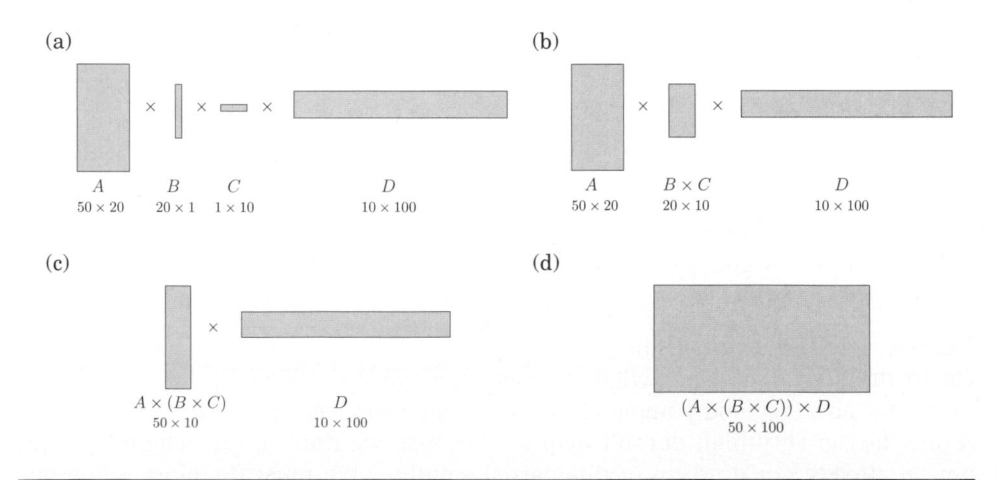

Memoization

In dynamic programming, we write out a recursive formula that expresses large problems in terms of smaller ones and then use it to fill out a table of solution values in a bottom-up manner, from smallest subproblem to largest.

The formula also suggests a recursive algorithm, but we saw earlier that naive recursion can be terribly inefficient, because it solves the same subproblems over and over again. What about a more intelligent recursive implementation, one that remembers its previous invocations and thereby avoids repeating them?

On the knapsack problem (with repetitions), such an algorithm would use a hash table (recall Section 1.5) to store the values of $K(\cdot)$ that had already been computed. At each recursive call requesting some $K(w)$, the algorithm would first check if the answer was already in the table and then would proceed to its calculation only if it wasn't. This trick is called *memoization*:

```
A hash table, initially empty, holds values of K(w) indexed by w

function knapsack(w)
if w is in hash table: return K(w)
K(w) = max{knapsack(w − w_i) + v_i : w_i ≤ w}
insert K(w) into hash table, with key w
return K(w)
```

Since this algorithm never repeats a subproblem, its running time is $O(nW)$, just like the dynamic program. However, the constant factor in this big-O notation is substantially larger because of the overhead of recursion.

In some cases, though, memoization pays off. Here's why: dynamic programming automatically solves every subproblem that could conceivably be needed, while memoization only ends up solving the ones that are actually used. For instance, suppose that W and all the weights w_i are multiples of 100. Then a subproblem $K(w)$ is useless if 100 does not divide w. The memoized recursive algorithm will never look at these extraneous table entries.

commutative (in general, $A \times B \neq B \times A$), but it is *associative*, which means for instance that $A \times (B \times C) = (A \times B) \times C$. Thus we can compute our product of four matrices in many different ways, depending on how we parenthesize it. Are some of these better than others?

Multiplying an $m \times n$ matrix by an $n \times p$ matrix takes mnp multiplications, to a good enough approximation. Using this formula, let's compare several different ways of evaluating $A \times B \times C \times D$:

Parenthesization	Cost computation	Cost
$A \times ((B \times C) \times D)$	$20 \cdot 1 \cdot 10 + 20 \cdot 10 \cdot 100 + 50 \cdot 20 \cdot 100$	120,200
$(A \times (B \times C)) \times D$	$20 \cdot 1 \cdot 10 + 50 \cdot 20 \cdot 10 + 50 \cdot 10 \cdot 100$	60,200
$(A \times B) \times (C \times D)$	$50 \cdot 20 \cdot 1 + 1 \cdot 10 \cdot 100 + 50 \cdot 1 \cdot 100$	7,000

As you can see, the order of multiplications makes a big difference in the final running time! Moreover, the natural *greedy* approach, to always perform the cheapest matrix multiplication available, leads to the second parenthesization shown here and is therefore a failure.

How do we determine the optimal order, if we want to compute $A_1 \times A_2 \times \cdots \times A_n$, where the A_i's are matrices with dimensions $m_0 \times m_1, m_1 \times m_2, \ldots, m_{n-1} \times m_n$, respectively? The first thing to notice is that a particular parenthesization can be represented very naturally by a binary tree in which the individual matrices correspond to the leaves, the root is the final product, and interior nodes are intermediate products (Figure 6.7). The possible orders in which to do the multiplication correspond to the various full binary trees with n leaves, whose number is exponential in n (Exercise 2.13). We certainly cannot try each tree, and with brute force thus ruled out, we turn to dynamic programming.

The binary trees of Figure 6.7 are suggestive: for a tree to be optimal, its subtrees must also be optimal. What are the subproblems corresponding to the subtrees? They are products of the form $A_i \times A_{i+1} \times \cdots \times A_j$. Let's see if this works: for $1 \le i \le j \le n$, define

$$C(i, j) = \text{minimum cost of multiplying } A_i \times A_{i+1} \times \cdots \times A_j.$$

The size of this subproblem is the number of matrix multiplications, $|j - i|$. The smallest subproblem is when $i = j$, in which case there's nothing to multiply, so $C(i, i) = 0$. For $j > i$, consider the optimal subtree for $C(i, j)$. The first branch in this subtree, the one at the top, will split the product in two pieces, of the form $A_i \times \cdots \times A_k$ and $A_{k+1} \times \cdots \times A_j$, for some k between i and j. The cost of the subtree is then the cost of these two partial products, plus the cost of combining them: $C(i, k) + C(k + 1, j) + m_{i-1} \cdot m_k \cdot m_j$. And we just need to find the splitting

Figure 6.7 (a) $((A \times B) \times C) \times D$; (b) $A \times ((B \times C) \times D)$; (c) $(A \times (B \times C)) \times D$.

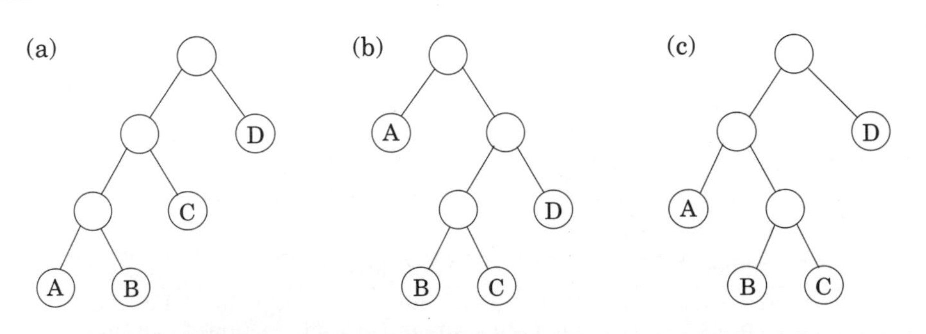

point k for which this is smallest:

$$C(i, j) = \min_{i \le k < j} \left\{ C(i, k) + C(k+1, j) + m_{i-1} \cdot m_k \cdot m_j \right\}.$$

We are ready to code! In the following, the variable s denotes subproblem size.

```
for i = 1 to n:  C(i, i) = 0
for s = 1 to n-1:
  for i = 1 to n-s:
    j = i + s
    C(i, j) = min{C(i, k) + C(k+1, j) + m_{i-1} · m_k · m_j : i ≤ k < j}
return C(1, n)
```

The subproblems constitute a two-dimensional table, each of whose entries takes $O(n)$ time to compute. The overall running time is thus $O(n^3)$.

6.6 Shortest paths

We started this chapter with a dynamic programming algorithm for the elementary task of finding the shortest path in a dag. We now turn to more sophisticated shortest-path problems and see how these too can be accommodated by our powerful algorithmic technique.

Shortest reliable paths

Life is complicated, and abstractions such as graphs, edge lengths, and shortest paths rarely capture the whole truth. In a communications network, for example, even if edge lengths faithfully reflect transmission delays, there may be other considerations involved in choosing a path. For instance, each extra edge in the path might be an extra "hop" fraught with uncertainties and dangers of packet loss. In such cases, we would like to avoid paths with too many edges. Figure 6.8 illustrates this problem with a graph in which the shortest path from S to T has four edges, while there is another path that is a little longer but uses only two edges. If four edges translate to prohibitive unreliability, we may have to choose the latter path.

Figure 6.8 We want a path from s to t that is both short *and* has few edges.

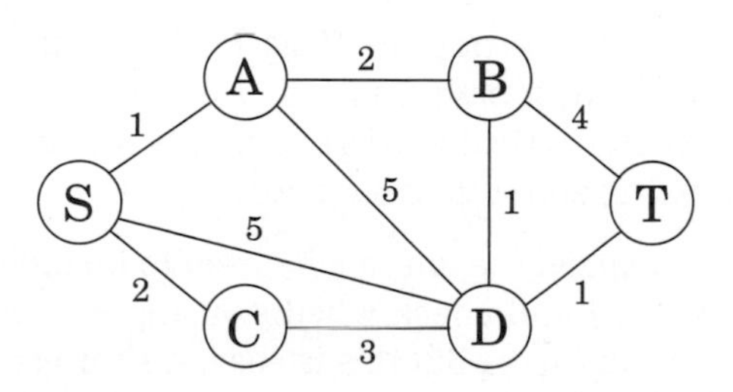

Suppose then that we are given a graph G with lengths on the edges, along with two nodes s and t and an integer k, and we want the shortest path from s to t *that uses at most k edges*.

Is there a quick way to adapt Dijkstra's algorithm to this new task? Not quite: that algorithm focuses on the length of each shortest path without "remembering" the number of hops in the path, which is now a crucial piece of information.

In dynamic programming, the trick is to choose subproblems so that all vital information is remembered and carried forward. In this case, let us define, for each vertex v and each integer $i \leq k$, $\texttt{dist}(v, i)$ to be *the length of the shortest path from s to v that uses i edges*. The starting values $\texttt{dist}(v, 0)$ are ∞ for all vertices except s, for which it is 0. And the general update equation is, naturally enough,

$$\texttt{dist}(v, i) = \min_{(u,v)\in E} \{\texttt{dist}(u, i-1) + \ell(u, v)\}.$$

Need we say more?

All-pairs shortest paths

What if we want to find the shortest path not just between s and t but between *all* pairs of vertices? One approach would be to execute our general shortest-path algorithm from Section 4.6.1 (since there may be negative edges) $|V|$ times, once for each starting node. The total running time would then be $O(|V|^2|E|)$. We'll now see a better alternative, the $O(|V|^3)$ dynamic programming-based *Floyd-Warshall algorithm*.

Is there a good subproblem for computing distances between all pairs of vertices in a graph? Simply solving the problem for more and more pairs or starting points is unhelpful, because it leads right back to the $O(|V|^2|E|)$ algorithm.

One idea comes to mind: the shortest path $u \rightarrow w_1 \rightarrow \cdots \rightarrow w_l \rightarrow v$ between u and v uses some number of intermediate nodes—possibly none. Suppose we disallow intermediate nodes altogether. Then we can solve all-pairs shortest paths at once: the shortest path from u to v is simply the direct edge (u, v), if it exists. What if we now gradually expand the *set of permissible intermediate nodes*? We can do this one node at a time, updating the shortest path lengths at each stage. Eventually this set grows to all of V, at which point all vertices are allowed to be on all paths, and we have found the true shortest paths between vertices of the graph!

More concretely, number the vertices in V as $\{1, 2, \ldots, n\}$, and let $\texttt{dist}(i, j, k)$ denote the length of the shortest path from i to j in which only nodes $\{1, 2, \ldots, k\}$ can be used as intermediates. Initially, $\texttt{dist}(i, j, 0)$ is the length of the direct edge between i and j, if it exists, and is ∞ otherwise.

What happens when we expand the intermediate set to include an extra node k? We must reexamine all pairs i, j and check whether using k as an intermediate point gives us a shorter path from i to j. But this is easy: a shortest path from i to j that uses k along with possibly other lower-numbered intermediate nodes goes through k just once (why? because we assume that there are no negative cycles). And we have already calculated the length of the shortest path from i to k and from k to j using only lower-numbered vertices:

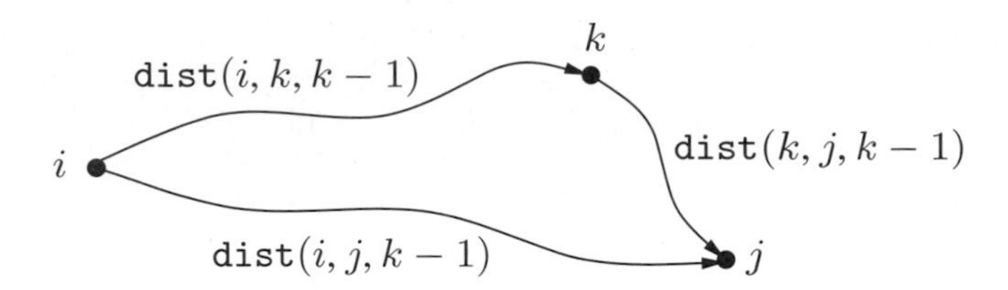

Thus, using k gives us a shorter path from i to j if and only if

$$\texttt{dist}(i, k, k-1) + \texttt{dist}(k, j, k-1) < \texttt{dist}(i, j, k-1),$$

in which case $\texttt{dist}(i, j, k)$ should be updated accordingly.

Here is the Floyd-Warshall algorithm—and as you can see, it takes $O(|V|^3)$ time.

```
for i = 1 to n:
   for j = 1 to n:
      dist(i, j, 0) = ∞
for all (i, j) ∈ E:
   dist(i, j, 0) = ℓ(i, j)
for k = 1 to n:
   for i = 1 to n:
      for j = 1 to n:
         dist(i, j, k) = min{dist(i, k, k − 1) + dist(k, j, k − 1),  dist(i, j, k − 1)}
```

The traveling salesman problem

A traveling salesman is getting ready for a big sales tour. Starting at his hometown, suitcase in hand, he will conduct a journey in which each of his target cities is visited exactly once before he returns home. Given the pairwise distances between cities, what is the best order in which to visit them, so as to minimize the overall distance traveled?

Denote the cities by $1, \ldots, n$, the salesman's hometown being 1, and let $D = (d_{ij})$ be the matrix of intercity distances. The goal is to design a tour that starts and ends at 1, includes all other cities exactly once, and has minimum total length. Figure 6.9 shows an example involving five cities. Can you spot the optimal tour? Even in this tiny example, it is tricky for a human to find the solution; imagine what happens when hundreds of cities are involved.

It turns out this problem is also difficult for computers. In fact, the traveling salesman problem (TSP) is one of the most notorious computational tasks. There is a long history of attempts at solving it, a long saga of failures and partial successes, and along the way, major advances in algorithms and complexity theory. The most basic

Figure 6.9 The optimal traveling salesman tour has length 10.

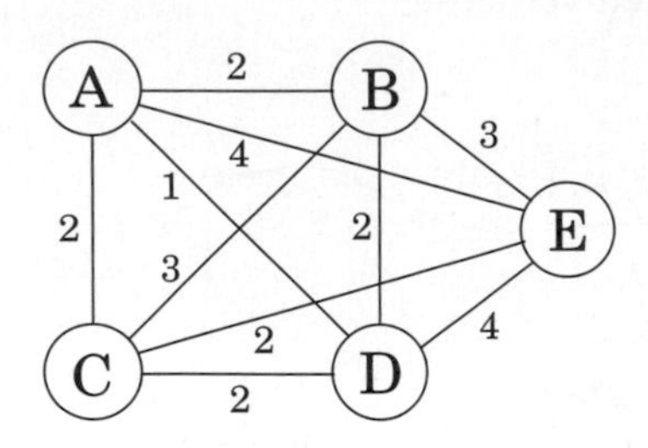

piece of bad news about the TSP, which we will better understand in Chapter 8, is that it is highly unlikely to be solvable in polynomial time.

How long does it take, then? Well, the brute-force approach is to evaluate every possible tour and return the best one. Since there are $(n-1)!$ possibilities, this strategy takes $O(n!)$ time. We will now see that dynamic programming yields a much faster solution, though not a polynomial one.

What is the appropriate subproblem for the TSP? Subproblems refer to partial solutions, and in this case the most obvious partial solution is the initial portion of a tour. Suppose we have started at city 1 as required, have visited a few cities, and are now in city j. What information do we need in order to extend this partial tour? We certainly need to know j, since this will determine which cities are most convenient to visit next. And we also need to know all the cities visited so far, so that we don't repeat any of them. Here, then, is an appropriate subproblem.

For a subset of cities $S \subseteq \{1, 2, \ldots, n\}$ that includes 1, and $j \in S$, let $C(S, j)$ be the length of the shortest path visiting each node in S exactly once, starting at 1 and ending at j.

When $|S| > 1$, we define $C(S, 1) = \infty$ since the path cannot both start and end at 1.

Now, let's express $C(S, j)$ in terms of smaller subproblems. We need to start at 1 and end at j; what should we pick as the second-to-last city? It has to be some $i \in S$, so the overall path length is the distance from 1 to i, namely, $C(S - \{j\}, i)$, plus the length of the final edge, d_{ij}. We must pick the best such i:

$$C(S, j) = \min_{i \in S : i \neq j} C(S - \{j\}, i) + d_{ij}.$$

On time and memory

The amount of time it takes to run a dynamic programming algorithm is easy to discern from the dag of subproblems: in many cases *it is just the total number of edges in the dag*! All we are really doing is visiting the nodes in linearized order, examining each node's inedges, and, most often, doing a constant amount of work per edge. By the end, each edge of the dag has been examined once.

But how much computer *memory* is required? There is no simple parameter of the dag characterizing this. It is certainly possible to do the job with an amount of memory proportional to the number of vertices (subproblems), but we can usually get away with much less. The reason is that the value of a particular subproblem only needs to be remembered until the larger subproblems depending on it have been solved. Thereafter, the memory it takes up can be released for reuse.

For example, in the Floyd-Warshall algorithm the value of dist(i, j, k) is not needed once the dist$(\cdot, \cdot, k+1)$ values have been computed. Therefore, we only need two $|V| \times |V|$ arrays to store the dist values, one for odd values of k and one for even values: when computing dist(i, j, k), we overwrite dist$(i, j, k-2)$.

(And let us not forget that, as always in dynamic programming, we also need one more array, prev(i, j), storing the next to last vertex in the current shortest path from i to j, a value that must be updated with dist(i, j, k). We omit this mundane but crucial bookkeeping step from our dynamic programming algorithms.)

Can you see why the edit distance dag in Figure 6.5 only needs memory proportional to the length of the shorter string?

The subproblems are ordered by $|S|$. Here's the code.

```
C({1}, 1) = 0
for s = 2 to n:
    for all subsets S ⊆ {1, 2, ..., n} of size s and containing 1:
        C(S, 1) = ∞
        for all j ∈ S, j ≠ 1:
            C(S, j) = min{C(S − {j}, i) + d_ij : i ∈ S, i ≠ j}
return min_j C({1, ..., n}, j) + d_j1
```

There are at most $2^n \cdot n$ subproblems, and each one takes linear time to solve. The total running time is therefore $O(n^2 2^n)$.

6.7 Independent sets in trees

A subset of nodes $S \subset V$ is an *independent set* of graph $G = (V, E)$ if there are no edges between them. For instance, in Figure 6.10 the nodes $\{1, 5\}$ form an independent set, but nodes $\{1, 4, 5\}$ do not, because of the edge between 4 and 5. The largest independent set is $\{2, 3, 6\}$.

Figure 6.10 The largest independent set in this graph has size 3.

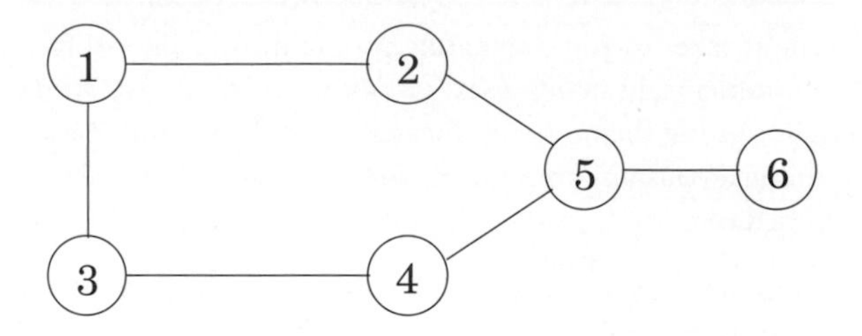

Like several other problems we have seen in this chapter (knapsack, traveling salesman), finding the largest independent set in a graph is believed to be intractable. However, when the graph happens to be a *tree*, the problem can be solved in linear time, using dynamic programming. And what are the appropriate subproblems? Already in the chain matrix multiplication problem we noticed that the layered structure of a tree provides a natural definition of a subproblem—as long as one node of the tree has been identified as a root.

So here's the algorithm: Start by rooting the tree at any node r. Now, each node defines a subtree—the one hanging from it. This immediately suggests subproblems:

$$I(u) = \text{size of largest independent set of subtree hanging from } u.$$

Our final goal is $I(r)$.

Dynamic programming proceeds as always from smaller subproblems to larger ones, that is to say, bottom-up in the rooted tree. Suppose we know the largest independent sets for all subtrees below a certain node u; in other words, suppose we know $I(w)$ for all descendants w of u. How can we compute $I(u)$? Let's split the computation into two cases: any independent set either includes u or it doesn't (Figure 6.11).

$$I(u) = \max\left\{1 + \sum_{\text{grandchildren } w \text{ of } u} I(w),\ \sum_{\text{children } w \text{ of } u} I(w)\right\}.$$

If the independent set includes u, then we get one point for it, but we aren't allowed to include the children of u—therefore we move on to the grandchildren. This is the first case in the formula. On the other hand, if we don't include u, then we don't get a point for it, but we can move on to its children.

The number of subproblems is exactly the number of vertices. With a little care, the running time can be made linear, $O(|V| + |E|)$.

Figure 6.11 $I(u)$ is the size of the largest independent set of the subtree rooted at u. Two cases: either u is in this independent set, or it isn't.

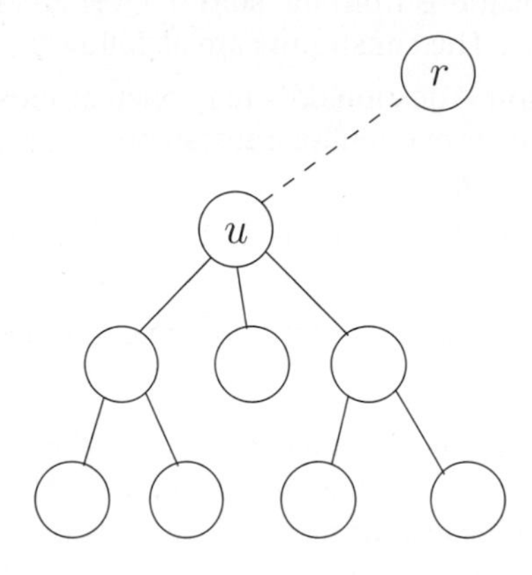

Exercises

6.1. A *contiguous subsequence* of a list S is a subsequence made up of consecutive elements of S. For instance, if S is

$$5, 15, -30, 10, -5, 40, 10,$$

then $15, -30, 10$ is a contiguous subsequence but $5, 15, 40$ is not. Give a linear-time algorithm for the following task:

Input: A list of numbers, $a_1, a_2, \ldots, a_n$.
Output: The contiguous subsequence of maximum sum (a subsequence of length zero has sum zero).

For the preceding example, the answer would be $10, -5, 40, 10$, with a sum of 55.

(*Hint:* For each $j \in \{1, 2, \ldots, n\}$, consider contiguous subsequences ending exactly at position j.)

6.2. You are going on a long trip. You start on the road at mile post 0. Along the way there are n hotels, at mile posts $a_1 < a_2 < \cdots < a_n$, where each a_i is measured from the starting point. The only places you are allowed to stop are at these hotels, but you can choose which of the hotels you stop at. You must stop at the final hotel (at distance a_n), which is your destination.

You'd ideally like to travel 200 miles a day, but this may not be possible (depending on the spacing of the hotels). If you travel x miles during a day, the *penalty* for that day is $(200 - x)^2$. You want to plan your trip so as to minimize the total penalty—that is, the sum, over all travel days, of the daily penalties. Give an efficient algorithm that determines the optimal sequence of hotels at which to stop.

6.3. Yuckdonald's is considering opening a series of restaurants along Quaint Valley Highway (QVH). The n possible locations are along a straight line, and the distances of these locations from the start of QVH are, in miles and in increasing order, $m_1, m_2, \ldots, m_n$. The constraints are as follows:

- At each location, Yuckdonald's may open at most one restaurant. The expected profit from opening a restaurant at location i is p_i, where $p_i > 0$ and $i = 1, 2, \ldots, n$.
- Any two restaurants should be at least k miles apart, where k is a positive integer.

Give an efficient algorithm to compute the maximum expected total profit subject to the given constraints.

6.4. You are given a string of n characters $s[1 \ldots n]$, which you believe to be a corrupted text document in which all punctuation has vanished (so that it looks something like "itwasthebestoftimes..."). You wish to reconstruct the document using a dictionary, which is available in the form of a Boolean function $\texttt{dict}(\cdot)$: for any string w,

$$\texttt{dict}(w) = \begin{cases} \texttt{true} & \text{if } w \text{ is a valid word} \\ \texttt{false} & \text{otherwise.} \end{cases}$$

(a) Give a dynamic programming algorithm that determines whether the string $s[\cdot]$ can be reconstituted as a sequence of valid words. The running time should be at most $O(n^2)$, assuming calls to dict take unit time.

(b) In the event that the string is valid, make your algorithm output the corresponding sequence of words.

6.5. *Pebbling a checkerboard.* We are given a checkerboard which has 4 rows and n columns, and has an integer written in each square. We are also given a set of $2n$ pebbles, and we want to place some or all of these on the checkerboard (each pebble can be placed on exactly one square) so as to maximize the sum of the integers in the squares that are covered by pebbles. There is one constraint: for a placement of pebbles to be legal, no two of them can be on horizontally or vertically adjacent squares (diagonal adjacency is fine).

(a) Determine the number of legal *patterns* that can occur in any column (in isolation, ignoring the pebbles in adjacent columns) and describe these patterns.

Call two patterns *compatible* if they can be placed on adjacent columns to form a legal placement. Let us consider subproblems consisting of the first k columns $1 \le k \le n$. Each subproblem can be assigned a *type*, which is the pattern occurring in the last column.

(b) Using the notions of compatibility and type, give an $O(n)$-time dynamic programming algorithm for computing an optimal placement.

6.6. Let us define a multiplication operation on three symbols a, b, c according to the following table; thus $ab = b$, $ba = c$, and so on. Notice that the multiplication operation defined by the table is neither associative nor commutative.

	a	b	c
a	b	b	a
b	c	b	a
c	a	c	c

Find an efficient algorithm that examines a string of these symbols, say *bbbbac*, and decides whether or not it is possible to parenthesize the string in such a way that the value of the resulting expression is a. For example, on input *bbbbac* your algorithm should return *yes* because $((b(bb))(ba))c = a$.

6.7. A subsequence is *palindromic* if it is the same whether read left to right or right to left. For instance, the sequence

$$A, C, G, T, G, T, C, A, A, A, A, T, C, G$$

has many palindromic subsequences, including A, C, G, C, A and A, A, A, A (on the other hand, the subsequence A, C, T is *not* palindromic). Devise an algorithm that takes a sequence $x[1 \ldots n]$ and returns the (length of the) longest palindromic subsequence. Its running time should be $O(n^2)$.

6.8. Given two strings $x = x_1 x_2 \cdots x_n$ and $y = y_1 y_2 \cdots y_m$, we wish to find the length of their *longest common substring*, that is, the largest k for which there are indices i and j with $x_i x_{i+1} \cdots x_{i+k-1} = y_j y_{j+1} \cdots y_{j+k-1}$. Show how to do this in time $O(mn)$.

6.9. A certain string-processing language offers a primitive operation which splits a string into two pieces. Since this operation involves copying the original string, it takes n units of time for a string of length n, regardless of the location of the cut. Suppose, now, that you want to break a string into many pieces. The order in which the breaks are made can affect the total running time. For example, if you want to cut a 20-character string at positions 3 and 10, then making the first cut at position 3 incurs a total cost of $20 + 17 = 37$, while doing position 10 first has a better cost of $20 + 10 = 30$.

Give a dynamic programming algorithm that, given the locations of m cuts in a string of length n, finds the minimum cost of breaking the string into $m + 1$ pieces.

6.10. *Counting heads.* Given integers n and k, along with $p_1, \ldots, p_n \in [0, 1]$, you want to determine the probability of obtaining exactly k heads when n biased coins are tossed independently at random, where p_i is the probability that the ith coin comes up heads. Give an $O(nk)$ algorithm for this task.[2] Assume you can multiply and add two numbers in $[0, 1]$ in $O(1)$ time.

6.11. Given two strings $x = x_1 x_2 \cdots x_n$ and $y = y_1 y_2 \cdots y_m$, we wish to find the length of their *longest common subsequence*, that is, the largest k for which there are indices $i_1 < i_2 < \cdots < i_k$ and $j_1 < j_2 < \cdots < j_k$ with $x_{i_1} x_{i_2} \cdots x_{i_k} = y_{j_1} y_{j_2} \cdots y_{j_k}$. Show how to do this in time $O(mn)$.

6.12. You are given a convex polygon P on n vertices in the plane (specified by their x and y coordinates). A *triangulation* of P is a collection of $n - 3$ diagonals of P such that no two diagonals intersect (except possibly at their endpoints). Notice that a triangulation splits the polygon's interior into $n - 2$ disjoint triangles. The cost of a triangulation is the sum of the lengths of the diagonals in it. Give an efficient algorithm for finding a triangulation of minimum cost. (*Hint:* Label the vertices of P by $1, \ldots, n$, starting from an arbitrary vertex and walking clockwise. For $1 \le i < j \le n$, let the subproblem $A(i, j)$ denote the minimum cost triangulation of the polygon spanned by vertices $i, i + 1, \ldots, j$.)

6.13. Consider the following game. A "dealer" produces a sequence $s_1 \cdots s_n$ of "cards," face up, where each card s_i has a value v_i. Then two players take turns picking a card from the sequence, but can only pick the first or the last card of the (remaining) sequence. The goal is to collect cards of largest total value. (For example, you can think of the cards as bills of different denominations.) Assume n is even.

(a) Show a sequence of cards such that it is not optimal for the first player to start by picking up the available card of larger value. That is, the natural *greedy* strategy is suboptimal.

(b) Give an $O(n^2)$ algorithm to compute an optimal strategy for the first player. Given the initial sequence, your algorithm should precompute in $O(n^2)$ time some information, and then the first player should be able to make each move optimally in $O(1)$ time by looking up the precomputed information.

6.14. *Cutting cloth.* You are given a rectangular piece of cloth with dimensions $X \times Y$, where X and Y are positive integers, and a list of n products that can be made using the cloth. For each product $i \in [1, n]$ you know that a rectangle of cloth of dimensions $a_i \times b_i$ is needed and that the final selling price of the product is c_i. Assume the a_i, b_i, and c_i are all positive integers. You have a machine that can cut any rectangular piece of cloth into two pieces either horizontally or vertically. Design an algorithm that determines the best return on the $X \times Y$ piece of cloth, that is, a strategy for cutting the cloth so that the products made from the resulting pieces give the maximum sum of selling prices. You are free to make as many copies of a given product as you wish, or none if desired.

6.15. Suppose two teams, A and B, are playing a match to see who is the first to win n games (for some particular n). We can suppose that A and B are equally

[2]In fact, there is also a $O(n \log^2 n)$ algorithm within your reach.

competent, so each has a 50% chance of winning any particular game. Suppose they have already played $i + j$ games, of which A has won i and B has won j. Give an efficient algorithm to compute the probability that A will go on to win the match. For example, if $i = n - 1$ and $j = n - 3$ then the probability that A will win the match is 7/8, since it must win any of the next three games.

6.16. The *garage sale problem*. On a given Sunday morning, there are n garage sales going on, $g_1, g_2, \ldots, g_n$. For each garage sale g_j, you have an estimate of its value to you, v_j. For any two garage sales you have an estimate of the transportation cost d_{ij} of getting from g_i to g_j. You are also given the costs d_{0j} and d_{j0} of going between your home and each garage sale. You want to find a tour of a *subset* of the given garage sales, starting and ending at home, that maximizes your total benefit minus your total transportation costs.

Give an algorithm that solves this problem in time $O(n^2 2^n)$. (*Hint:* This is closely related to the traveling salesman problem.)

6.17. Given an unlimited supply of coins of denominations $x_1, x_2, \ldots, x_n$, we wish to make change for a value v; that is, we wish to find a set of coins whose total value is v. This might not be possible: for instance, if the denominations are 5 and 10 then we can make change for 15 but not for 12. Give an $O(nv)$ dynamic-programming algorithm for the following problem.

Input: $x_1, \ldots, x_n; v$.
Question: Is it possible to make change for v using coins of denominations $x_1, \ldots, x_n$?

6.18. Consider the following variation on the change-making problem (Exercise 6.17): you are given denominations $x_1, x_2, \ldots, x_n$, and you want to make change for a value v, but you are allowed to use each denomination *at most once*. For instance, if the denominations are 1, 5, 10, 20, then you can make change for $16 = 1 + 15$ and for $31 = 1 + 10 + 20$ but not for 40 (because you can't use 20 twice).

Input: Positive integers $x_1, x_2, \ldots, x_n$; another integer v.
Output: Can you make change for v, using each denomination x_i at most once?

Show how to solve this problem in time $O(nv)$.

6.19. Here is yet another variation on the change-making problem (Exercise 6.17).

Given an unlimited supply of coins of denominations $x_1, x_2, \ldots, x_n$, we wish to make change for a value v using at most k coins; that is, we wish to find a set of $\leq k$ coins whose total value is v. This might not be possible: for instance, if the denominations are 5 and 10 and $k = 6$, then we can make change for 55 but not for 65. Give an efficient dynamic-programming algorithm for the following problem.

Input: $x_1, \ldots, x_n$; k; v.
Question: Is it possible to make change for v using at most k coins, of denominations $x_1, \ldots, x_n$?

Figure 6.12 Two binary search trees for the keywords of a programming language.

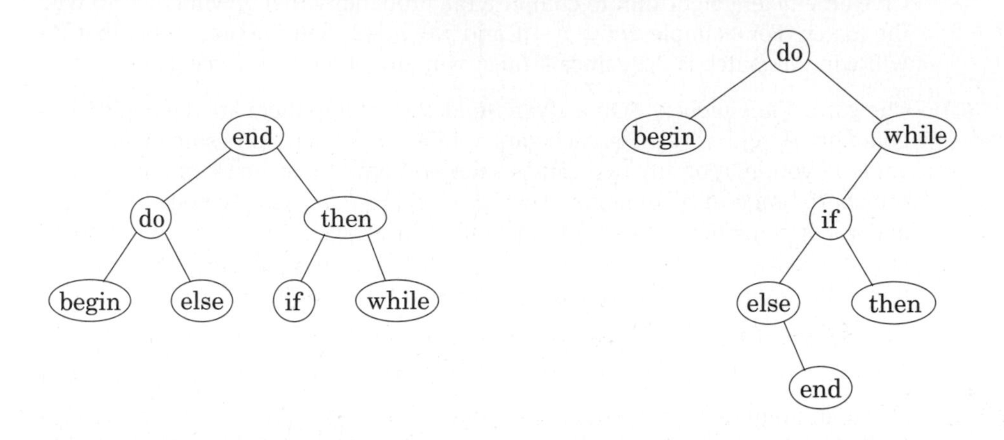

6.20. *Optimal binary search trees.* Suppose we know the frequency with which keywords occur in programs of a certain language, for instance:

begin	5%
do	40%
else	8%
end	4%
if	10%
then	10%
while	23%

We want to organize them in a *binary search tree*, so that the keyword in the root is alphabetically bigger than all the keywords in the left subtree and smaller than all the keywords in the right subtree (and this holds for all nodes).

Figure 6.12 has a nicely-balanced example on the left. In this case, when a keyword is being looked up, the number of comparisons needed is at most three: for instance, in finding "while", only the three nodes "end", "then", and "while" get examined. But since we know the frequency with which keywords are accessed, we can use an even more fine-tuned cost function, the *average number of comparisons* to look up a word. For the search tree on the left, it is

$$\text{cost} = 1(0.04) + 2(0.40 + 0.10) + 3(0.05 + 0.08 + 0.10 + 0.23) = 2.42.$$

By this measure, the best search tree is the one on the right, which has a cost of 2.18.

Give an efficient algorithm for the following task.

Input: n words (in sorted order); frequencies of these words: $p_1, p_2, \ldots, p_n$.

Output: The binary search tree of lowest cost (defined above as the expected number of comparisons in looking up a word).

6.21. A *vertex cover* of a graph $G = (V, E)$ is a subset of vertices $S \subseteq V$ that includes at least one endpoint of every edge in E. Give a linear-time algorithm for the following task.

Input: An undirected tree $T = (V, E)$.
Output: The size of the smallest vertex cover of T.

For instance, in the following tree, possible vertex covers include $\{A, B, C, D, E, F, G\}$ and $\{A, C, D, F\}$ but not $\{C, E, F\}$. The smallest vertex cover has size 3: $\{B, E, G\}$.

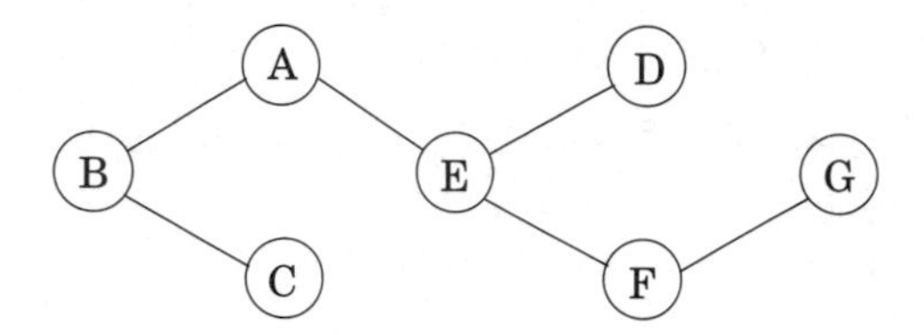

6.22. Give an $O(nt)$ algorithm for the following task.

Input: A list of n positive integers $a_1, a_2, \ldots, a_n$; a positive integer t.
Question: Does some subset of the a_i's add up to t? (You can use each a_i at most once.)

6.23. A mission-critical production system has n stages that have to be performed sequentially; stage i is performed by machine M_i. Each machine M_i has a probability r_i of functioning reliably and a probability $1 - r_i$ of failing (and the failures are independent). Therefore, if we implement each stage with a single machine, the probability that the whole system works is $r_1 \cdot r_2 \cdots r_n$. To improve this probability we add redundancy, by having m_i copies of the machine M_i that performs stage i. The probability that all m_i copies fail simultaneously is only $(1 - r_i)^{m_i}$, so the probability that stage i is completed correctly is $1 - (1 - r_i)^{m_i}$ and the probability that the whole system works is $\prod_{i=1}^{n}(1 - (1 - r_i)^{m_i})$. Each machine M_i has a cost c_i, and there is a total budget B to buy machines. (Assume that B and c_i are positive integers.)

Given the probabilities $r_1, \ldots, r_n$, the costs $c_1, \ldots, c_n$, and the budget B, find the redundancies $m_1, \ldots, m_n$ that are within the available budget and that maximize the probability that the system works correctly.

6.24. *Time and space complexity of dynamic programming.* Our dynamic programming algorithm for computing the edit distance between strings of length m and n creates a table of size $n \times m$ and therefore needs $O(mn)$ time and space. In practice, it will run out of space long before it runs out of time. How can this space requirement be reduced?

(a) Show that if we just want to compute the value of the edit distance (rather than the optimal sequence of edits), then only $O(n)$ space is needed, because only a small portion of the table needs to be maintained at any given time.

(b) Now suppose that we also want the optimal sequence of edits. As we saw earlier, this problem can be recast in terms of a corresponding grid-shaped dag, in which the goal is to find the optimal path from node $(0, 0)$ to node (n, m). It will be convenient to work with this formulation, and while we're talking about convenience, we might as well also assume that m is a power of 2.

Let's start with a small addition to the edit distance algorithm that will turn out to be very useful. The optimal path in the dag must pass through an intermediate node $(k, m/2)$ for some k; show how the algorithm can be modified to also return this value k.

(c) Now consider a recursive scheme:

```
procedure find-path((0,0) → (n,m))
compute the value k above
find-path((0,0) → (k,m/2))
find-path((k,m/2) → (n,m))
concatenate these two paths, with k in the middle
```

Show that this scheme can be made to run in $O(mn)$ time and $O(n)$ space.

6.25. Consider the following 3-PARTITION problem. Given integers $a_1, \ldots, a_n$, we want to determine whether it is possible to partition $\{1, \ldots, n\}$ into three disjoint subsets I, J, K such that

$$\sum_{i \in I} a_i = \sum_{j \in J} a_j = \sum_{k \in K} a_k = \frac{1}{3} \sum_{i=1}^{n} a_i$$

For example, for input $(1, 2, 3, 4, 4, 5, 8)$ the answer is *yes*, because there is the partition $(1, 8)$, $(4, 5)$, $(2, 3, 4)$. On the other hand, for input $(2, 2, 3, 5)$ the answer is *no*.

Devise and analyze a dynamic programming algorithm for 3-PARTITION that runs in time polynomial in n and in $\sum_i a_i$.

6.26. *Sequence alignment.* When a new gene is discovered, a standard approach to understanding its function is to look through a database of known genes and find close matches. The closeness of two genes is measured by the extent to which they are *aligned*. To formalize this, think of a gene as being a long string over an alphabet $\Sigma = \{A, C, G, T\}$. Consider two genes (strings) $x = ATGCC$ and $y = TACGCA$. An alignment of x and y is a way of matching up these two strings by writing them in columns, for instance:

$$\begin{array}{ccccccc} - & A & T & - & G & C & C \\ T & A & - & C & G & C & A \end{array}$$

Here the "$-$" indicates a "gap." The characters of each string must appear in order, and each column must contain a character from at least one of the strings. The score of an alignment is specified by a scoring matrix δ of size $(|\Sigma|+1) \times (|\Sigma|+1)$, where the extra row and column are to accommodate gaps. For instance the preceding alignment has the following score:

$$\delta(-, T) + \delta(A, A) + \delta(T, -) + \delta(-, C) + \delta(G, G) + \delta(C, C) + \delta(C, A).$$

Give a dynamic programming algorithm that takes as input two strings $x[1 \ldots n]$ and $y[1 \ldots m]$ and a scoring matrix δ, and returns the highest-scoring alignment. The running time should be $O(mn)$.

6.27. *Alignment with gap penalties.* The alignment algorithm of Exercise 6.26 helps to identify DNA sequences that are close to one another. The discrepancies between these closely matched sequences are often caused by errors in DNA replication. However, a closer look at the biological replication process reveals that the scoring function we considered earlier has a qualitative problem: nature often inserts or removes entire substrings of nucleotides (creating long gaps), rather than editing just one position at a time. Therefore, the penalty for a gap of length 10 should not be 10 times the penalty for a gap of length 1, but something significantly smaller.

Repeat Exercise 6.26, but this time use a modified scoring function in which the penalty for a gap of length k is $c_0 + c_1 k$, where c_0 and c_1 are given constants (and c_0 is larger than c_1).

6.28. *Local sequence alignment.* Often two DNA sequences are significantly different, but contain regions that are very similar and are *highly conserved*. Design an algorithm that takes an input two strings $x[1 \ldots n]$ and $y[1 \ldots m]$ and a scoring matrix δ (as defined in Exercise 6.26), and outputs substrings x' and y' of x and y, respectively, that have the highest-scoring alignment over all pairs of such substrings. Your algorithm should take time $O(mn)$.

6.29. *Exon chaining.* Each gene corresponds to a subregion of the overall genome (the DNA sequence); however, part of this region might be "junk DNA." Frequently, a gene consists of several pieces called exons, which are separated by junk fragments called introns. This complicates the process of identifying genes in a newly sequenced genome.

Suppose we have a new DNA sequence and we want to check whether a certain gene (a string) is present in it. Because we cannot hope that the gene will be a contiguous subsequence, we look for partial matches—fragments of the DNA that are also present in the gene (actually, even these partial matches will be approximate, not perfect). We then attempt to assemble these fragments.

Let $x[1 \ldots n]$ denote the DNA sequence. Each partial match can be represented by a triple (l_i, r_i, w_i), where $x[l_i \ldots r_i]$ is the fragment and w_i is a weight representing the strength of the match (it might be a local alignment score or some other statistical quantity). Many of these potential matches could be false, so the goal is to find a subset of the triples that are consistent (nonoverlapping) and have a maximum total weight.

Show how to do this efficiently.

6.30. *Reconstructing evolutionary trees by maximum parsimony.* Suppose we manage to sequence a particular gene across a whole bunch of different species. For concreteness, say there are n species, and the sequences are strings of length k over alphabet $\Sigma = \{A, C, G, T\}$. How can we use this information to reconstruct the evolutionary history of these species?

Evolutionary history is commonly represented by a tree whose leaves are the different species, whose root is their common ancestor, and whose internal branches represent speciation events (that is, moments when a new species broke off from an existing one). Thus we need to find the following:

- A (binary) evolutionary tree with the given species at the leaves.
- For each internal node, a string of length k: the gene sequence for that particular ancestor.

For each possible tree T, annotated with sequences $s(u) \in \Sigma^k$ at each of its nodes u, we can assign a score based on the principle of *parsimony*: fewer mutations are more likely.

$$\text{score}(T) = \sum_{(u,v)\in E(T)} (\text{number of positions on which } s(u) \text{ and } s(v) \text{ agree}).$$

Finding the highest-score tree is a difficult problem. Here we will consider just a small part of it: suppose we know the structure of the tree, and we want to fill in the sequences $s(u)$ of the internal nodes u. Here's an example with $k = 4$ and $n = 5$:

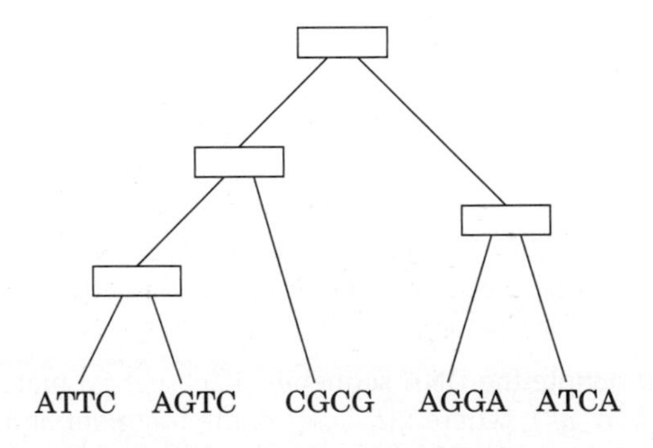

(a) In this particular example, there are several maximum parsimony reconstructions of the internal node sequences. Find one of them.

(b) Give an efficient (in terms of n and k) algorithm for this task. (*Hint:* Even though the sequences might be long, you can do just one position at a time.)

Chapter 7

Linear programming and reductions

Many of the problems for which we want algorithms are *optimization* tasks: the *shortest* path, the *cheapest* spanning tree, the *longest* increasing subsequence, and so on. In such cases, we seek a solution that (1) satisfies certain constraints (for instance, the path must use edges of the graph and lead from s to t, the tree must touch all nodes, the subsequence must be increasing); and (2) is the best possible, with respect to some well-defined criterion, among all solutions that satisfy these constraints.

Linear programming describes a broad class of optimization tasks in which both the constraints and the optimization criterion are *linear functions*. It turns out an enormous number of problems can be expressed in this way.

Given the vastness of its topic, this chapter is divided into several parts, which can be read separately subject to the following dependencies.

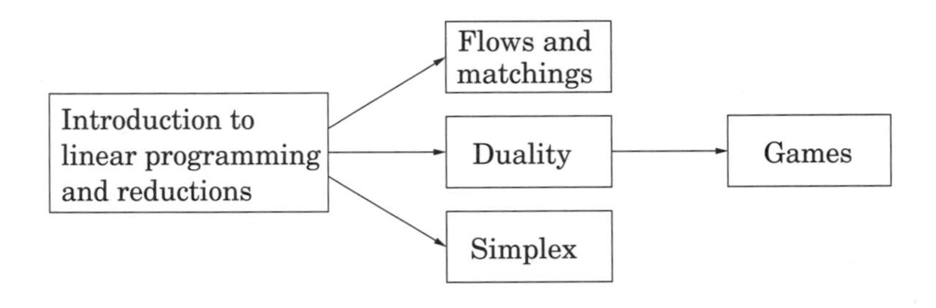

7.1 An introduction to linear programming

In a linear programming problem we are given a set of variables, and we want to assign real values to them so as to (1) satisfy a set of linear equations and/or linear inequalities involving these variables and (2) maximize or minimize a given linear objective function.

7.1.1 Example: profit maximization

A boutique chocolatier has two products: its flagship assortment of triangular chocolates, called *Pyramide*, and the more decadent and deluxe *Pyramide Nuit*. How much of each should it produce to maximize profits? Let's say it makes x_1 boxes of Pyramide per day, at a profit of \$1 each, and x_2 boxes of Nuit, at a more substantial profit of \$6 apiece; x_1 and x_2 are unknown values that we wish to determine. But this is not all; there are also some constraints on x_1 and x_2 that must be accommodated (besides the obvious one, $x_1, x_2 \geq 0$). First, the daily demand for these exclusive chocolates is limited to at most 200 boxes of Pyramide and 300 boxes of Nuit. Also, the current workforce can produce a total of at most 400 boxes of chocolate per day. What are the optimal levels of production?

We represent the situation by a *linear program*, as follows.

$$\begin{array}{ll} \text{Objective function} & \max x_1 + 6x_2 \\ \text{Constraints} & x_1 \leq 200 \\ & x_2 \leq 300 \\ & x_1 + x_2 \leq 400 \\ & x_1, x_2 \geq 0 \end{array}$$

A linear equation in x_1 and x_2 defines a line in the two-dimensional (2D) plane, and a linear inequality designates a *half-space*, the region on one side of the line. Thus the set of all *feasible solutions* of this linear program, that is, the points (x_1, x_2) which satisfy all constraints, is the intersection of five half-spaces. It is a convex polygon, shown in Figure 7.1.

We want to find the point in this polygon at which the objective function—the profit—is maximized. The points with a profit of c dollars lie on the line $x_1 + 6x_2 = c$, which has a slope of $-1/6$ and is shown in Figure 7.1 for selected values of c. As c increases, this "profit line" moves parallel to itself, up and to the right. Since the goal

Figure 7.1 (a) The feasible region for a linear program. (b) Contour lines of the objective function: $x_1 + 6x_2 = c$ for different values of the profit c.

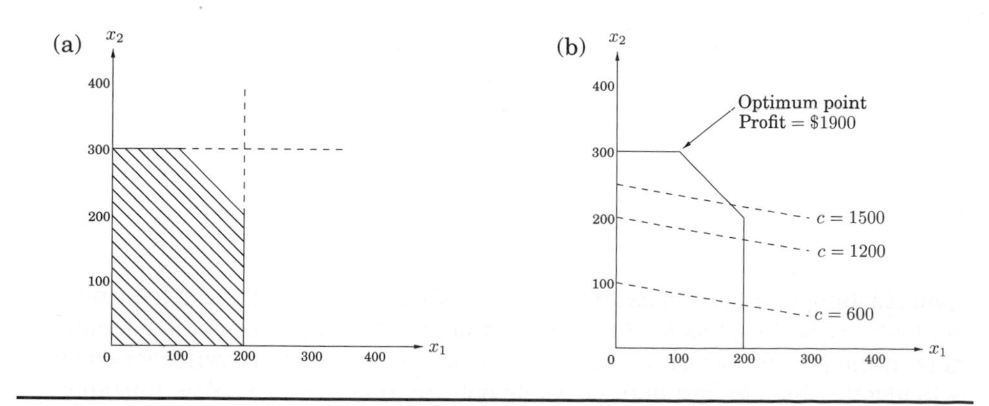

is to maximize c, we must move the line as far up as possible, while still touching the feasible region. The optimum solution will be the very last feasible point that the profit line sees and must therefore be a vertex of the polygon, as shown in the figure. If the slope of the profit line were different, then its last contact with the polygon could be an entire edge rather than a single vertex. In this case, the optimum solution would not be unique, but there would certainly be an optimum vertex.

It is a general rule of linear programs that the optimum is achieved at a vertex of the feasible region. The only exceptions are cases in which there is no optimum; this can happen in two ways:

1. The linear program is *infeasible*; that is, the constraints are so tight that it is impossible to satisfy all of them. For instance,

$$x \leq 1, \quad x \geq 2.$$

2. The constraints are so loose that the feasible region is *unbounded*, and it is possible to achieve arbitrarily high objective values. For instance,

$$\max \; x_1 + x_2$$
$$x_1, x_2 \geq 0.$$

Solving linear programs

Linear programs (LPs) can be solved by the *simplex method*, devised by George Dantzig in 1947. We shall explain it in more detail in Section 7.6, but briefly, this algorithm starts at a vertex, in our case perhaps (0, 0), and repeatedly looks for an adjacent vertex (connected by an edge of the feasible region) of better objective value. In this way it does *hill-climbing* on the vertices of the polygon, walking from neighbor to neighbor so as to steadily increase profit along the way. Here's a possible trajectory.

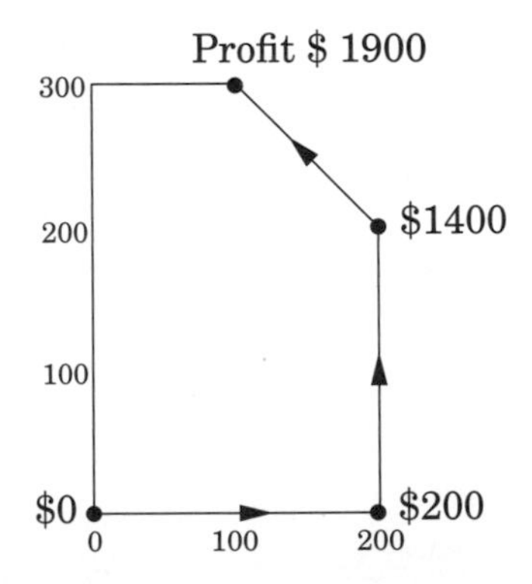

Upon reaching a vertex that has no better neighbor, simplex declares it to be optimal and halts. Why does this *local* test imply *global* optimality? By simple geometry—think of the profit line passing through this vertex. Since all the vertex's neighbors lie below the line, the rest of the feasible polygon must also lie below this line.

More products
Encouraged by consumer demand, the chocolatier decides to introduce a third and even more exclusive line of chocolates, called *Pyramide Luxe*. One box of these will bring in a profit of \$13. Let x_1, x_2, x_3 denote the number of boxes of each chocolate produced daily, with x_3 referring to Luxe. The old constraints on x_1 and x_2 persist, although the labor restriction now extends to x_3 as well: the sum of all three variables can be at most 400. What's more, it turns out that Nuit and Luxe require the same packaging machinery, except that Luxe uses it three times as much, which imposes another constraint $x_2 + 3x_3 \leq 600$. What are the best possible levels of production?

Here is the updated linear program.

$$\begin{aligned} \max \quad x_1 + 6x_2 + 13x_3 & \\ x_1 &\leq 200 \\ x_2 &\leq 300 \\ x_1 + x_2 + x_3 &\leq 400 \\ x_2 + 3x_3 &\leq 600 \\ x_1, x_2, x_3 &\geq 0 \end{aligned}$$

Figure 7.2 The feasible polyhedron for a three-variable linear program.

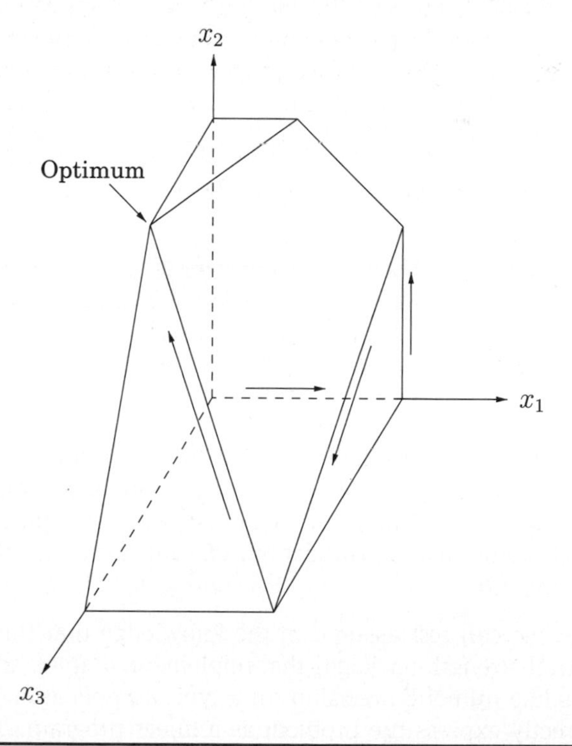

The space of solutions is now three-dimensional. Each linear equation defines a 3D plane, and each inequality a half-space on one side of the plane. The feasible region is an intersection of seven half-spaces, a polyhedron (Figure 7.2). Looking at the figure, can you decipher which inequality corresponds to each face of the polyhedron?

A profit of c corresponds to the plane $x_1 + 6x_2 + 13x_3 = c$. As c increases, this profit-plane moves parallel to itself, further and further into the positive orthant until it no longer touches the feasible region. The point of final contact is the optimal vertex: $(0, 300, 100)$, with total profit \$3100.

How would the simplex algorithm behave on this modified problem? As before, it would move from vertex to vertex, along edges of the polyhedron, increasing profit steadily. A possible trajectory is shown in Figure 7.2, corresponding to the following sequence of vertices and profits:

$$\begin{array}{c} (0,0,0) \\ \$0 \end{array} \longrightarrow \begin{array}{c} (200,0,0) \\ \$200 \end{array} \longrightarrow \begin{array}{c} (200,200,0) \\ \$1400 \end{array} \longrightarrow \begin{array}{c} (200,0,200) \\ \$2800 \end{array} \longrightarrow \begin{array}{c} (0,300,100) \\ \$3100 \end{array}$$

Finally, upon reaching a vertex with no better neighbor, it would stop and declare this to be the optimal point. Once again by basic geometry, if all the vertex's neighbors lie on one side of the profit-plane, then so must the entire polyhedron.

A magic trick called duality

Here is why you should believe that $(0, 300, 100)$, with a total profit of \$3100, is the optimum: Look back at the linear program. Add the second inequality to the third, and add to them the fourth multiplied by 4. The result is the inequality $x_1 + 6x_2 + 13x_3 \leq 3100$.

Do you see? This inequality says that no feasible solution (values x_1, x_2, x_3 satisfying the constraints) can possibly have a profit greater than 3100. So we must indeed have found the optimum! The only question is, where did we get these mysterious multipliers $(0, 1, 1, 4)$ for the four inequalities?

In Section 7.4 we'll see that it is always possible to come up with such multipliers by solving another LP! Except that (it gets even better) we do not even need to solve this other LP, because it is in fact so intimately connected to the original one—it is called the *dual*—that solving the original LP solves the dual as well! But we are getting far ahead of our story.

What if we add a fourth line of chocolates, or hundreds more of them? Then the problem becomes high-dimensional, and hard to visualize. Simplex continues to work in this general setting, although we can no longer rely upon simple geometric intuitions for its description and justification. We will study the full-fledged simplex algorithm in Section 7.6.

In the meantime, we can rest assured in the knowledge that there are many professional, industrial-strength packages that implement simplex and take care of all the tricky details like numeric precision. In a typical application, the main task is therefore to correctly express the problem as a linear program. The package then takes care of the rest.

With this in mind, let's look at a high-dimensional application.

7.1.2 Example: production planning

This time, our company makes handwoven carpets, a product for which the demand is extremely seasonal. Our analyst has just obtained demand estimates for all months of the next calendar year: $d_1, d_2, \ldots, d_{12}$. As feared, they are very uneven, ranging from 440 to 920.

Here's a quick snapshot of the company. We currently have 30 employees, each of whom makes 20 carpets per month and gets a monthly salary of $2,000. We have no initial surplus of carpets.

How can we handle the fluctuations in demand? There are three ways:

1. *Overtime*, but this is expensive since overtime pay is 80% more than regular pay. Also, workers can put in at most 30% overtime.
2. *Hiring and firing*, but these cost $320 and $400, respectively, per worker.
3. *Storing surplus production*, but this costs $8 per carpet per month. We currently have no stored carpets on hand, and we must end the year without any carpets stored.

This rather involved problem can be formulated and solved as a linear program!

A crucial first step is defining the variables.

$$
\begin{aligned}
w_i &= \text{number of workers during } i\text{th month; } w_0 = 30. \\
x_i &= \text{number of carpets made during } i\text{th month.} \\
o_i &= \text{number of carpets made by overtime in month } i. \\
h_i,\ f_i &= \text{number of workers hired and fired, respectively, at beginning of month } i. \\
s_i &= \text{number of carpets stored at end of month } i\text{; } s_0 = 0.
\end{aligned}
$$

All in all, there are 72 variables (74 if you count w_0 and s_0).

We now write the constraints. First, all variables must be nonnegative:

$$w_i, x_i, o_i, h_i, f_i, s_i \geq 0, \quad i = 1, \ldots, 12.$$

The total number of carpets made per month consists of regular production plus overtime:

$$x_i = 20w_i + o_i$$

(one constraint for each $i = 1, \ldots, 12$). The number of workers can potentially change at the start of each month:

$$w_i = w_{i-1} + h_i - f_i.$$

The number of carpets stored at the end of each month is what we started with, plus the number we made, minus the demand for the month:

$$s_i = s_{i-1} + x_i - d_i.$$

And overtime is limited:

$$o_i \leq 6w_i.$$

Finally, what is the objective function? It is to minimize the total cost:

$$\min \ 2000\sum_i w_i + 320\sum_i h_i + 400\sum_i f_i + 8\sum_i s_i + 180\sum_i o_i,$$

a linear function of the variables. Solving this linear program by simplex should take less than a second and will give us the optimum business strategy for our company.

Well, almost. The optimum solution might turn out to be *fractional*; for instance, it might involve hiring 10.6 workers in the month of March. This number would have to be rounded to either 10 or 11 in order to make sense, and the overall cost would then increase correspondingly. In the present example, most of the variables take on fairly large (double-digit) values, and thus rounding is unlikely to affect things too much. There are other LPs, however, in which rounding decisions have to be made very carefully in order to end up with an integer solution of reasonable quality.

In general, there is a tension in linear programming between the ease of obtaining fractional solutions and the desirability of integer ones. As we shall see in Chapter 8, finding the optimum integer solution of an LP is an important but very hard problem, called *integer linear programming*.

7.1.3 Example: optimum bandwidth allocation

Next we turn to a miniaturized version of the kind of problem a network service provider might face.

Suppose we are managing a network whose lines have the bandwidths shown in Figure 7.3, and we need to establish three connections: between users A and B, between B and C, and between A and C. Each connection requires at least two units of bandwidth, but can be assigned more. Connection A–B pays \$3 per unit of bandwidth, and connections B–C and A–C pay \$2 and \$4, respectively.

Each connection can be routed in two ways, a long path and a short path, or by a combination: for instance, two units of bandwidth via the short route, one via the long route. How do we route these connections to maximize our network's revenue?

This is a linear program. We have variables for each connection and each path (long or short); for example, x_{AB} is the short-path bandwidth allocated to the connection between A and B, and x'_{AB} the long-path bandwidth for this same connection. We demand that no edge's bandwidth is exceeded and that each connection gets a

Figure 7.3 A communications network between three users A, B, and C. Bandwidths are shown.

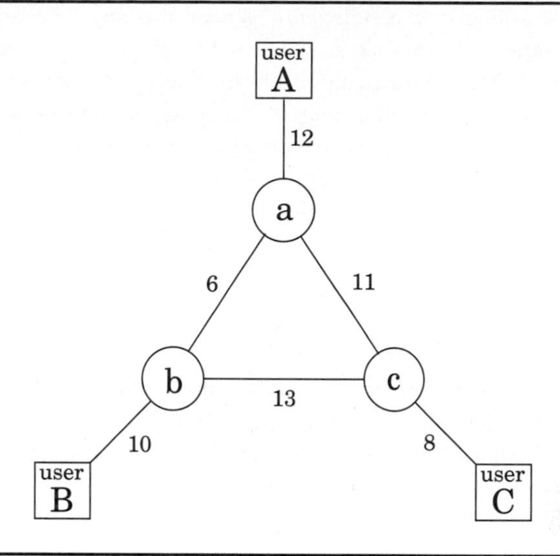

bandwidth of at least 2 units.

$$
\begin{aligned}
\max \quad 3x_{AB} + 3x'_{AB} + 2x_{BC} + 2x'_{BC} + 4x_{AC} + 4x'_{AC} & \\
x_{AB} + x'_{AB} + x_{BC} + x'_{BC} \le 10 & \quad \text{[edge } (b, B)] \\
x_{AB} + x'_{AB} + x_{AC} + x'_{AC} \le 12 & \quad \text{[edge } (a, A)] \\
x_{BC} + x'_{BC} + x_{AC} + x'_{AC} \le 8 & \quad \text{[edge } (c, C)] \\
x_{AB} + x'_{BC} + x'_{AC} \le 6 & \quad \text{[edge } (a, b)] \\
x'_{AB} + x_{BC} + x'_{AC} \le 13 & \quad \text{[edge } (b, c)] \\
x'_{AB} + x'_{BC} + x_{AC} \le 11 & \quad \text{[edge } (a, c)] \\
x_{AB} + x'_{AB} \ge 2 & \\
x_{BC} + x'_{BC} \ge 2 & \\
x_{AC} + x'_{AC} \ge 2 & \\
x_{AB}, x'_{AB}, x_{BC}, x'_{BC}, x_{AC}, x'_{AC} \ge 0 &
\end{aligned}
$$

Even a tiny example like this one is hard to solve on one's own (try it!), and yet the optimal solution is obtained instantaneously via simplex:

$$x_{AB} = 0,\ x'_{AB} = 7,\ x_{BC} = x'_{BC} = 1.5,\ x_{AC} = 0.5,\ x'_{AC} = 4.5.$$

This solution is not integral, but in the present application we don't need it to be, and thus no rounding is required. Looking back at the original network, we see that every edge except a–c is used at full capacity.

Reductions

Sometimes a computational task is sufficiently general that any subroutine for it can also be used to solve a variety of other tasks, which at first glance might seem unrelated. For instance, we saw in Chapter 6 how an algorithm for finding the longest path in a dag can, surprisingly, also be used for finding longest increasing subsequences. We describe this phenomenon by saying that the longest increasing subsequence problem *reduces to* the longest path problem in a dag. In turn, the longest path in a dag reduces to the shortest path in a dag; here's how a subroutine for the latter can be used to solve the former:

```
function LONGEST PATH(G)
  negate all edge weights of G
  return SHORTEST PATH(G)
```

Let's step back and take a slightly more formal view of reductions. If any subroutine for task Q can also be used to solve P, we say P *reduces to* Q. Often, P is solvable by a single call to Q's subroutine, which means any instance x of P can be transformed into an instance y of Q such that $P(x)$ can be deduced from $Q(y)$:

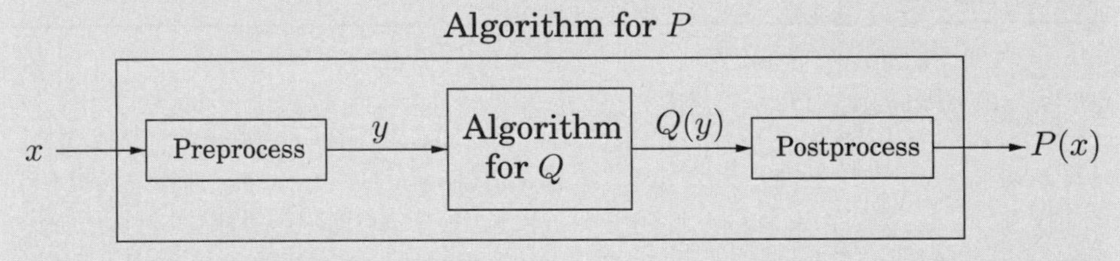

(Do you see that the reduction from $P =$ LONGEST PATH to $Q =$ SHORTEST PATH follows this schema?) If the pre- and postprocessing procedures are efficiently computable then this creates an efficient algorithm for P out of *any* efficient algorithm for Q!

Reductions enhance the power of algorithms: Once we have an algorithm for problem Q (which could be shortest path, for example) we can use it to solve other problems. In fact, most of the computational tasks we study in this book are considered core computer science problems precisely because they arise in so many different applications, which is another way of saying that many problems reduce to them. This is especially true of linear programming.

One cautionary observation: our LP has one variable for every possible path between the users. In a larger network, there could easily be exponentially many such paths, and therefore this particular way of translating the network problem into an LP will not scale well. We will see a cleverer and more scalable formulation in Section 7.2.

Here's a parting question for you to consider. Suppose we removed the constraint that each connection should receive at least two units of bandwidth. Would the optimum change?

7.1.4 Variants of linear programming

As evidenced in our examples, a general linear program has many degrees of freedom.

1. It can be either a maximization or a minimization problem.
2. Its constraints can be equations and/or inequalities.
3. The variables are often restricted to be nonnegative, but they can also be unrestricted in sign.

We will now show that these various LP options *can all be reduced to one another* via simple transformations. Here's how.

1. To turn a maximization problem into a minimization (or vice versa), just multiply the coefficients of the objective function by -1.

2a. To turn an inequality constraint like $\sum_{i=1}^{n} a_i x_i \leq b$ into an equation, introduce a new variable s and use

$$\begin{aligned} \sum_{i=1}^{n} a_i x_i + s &= b \\ s &\geq 0. \end{aligned}$$

This s is called the *slack variable* for the inequality. As justification, observe that a vector $(x_1, \ldots, x_n)$ satisfies the original inequality constraint if and only if there is some $s \geq 0$ for which it satisfies the new equality constraint.

2b. To change an equality constraint into inequalities is easy: rewrite $ax = b$ as the equivalent pair of constraints $ax \leq b$ and $ax \geq b$.

3. Finally, to deal with a variable x that is unrestricted in sign, do the following:
   - Introduce two nonnegative variables, $x^+, x^- \geq 0$.
   - Replace x, wherever it occurs in the constraints or the objective function, by $x^+ - x^-$.

   This way, x can take on any real value by appropriately adjusting the new variables. More precisely, any feasible solution to the original LP involving x can be mapped to a feasible solution of the new LP involving x^+, x^-, and vice versa.

By applying these transformations we can reduce any LP (maximization or minimization, with both inequalities and equations, and with both nonnegative and unrestricted variables) into an LP of a much more constrained kind that we call the *standard form*, in which the variables are all nonnegative, the constraints are all equations, and the objective function is to be minimized.

Matrix-vector notation

A linear function like $x_1 + 6x_2$ can be written as the dot product of two vectors

$$\mathbf{c} = \begin{pmatrix} 1 \\ 6 \end{pmatrix} \text{ and } \mathbf{x} = \begin{pmatrix} x_1 \\ x_2 \end{pmatrix},$$

denoted $\mathbf{c} \cdot \mathbf{x}$ or $\mathbf{c}^T\mathbf{x}$. Similarly, linear constraints can be compiled into matrix-vector form:

$$\begin{aligned} x_1 &\le 200 \\ x_2 &\le 300 \\ x_1 + x_2 &\le 400 \end{aligned} \quad \Longrightarrow \quad \underbrace{\begin{pmatrix} 1 & 0 \\ 0 & 1 \\ 1 & 1 \end{pmatrix}}_{\mathbf{A}} \underbrace{\begin{pmatrix} x_1 \\ x_2 \end{pmatrix}}_{\mathbf{x}} \le \underbrace{\begin{pmatrix} 200 \\ 300 \\ 400 \end{pmatrix}}_{\mathbf{b}}.$$

Here each row of matrix $\mathbf{A}$ corresponds to one constraint: its dot product with $\mathbf{x}$ is at most the value in the corresponding row of $\mathbf{b}$. In other words, if the rows of $\mathbf{A}$ are the vectors $\mathbf{a}_1, \ldots, \mathbf{a}_m$, then the statement $\mathbf{Ax} \le \mathbf{b}$ is equivalent to

$$\mathbf{a}_i \cdot \mathbf{x} \le b_i \text{ for all } i = 1, \ldots, m.$$

With these notational conveniences, a generic LP can be expressed simply as

$$\begin{aligned} &\max\ \mathbf{c}^T\mathbf{x} \\ &\mathbf{Ax} \le \mathbf{b} \\ &\mathbf{x} \ge 0. \end{aligned}$$

For example, our first linear program gets rewritten thus:

$$\begin{aligned} \max\ x_1 + 6x_2 & \\ x_1 &\le 200 \\ x_2 &\le 300 \\ x_1 + x_2 &\le 400 \\ x_1, x_2 &\ge 0 \end{aligned} \quad \Longrightarrow \quad \begin{aligned} \min\ -x_1 - 6x_2 & \\ x_1 + s_1 &= 200 \\ x_2 + s_2 &= 300 \\ x_1 + x_2 + s_3 &= 400 \\ x_1, x_2, s_1, s_2, s_3 &\ge 0 \end{aligned}$$

The original was also in a useful form: maximize an objective subject to certain inequalities. Any LP can likewise be recast in this way, using the reductions given earlier.

7.2 Flows in networks

7.2.1 Shipping oil

Figure 7.4(a) shows a directed graph representing a network of pipelines along which oil can be sent. The goal is to ship as much oil as possible from the *source* s to the *sink* t. Each pipeline has a maximum *capacity* it can handle, and there are

Figure 7.4 (a) A network with edge capacities. (b) A *flow* in the network.

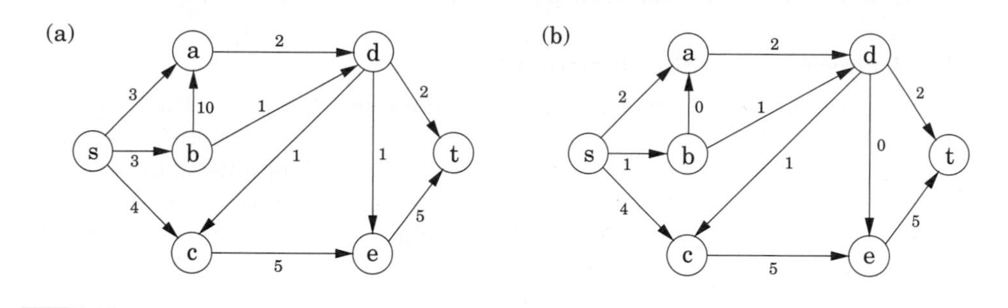

no opportunities for storing oil en route. Figure 7.4(b) shows a possible *flow* from s to t, which ships 7 units in all. Is this the best that can be done?

7.2.2 Maximizing flow

The networks we are dealing with consist of a directed graph $G = (V, E)$; two special nodes $s, t \in V$, which are, respectively, a source and sink of G; and *capacities* $c_e > 0$ on the edges.

We would like to send as much oil as possible from s to t without exceeding the capacities of any of the edges. A particular shipping scheme is called a *flow* and consists of a variable f_e for each edge e of the network, satisfying the following two properties:

1. It doesn't violate edge capacities: $0 \leq f_e \leq c_e$ for all $e \in E$.
2. For all nodes u except s and t, the amount of flow entering u equals the amount leaving u:

$$\sum_{(w,u)\in E} f_{wu} = \sum_{(u,z)\in E} f_{uz}.$$

   In other words, flow is *conserved*.

The *size* of a flow is the total quantity sent from s to t and, by the conservation principle, is equal to the quantity leaving s:

$$\text{size}(f) = \sum_{(s,u)\in E} f_{su}.$$

In short, our goal is to assign values to $\{f_e : e \in E\}$ that will satisfy a set of linear constraints and maximize a linear objective function. But this is a linear program! *The maximum-flow problem reduces to linear programming.*

For example, for the network of Figure 7.4 the LP has 11 variables, one per edge. It seeks to maximize $f_{sa} + f_{sb} + f_{sc}$ subject to a total of 27 constraints: 11 for nonnegativity (such as $f_{sa} \geq 0$), 11 for capacity (such as $f_{sa} \leq 3$), and 5 for flow conservation (one for each node of the graph other than s and t, such as $f_{sc} + f_{dc} = f_{ce}$). Simplex would take no time at all to correctly solve the problem and to confirm that, in our example, a flow of 7 is in fact optimal.

7.2.3 A closer look at the algorithm

All we know so far of the simplex algorithm is the vague geometric intuition that it keeps making local moves on the surface of a convex feasible region, successively improving the objective function until it finally reaches the optimal solution. Once we have studied it in more detail (Section 7.6), we will be in a position to understand exactly how it handles flow LPs, which is useful as a source of inspiration for designing *direct* max-flow algorithms.

It turns out that in fact the behavior of simplex has an elementary interpretation:

> Start with zero flow.
>
> *Repeat:* choose an appropriate path from s to t, and increase flow along the edges of this path as much as possible.

Figure 7.5(a)–(d) shows a small example in which simplex halts after two iterations. The final flow has size 2, which is easily seen to be optimal.

There is just one complication. What if we had initially chosen a different path, the one in Figure 7.5(e)? This gives only one unit of flow and yet seems to block all other paths. Simplex gets around this problem by also allowing paths to *cancel existing flow*. In this particular case, it would subsequently choose the path of Figure 7.5(f). Edge (b, a) of this path isn't in the original network and has the effect of canceling flow previously assigned to edge (a, b).

To summarize, in each iteration simplex looks for an $s - t$ path whose edges (u, v) can be of two types:

1. (u, v) is in the original network, and is not yet at full capacity.
2. The reverse edge (v, u) is in the original network, and there is some flow along it.

If the current flow is f, then in the first case, edge (u, v) can handle up to $c_{uv} - f_{uv}$ additional units of flow, and in the second case, up to f_{vu} additional units (canceling all or part of the existing flow on (v, u)). These flow-increasing opportunities can be captured in a *residual network* $G^f = (V, E^f)$, which has exactly the two types of edges listed, with residual capacities c^f:

$$\begin{cases} c_{uv} - f_{uv} & \text{if } (u, v) \in E \text{ and } f_{uv} < c_{uv} \\ f_{vu} & \text{if } (v, u) \in E \text{ and } f_{vu} > 0. \end{cases}$$

Thus we can equivalently think of simplex as choosing an $s - t$ path in the residual network.

By simulating the behavior of simplex, we get a direct algorithm for solving max-flow. It proceeds in iterations, each time explicitly constructing G^f, finding a suitable

Figure 7.5 An illustration of the max-flow algorithm. (a) A toy network. (b) The first path chosen. (c) The second path chosen. (d) The final flow. (e) We could have chosen this path first. (f) In which case, we would have to allow this second path.

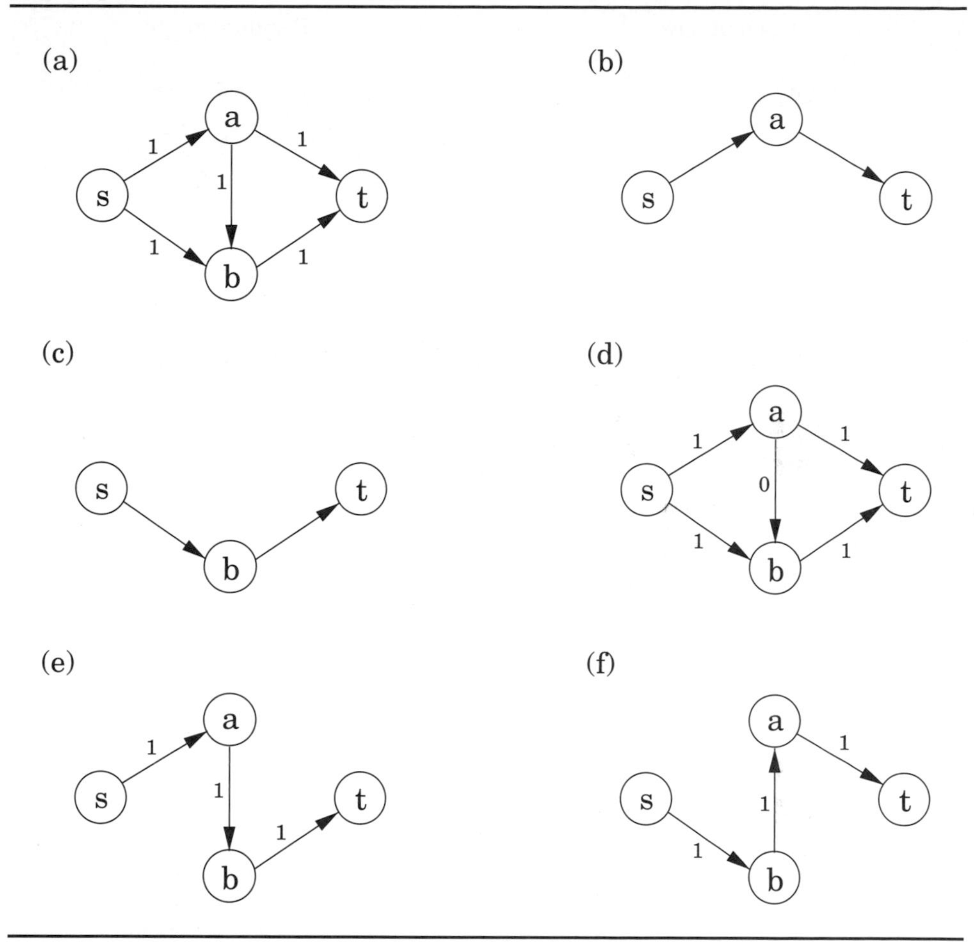

$s - t$ path in G^f by using, say, a linear-time breadth-first search, and halting if there is no longer any such path along which flow can be increased.

Figure 7.6 illustrates the algorithm on our oil example.

7.2.4 A certificate of optimality

Now for a truly remarkable fact: not only does simplex correctly compute a maximum flow, but it also generates a short proof of the optimality of this flow!

Let's see an example of what this means. Partition the nodes of the oil network (Figure 7.4) into two groups, $L = \{s, a, b\}$ and $R = \{c, d, e, t\}$:

Figure 7.6 The max-flow algorithm applied to the network of Figure 7.4. At each iteration, the current flow is shown on the left and the residual network on the right. The paths chosen are shown in bold.

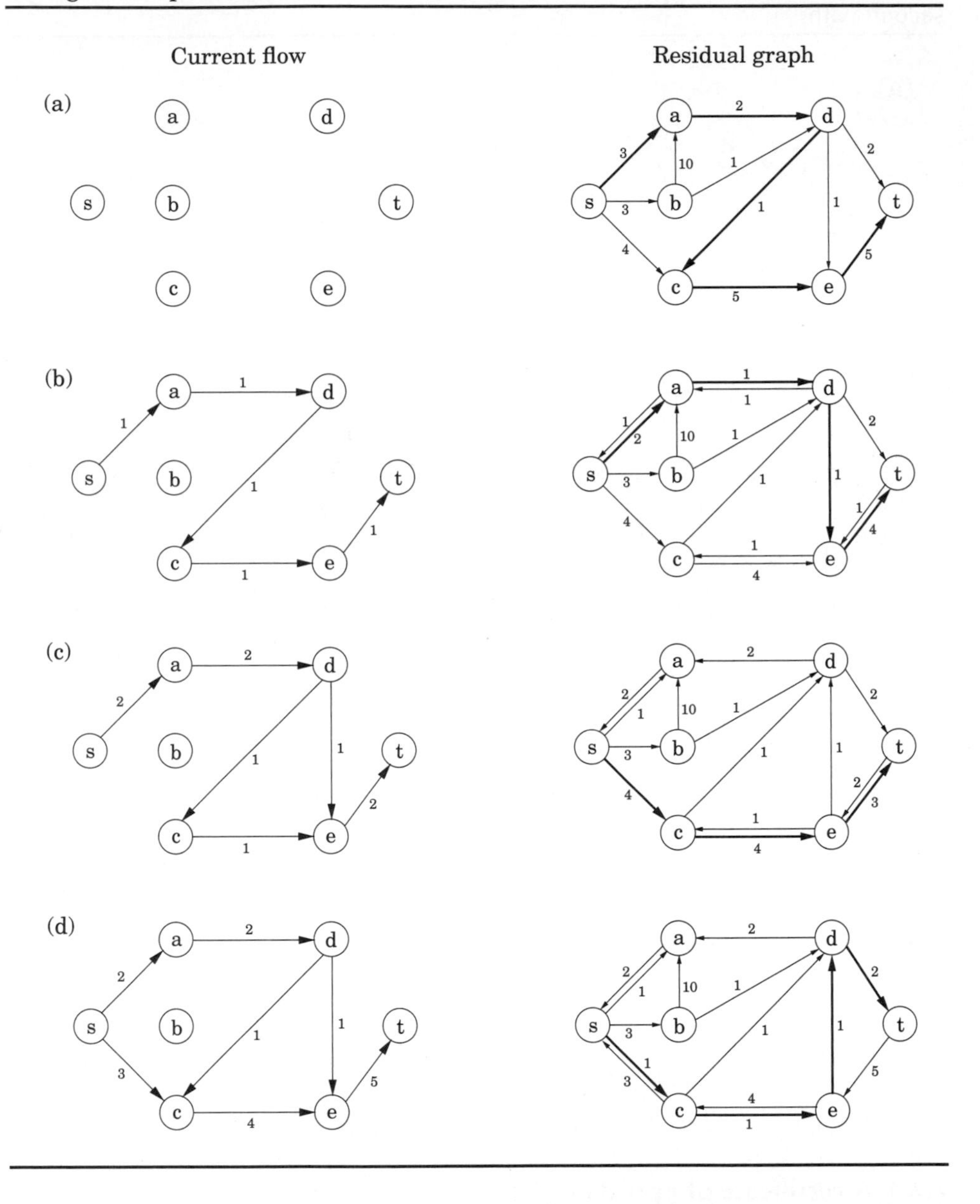

Figure 7.6 *Continued*

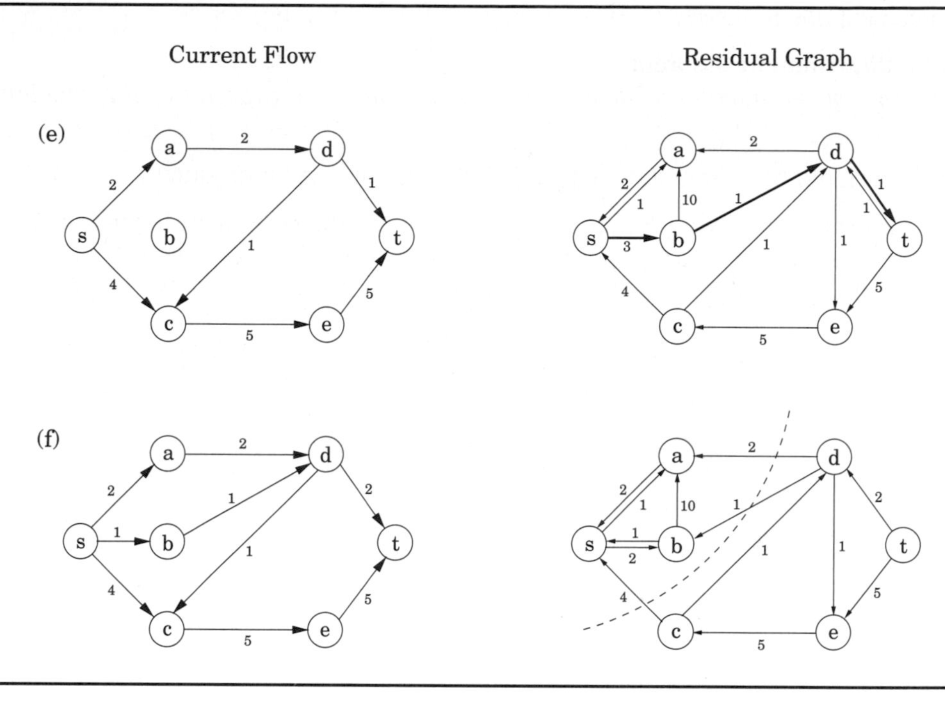

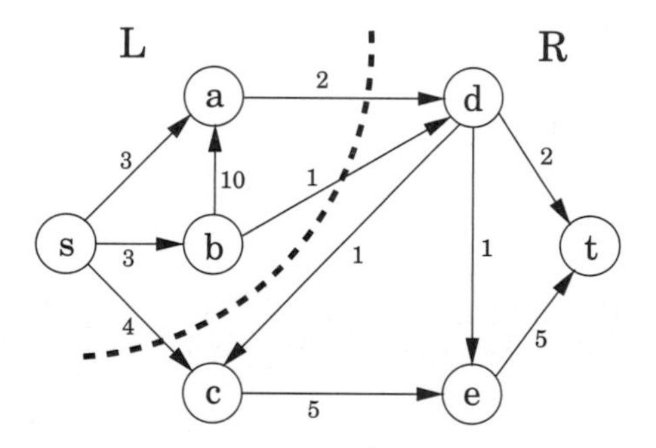

Any oil transmitted must pass from L to R. Therefore, no flow can possibly exceed the total capacity of the edges from L to R, which is 7. But this means that the flow we found earlier, of size 7, must be optimal!

More generally, an (s, t)-*cut* partitions the vertices into two disjoint groups L and R such that s is in L and t is in R. Its *capacity* is the total capacity of the edges from L to R, and as argued previously, is an upper bound on *any* flow:

Pick any flow f and any (s, t)-cut (L, R). Then size $(f) \leq$ capacity(L, R).

Some cuts are large and give loose upper bounds—cut $(\{s, b, c\}, \{a, d, e, t\})$ has a capacity of 19. But there is also a cut of capacity 7, which is effectively a *certificate of optimality* of the maximum flow. This isn't just a lucky property of our oil network; such a cut *always* exists.

Max-flow min-cut theorem
The size of the maximum flow in a network equals the capacity of the smallest (s, t)-cut.

Moreover, our algorithm automatically finds this cut as a by-product!

Let's see why this is true. Suppose f is the final flow when the algorithm terminates. We know that node t is no longer reachable from s in the residual network G^f. Let L be the nodes that *are* reachable from s in G^f, and let $R = V - L$ be the rest of the nodes. Then (L, R) is a cut in the graph G:

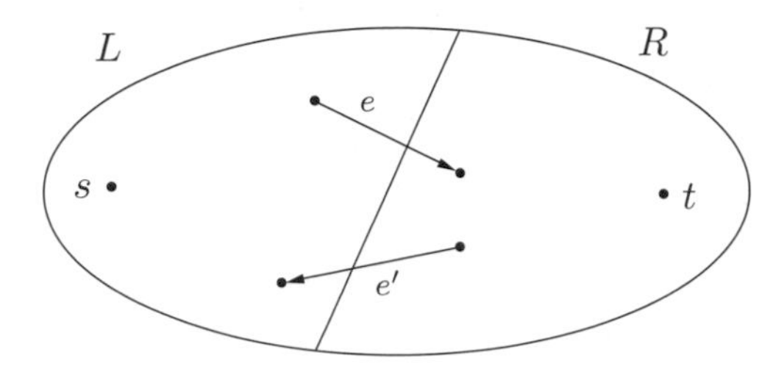

We claim that

$$\text{size}(f) = \text{capacity}(L, R).$$

To see this, observe that by the way L is defined, any edge going from L to R must be at full capacity (in the current flow f), and any edge from R to L must have zero flow. (So, in the figure, $f_e = c_e$ and $f_{e'} = 0$.) Therefore the net flow across (L, R) is exactly the capacity of the cut.

7.2.5 Efficiency

Each iteration of our maximum-flow algorithm is efficient, requiring $O(|E|)$ time if a depth-first or breadth-first search is used to find an $s - t$ path. But how many iterations are there?

Suppose all edges in the original network have *integer* capacities $\leq C$. Then an inductive argument shows that on each iteration of the algorithm, the flow is always an integer and increases by an integer amount. Therefore, since the maximum flow is at most $C|E|$ (why?), it follows that the number of iterations is at most this much. But this is hardly a reassuring bound: what if C is in the millions?

We examine this issue further in Exercise 7.31. It turns out that it is indeed possible to construct bad examples in which the number of iterations is proportional to C, *if* $s - t$ paths are not carefully chosen. However, if paths are chosen in a sensible manner—in particular, by using a breadth-first search, which finds the path with the fewest edges—then the number of iterations is at most $O(|V| \cdot |E|)$, no matter what the capacities are. This latter bound gives an overall running time of $O(|V| \cdot |E|^2)$ for maximum flow.

Figure 7.7 An edge between two people means they like each other. Is it possible to pair everyone up happily?

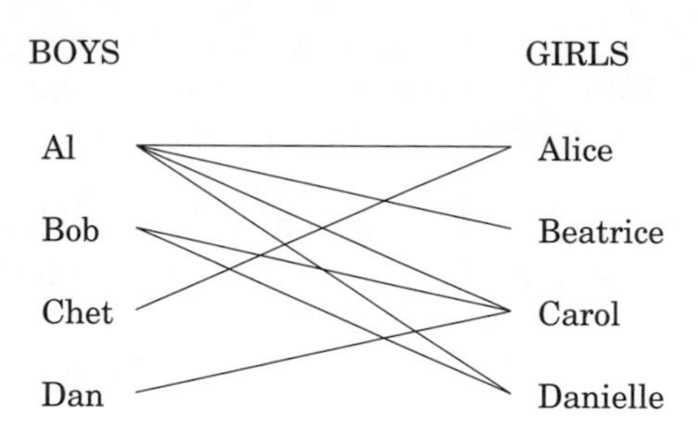

7.3 Bipartite matching

Figure 7.7 shows a graph with four nodes on the left representing boys and four nodes on the right representing girls.[1] There is an edge between a boy and girl if they like each other (for instance, Al likes all the girls). Is it possible to choose couples so that everyone has exactly one partner, and it is someone they like? In graph-theoretic jargon, is there a *perfect matching*?

This matchmaking game can be reduced to the maximum-flow problem, and thereby to linear programming! Create a new source node, s, with outgoing edges to all the boys; a new sink node, t, with incoming edges from all the girls; and direct all the edges in the original bipartite graph from boy to girl (Figure 7.8). Finally, give every edge a capacity of 1. Then there is a perfect matching if and only if this network has a flow whose size equals the number of couples. Can you find such a flow in the example?

Figure 7.8 A matchmaking network. Each edge has a capacity of one.

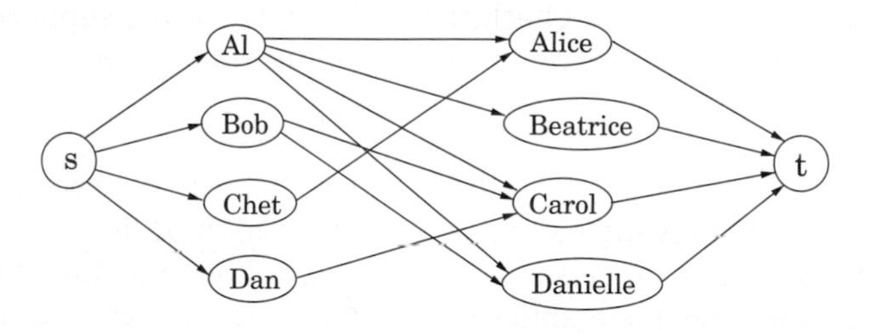

Actually, the situation is slightly more complicated than just stated: what is easy to see is that the optimum *integer-valued* flow corresponds to the optimum matching.

[1]This kind of graph, in which the nodes can be partitioned into two groups such that all edges are *between* the groups, is called *bipartite*.

We would be at a bit of a loss interpreting a flow that ships 0.7 units along the edge Al–Carol, for instance! Fortunately, the maximum-flow problem has the following

Property: *if all edge capacities are integers, then the optimal flow found by our algorithm is integral. We can see this directly from the algorithm, which in such cases would increment the flow by an integer amount on each iteration.*

Hence integrality comes for free in the maximum-flow problem. Unfortunately, this is the exception rather than the rule: as we will see in Chapter 8, it is a very difficult problem to find the optimum solution (or for that matter, *any* solution) of a general linear program, if we also demand that the variables be integers.

7.4 Duality

We have seen that in networks, flows are smaller than cuts, but the maximum flow and minimum cut exactly coincide and each is therefore a certificate of the other's optimality. Remarkable as this phenomenon is, we now generalize it from maximum flow to *any* problem that can be solved by linear programming! It turns out that every linear maximization problem has a *dual* minimization problem, and they relate to each other in much the same way as flows and cuts.

To understand what duality is about, recall our introductory LP with the two types of chocolate:

$$\begin{aligned} \max\ x_1 + 6x_2 & \\ x_1 &\leq 200 \\ x_2 &\leq 300 \\ x_1 + x_2 &\leq 400 \\ x_1, x_2 &\geq 0. \end{aligned}$$

Simplex declares the optimum solution to be $(x_1, x_2) = (100, 300)$, with objective value 1900. Can this answer be checked somehow? Let's see: suppose we take the first inequality and add it to six times the second inequality. We get

$$x_1 + 6x_2 \ \leq\ 2000.$$

This is interesting, because it tells us that it is impossible to achieve a profit of more than 2000. Can we add together some other combination of the LP constraints and bring this upper bound even closer to 1900? After a little experimentation, we find that multiplying the three inequalities by 0, 5, and 1, respectively, and adding them up yields

$$x_1 + 6x_2 \ \leq\ 1900.$$

So 1900 must indeed be the best possible value! The multipliers (0, 5, 1) magically constitute a *certificate of optimality*! It is remarkable that such a certificate exists for this LP—and even if we knew there were one, how would we systematically go about finding it?

Let's investigate the issue by describing what we expect of these three multipliers, call them y_1, y_2, y_3.

Multiplier	Inequality
y_1	$x_1 \leq 200$
y_2	$x_2 \leq 300$
y_3	$x_1 + x_2 \leq 400$

To start with, these y_i's must be nonnegative, for otherwise they are unqualified to multiply inequalities (multiplying an inequality by a negative number would flip the $\leq$ to $\geq$). After the multiplication and addition steps, we get the bound:

$$(y_1 + y_3)x_1 + (y_2 + y_3)x_2 \leq 200y_1 + 300y_2 + 400y_3.$$

We want the left-hand side to look like our objective function $x_1 + 6x_2$ so that the right-hand side is an upper bound on the optimum solution. For this we need $y_1 + y_3$ to be 1 and $y_2 + y_3$ to be 6. Come to think of it, it would be fine if $y_1 + y_3$ were larger than 1—the resulting certificate would be all the more convincing. Thus, we get an upper bound

$$x_1 + 6x_2 \leq 200y_1 + 300y_2 + 400y_3 \quad \text{if} \quad \left\{ \begin{array}{c} y_1, y_2, y_3 \geq 0 \\ y_1 + y_3 \geq 1 \\ y_2 + y_3 \geq 6 \end{array} \right\}.$$

We can easily find y's that satisfy the inequalities on the right by simply making them large enough, for example $(y_1, y_2, y_3) = (5, 3, 6)$. But these particular multipliers would tell us that the optimum solution of the LP is at most $200 \cdot 5 + 300 \cdot 3 + 400 \cdot 6 = 4300$, a bound that is far too loose to be of interest. What we want is a bound that is as tight as possible, so we should minimize $200y_1 + 300y_2 + 400y_3$ subject to the preceding inequalities. *And this is a new linear program!*

Therefore, finding the set of multipliers that gives the best upper bound on our original LP is tantamount to solving a new LP:

$$\begin{aligned} \min\ 200y_1 + 300y_2 + 400y_3 \\ y_1 + y_3 \geq 1 \\ y_2 + y_3 \geq 6 \\ y_1, y_2, y_3 \geq 0 \end{aligned}$$

By design, any feasible value of this *dual* LP is an upper bound on the original *primal* LP. So if we somehow find a pair of primal and dual feasible values that are equal, then they must both be optimal. Here is just such a pair:

$$\text{Primal} : (x_1, x_2) = (100, 300); \quad \text{Dual} : (y_1, y_2, y_3) = (0, 5, 1).$$

They both have value 1900, and therefore they certify each other's optimality (Figure 7.9).

Amazingly, this is not just a lucky example, but a general phenomenon. To start with, the preceding construction—creating a multiplier for each primal constraint;

Figure 7.9 By design, dual feasible values $\geq$ primal feasible values. The duality theorem tells us that moreover their optima coincide.

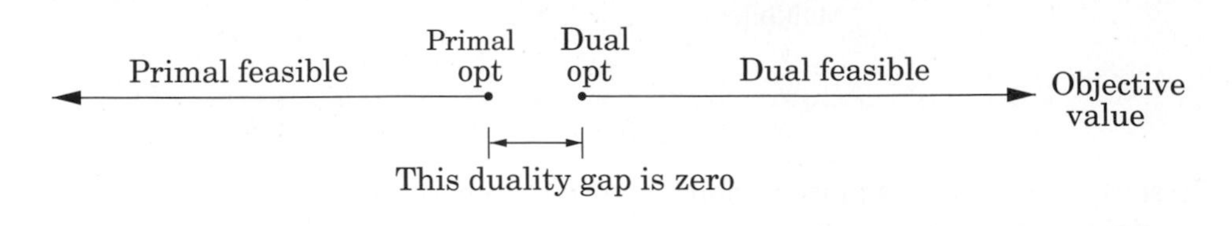

writing a constraint in the dual for every variable of the primal, in which the sum is required to be above the objective coefficient of the corresponding primal variable; and optimizing the sum of the multipliers weighted by the primal right-hand sides—can be carried out for any LP, as shown in Figure 7.10, and in even greater generality in Figure 7.11. The second figure has one noteworthy addition: if the primal has an equality constraint, then the corresponding multiplier (or *dual variable*) need not be nonnegative, because the validity of equations is preserved when multiplied by negative numbers. So, the multipliers of equations are unrestricted variables. Notice also the simple symmetry between the two LPs, in that the matrix $A = (a_{ij})$ defines one primal constraint with each of its *rows*, and one dual constraint with each of its *columns*.

By construction, any feasible solution of the dual is an upper bound on any feasible solution of the primal. But moreover, their optima coincide!

Duality theorem: *If a linear program has a bounded optimum, then so does its dual, and the two optimum values coincide.*

When the primal is the LP that expresses the max-flow problem, it is possible to assign interpretations to the dual variables that show the dual to be none other than the minimum-cut problem (Exercise 7.25). The relation between flows and cuts is therefore just a specific instance of the duality theorem. And in fact, the proof of this theorem falls out of the simplex algorithm, in much the same way as the max-flow min-cut theorem fell out of the analysis of the max-flow algorithm.

Figure 7.10 A generic primal LP in matrix-vector form, and its dual.

Primal LP:

$$\max \mathbf{c}^T\mathbf{x}$$
$$\mathbf{A}\mathbf{x} \leq \mathbf{b}$$
$$\mathbf{x} \geq 0$$

Dual LP:

$$\min \mathbf{y}^T\mathbf{b}$$
$$\mathbf{y}^T\mathbf{A} \geq \mathbf{c}^T$$
$$\mathbf{y} \geq 0$$

Figure 7.11 In the most general case of linear programming, we have a set I of inequalities and a set E of equalities (a total of $m = |I| + |E|$ constraints) over n variables, of which a subset N are constrained to be nonnegative. The dual has $m = |I| + |E|$ variables, of which only those corresponding to I have nonnegativity constraints.

Primal LP:

$$\begin{aligned} \max\ c_1x_1 + \cdots + c_nx_n \\ a_{i1}x_1 + \cdots + a_{in}x_n \leq b_i \quad \text{for } i \in I \\ a_{i1}x_1 + \cdots + a_{in}x_n = b_i \quad \text{for } i \in E \\ x_j \geq 0 \quad \text{for } j \in N \end{aligned}$$

Dual LP:

$$\begin{aligned} \min\ b_1y_1 + \cdots + b_my_m \\ a_{1j}y_1 + \cdots + a_{mj}y_m \geq c_j \quad \text{for } j \in N \\ a_{1j}y_1 + \cdots + a_{mj}y_m = c_j \quad \text{for } j \notin N \\ y_i \geq 0 \quad \text{for } i \in I \end{aligned}$$

Visualizing duality

One can solve the shortest-path problem by the following "analog" device: Given a weighted undirected graph, build a *physical model* of it in which each edge is a string of length equal to the edge's weight, and each node is a knot at which the appropriate endpoints of strings are tied together. Then to find the shortest path from s to t, just *pull* s away from t until the gadget is taut. It is intuitively clear that this finds the shortest path from s to t.

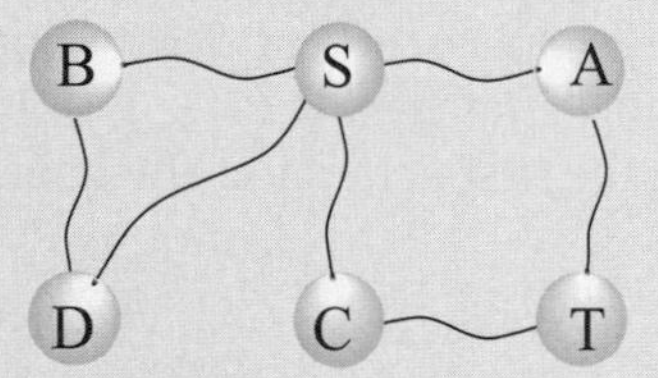

There is nothing remarkable or surprising about all this until we notice the following: the shortest-path problem is a *minimization* problem, right? Then why are we *pulling* s away from t, an act whose purpose is, obviously, *maximization*? Answer: By pulling s away from t we solve *the dual* of the shortest-path problem! This dual has a very simple form (Exercise 7.28), with one variable x_u for each node u:

$$\max\ x_s - x_t$$
$$|x_u - x_v| \leq w_{uv} \quad \text{for all edges } \{u, v\}.$$

In words, the dual problem is to stretch s and t as far apart as possible, subject to the constraint that the endpoints of any edge $\{u, v\}$ are separated by a distance of at most w_{uv}.

7.5 Zero-sum games

We can represent various conflict situations in life by *matrix games*. For example, the schoolyard *rock-paper-scissors* game is specified by the *payoff matrix* illustrated here. There are two players, called Row and Column, and they each pick a move from $\{r, p, s\}$. They then look up the matrix entry corresponding to their moves, and Column pays Row this amount. It is Row's gain and Column's loss.

$$G = \begin{array}{c|c|ccc} & & & \text{Column} & \\ & & r & p & s \\ \hline & r & 0 & -1 & 1 \\ \text{Row} & p & 1 & 0 & -1 \\ & s & -1 & 1 & 0 \end{array}$$

Now suppose the two of them play this game repeatedly. If Row always makes the same move, Column will quickly catch on and will always play the countermove, winning every time. Therefore Row should mix things up: we can model this by allowing Row to have a *mixed strategy*, in which on each turn she plays r with probability x_1, p with probability x_2, and s with probability x_3. This strategy is specified by the vector $\mathbf{x} = (x_1, x_2, x_3)$, positive numbers that add up to 1. Similarly, Column's mixed strategy is some $\mathbf{y} = (y_1, y_2, y_3)$.[2]

On any given round of the game, there is an $x_i y_j$ chance that Row and Column will play the ith and jth moves, respectively. Therefore the *expected* (average) payoff is

$$\sum_{i,j} G_{ij} \cdot \text{Prob[Row plays } i\text{, Column plays } j] = \sum_{i,j} G_{ij} x_i y_j.$$

Row wants to *maximize* this, while Column wants to *minimize* it. What payoffs can they hope to achieve in rock-paper-scissors? Well, suppose for instance that Row plays the "completely random" strategy $\mathbf{x} = (1/3, 1/3, 1/3)$. If Column plays r, then the average payoff (reading the first column of the game matrix) will be

$$\frac{1}{3} \cdot 0 + \frac{1}{3} \cdot 1 + \frac{1}{3} \cdot -1 = 0.$$

This is also true if Column plays p, or s. And since the payoff of any mixed strategy (y_1, y_2, y_3) is just a weighted average of the individual payoffs for playing r, p, and s, it must also be zero. This can be seen directly from the preceding formula,

$$\sum_{i,j} G_{ij} x_i y_j = \sum_{i,j} G_{ij} \cdot \frac{1}{3} y_j = \sum_j y_j \left(\sum_i \frac{1}{3} G_{ij} \right) = \sum_j y_j \cdot 0 = 0,$$

where the second-to-last equality is the observation that every column of G adds up to zero. Thus by playing the "completely random" strategy, Row forces an expected payoff of zero, *no matter what Column does*. This means that Column cannot hope for a negative (expected) payoff (remember that he wants the payoff to be as small as possible). But symmetrically, if Column plays the completely random strategy, he also forces an expected payoff of zero, and thus Row cannot hope for a positive (expected) payoff. In short, the best each player can do is to play completely

[2] Also of interest are scenarios in which players alter their strategies from round to round, but these can get very complicated and are a vast subject unto themselves.

randomly, with an expected payoff of zero. We have mathematically confirmed what you knew all along about rock-paper-scissors!

Let's think about this in a slightly different way, by considering two scenarios:

1. First Row announces her strategy, and then Column picks his.
2. First Column announces his strategy, and then Row chooses hers.

We've seen that the average payoff is the same (zero) in either case if both parties play optimally. But this might well be due to the high level of symmetry in rock-paper-scissors. In general games, we'd expect the first option to favor Column, since he knows Row's strategy and can fully exploit it while choosing his own. Likewise, we'd expect the second option to favor Row. Amazingly, this is not the case: if both play optimally, then it doesn't hurt a player to announce his or her strategy in advance! What's more, this remarkable property is a consequence of—and in fact equivalent to—linear programming duality.

Let's investigate this with a nonsymmetric game. Imagine a *presidential election* scenario in which there are two candidates for office, and the moves they make correspond to campaign issues on which they can focus (the initials stand for *economy*, *society*, *morality*, and *tax cut*). The payoff entries are millions of votes lost by Column.

$G =$

	m	t
e	3	−1
s	−2	1

Suppose Row announces that she will play the mixed strategy $\mathrm{x} = (1/2, 1/2)$. What should Column do? Move m will incur an expected loss of 1/2, while t will incur an expected loss of 0. The best response of Column is therefore the *pure* strategy $\mathrm{y} = (0, 1)$.

More generally, once Row's strategy $\mathrm{x} = (x_1, x_2)$ is fixed, there is always a *pure* strategy that is optimal for Column: either move m, with payoff $3x_1 - 2x_2$, or t, with payoff $-x_1 + x_2$, whichever is smaller. After all, any mixed strategy y is a weighted average of these two pure strategies and thus cannot beat the better of the two.

Therefore, if Row is forced to announce $\mathbf{x}$ before Column plays, she knows that his best response will achieve an expected payoff of $\min\{3x_1 - 2x_2, -x_1 + x_2\}$. She should choose $\mathbf{x}$ *defensively* to maximize her payoff against this best response:

$$\text{Pick } (x_1, x_2) \text{ that maximizes } \quad \underbrace{\min\{3x_1 - 2x_2, -x_1 + x_2\}}_{\text{payoff from Column's best response to } \mathbf{x}}$$

This choice of x_i's gives Row the best possible *guarantee* about her expected payoff. And we will now see that it can be found by an LP! The main trick is to notice that for *fixed* x_1 and x_2 the following are equivalent:

$$z = \min\{3x_1 - 2x_2, -x_1 + x_2\} \qquad \begin{array}{c} \max z \\ z \le 3x_1 - 2x_2 \\ z \le -x_1 + x_2 \end{array}$$

And Row needs to choose x_1 and x_2 to maximize this z.

$$\begin{array}{rcrcrcl} & & \max & & z & & \\ -3x_1 & + & 2x_2 & + & z & \le & 0 \\ x_1 & - & x_2 & + & z & \le & 0 \\ x_1 & + & x_2 & & & = & 1 \\ & & & & x_1, x_2 & \ge & 0 \end{array}$$

Symmetrically, if Column has to announce his strategy first, his best bet is to choose the mixed strategy y that minimizes his loss under Row's best response, in other words,

$$\text{Pick } (y_1, y_2) \text{ that minimizes} \quad \underbrace{\max\{3y_1 - y_2, -2y_1 + y_2\}}_{\text{outcome of Row's best response to } \mathbf{y}}$$

In LP form, this is

$$\begin{array}{rcrcrcl} & & \min & & w & & \\ -3y_1 & + & y_2 & + & w & \ge & 0 \\ 2y_1 & - & y_2 & + & w & \ge & 0 \\ y_1 & + & y_2 & & & = & 1 \\ & & & & y_1, y_2 & \ge & 0 \end{array}$$

The crucial observation now is that *these two LPs are dual to each other* (see Figure 7.11)! Hence, they have the same optimum, call it V.

Let us summarize. By solving an LP, Row (the maximizer) can determine a strategy for herself that guarantees an expected outcome of at least V no matter what Column does. And by solving the dual LP, Column (the minimizer) can guarantee an expected outcome of at most V, no matter what Row does. It follows that this is the uniquely defined optimal play: a priori it wasn't even certain that such a play existed. V is known as the *value* of the game. In our example, it is $1/7$ and is realized when Row plays her optimum mixed strategy $(3/7, 4/7)$ and Column plays his optimum mixed strategy $(2/7, 5/7)$.

This example is easily generalized to arbitrary games and shows the existence of mixed strategies that are optimal for both players and achieve the same value—a fundamental result of game theory called the *min-max theorem*. It can be written in equation form as follows:

$$\max_{\mathbf{x}} \min_{\mathbf{y}} \sum_{i,j} G_{ij} x_i y_j = \min_{\mathbf{y}} \max_{\mathbf{x}} \sum_{i,j} G_{ij} x_i y_j .$$

This is surprising, because the left-hand side, in which Row has to announce her strategy first, should presumably be better for Column than the right-hand side, in which he has to go first. Duality equalizes the two, as it did with maximum flows and minimum cuts.

7.6 The simplex algorithm

The extraordinary power and expressiveness of linear programs would be little consolation if we did not have a way to solve them efficiently. This is the role of the simplex algorithm.

At a high level, the simplex algorithm takes a set of linear inequalities and a linear objective function and finds the optimal feasible point by the following strategy:

```
let v be any vertex of the feasible region
while there is a neighbor v' of v with better objective value:
  set v = v'
```

In our 2D and 3D examples (Figure 7.1 and Figure 7.2), this was simple to visualize and made intuitive sense. But what if there are n variables, $x_1, \ldots, x_n$?

Any setting of the x_i's can be represented by an n-tuple of real numbers and plotted in n-dimensional space. A linear equation involving the x_i's defines a *hyperplane* in this same space $\mathbb{R}^n$, and the corresponding linear inequality defines a *half-space*, all points that are either precisely on the hyperplane or lie on one particular side of it. Finally, the feasible region of the linear program is specified by a set of inequalities and is therefore the intersection of the corresponding half-spaces, a convex polyhedron.

But what do the concepts of *vertex* and *neighbor* mean in this general context?

Figure 7.12 A polyhedron defined by seven inequalities.

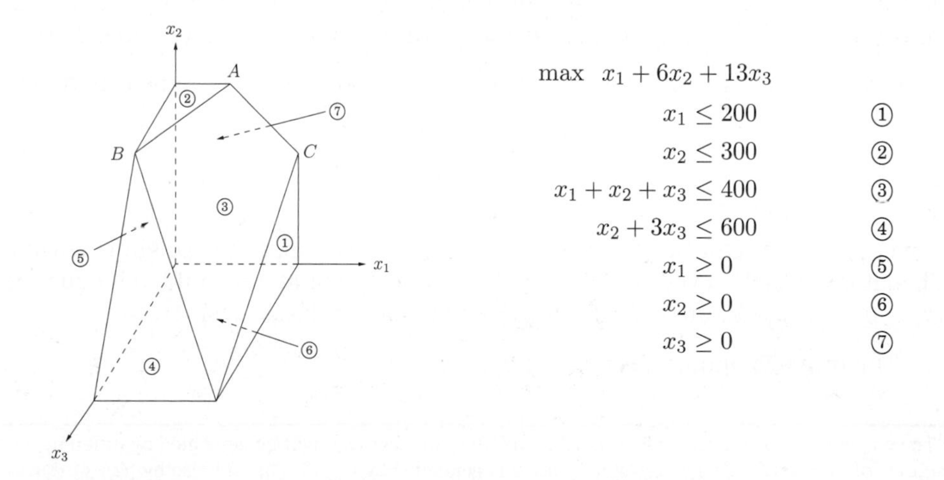

7.6.1 Vertices and neighbors in n-dimensional space

Figure 7.12 recalls an earlier example. Looking at it closely, we see that *each vertex is the unique point at which some subset of hyperplanes meet*. Vertex A, for instance, is the sole point at which constraints ②, ③, and ⑦ are satisfied with equality. On the other hand, the hyperplanes corresponding to inequalities ④ and ⑥ do not define a vertex, because their intersection is not just a single point but an entire line.

Let's make this definition precise.

> *Pick a subset of the inequalities. If there is a unique point that satisfies them with equality, and this point happens to be feasible, then it is a* vertex.

How many equations are needed to uniquely identify a point? When there are n variables, we need at least n linear equations if we want a unique solution. On the other hand, having more than n equations is redundant: at least one of them can be rewritten as a linear combination of the others and can therefore be disregarded. In short,

> Each vertex is specified by a set of n inequalities.[3]

A notion of *neighbor* now follows naturally.

> Two vertices are *neighbors* if they have $n - 1$ defining inequalities in common.

In Figure 7.12, for instance, vertices A and C share the two defining inequalities {③, ⑦} and are thus neighbors.

7.6.2 The algorithm

On each iteration, simplex has two tasks:

1. Check whether the current vertex is optimal (and if so, halt).
2. Determine where to move next.

As we will see, both tasks are easy if the vertex happens to be at the origin. And if the vertex is elsewhere, we will transform the coordinate system to move it to the origin!

First let's see why the origin is so convenient. Suppose we have some generic LP

$$\max \mathbf{c}^T\mathbf{x}$$
$$\mathbf{A}\mathbf{x} \leq \mathbf{b}$$
$$\mathbf{x} \geq 0$$

where x is the vector of variables, $\mathrm{x} = (x_1, \ldots, x_n)$. Suppose the origin is feasible. Then it is certainly a vertex, since it is the unique point at which the n inequalities $\{x_1 \geq 0, \ldots, x_n \geq 0\}$ are *tight*. Now let's solve our two tasks. Task 1:

> The origin is optimal if and only if all $c_i \leq 0$.

---

[3]There is one tricky issue here. It is possible that the same vertex might be generated by different subsets of inequalities. In Figure 7.12, vertex B is generated by {②, ③, ④}, but also by {②, ④, ⑤}. Such vertices are called *degenerate* and require special consideration. Let's assume for the time being that they don't exist, and we'll return to them later.

If all $c_i \leq 0$, then considering the constraints $\mathbf{x} \geq 0$, we can't hope for a better objective value. Conversely, if some $c_i > 0$, then the origin is not optimal, since we can increase the objective function by raising x_i.

Thus, for task 2, we can move by increasing some x_i for which $c_i > 0$. How much can we increase it? *Until we hit some other constraint.* That is, we release the tight constraint $x_i \geq 0$ and increase x_i until some other inequality, previously loose, now becomes tight. At that point, we again have exactly n tight inequalities, so we are at a new vertex.

For instance, suppose we're dealing with the following linear program.

$$\max\ 2x_1 + 5x_2$$

$$\begin{aligned} 2x_1 - x_2 &\leq 4 & \textcircled{1} \\ x_1 + 2x_2 &\leq 9 & \textcircled{2} \\ -x_1 + x_2 &\leq 3 & \textcircled{3} \\ x_1 &\geq 0 & \textcircled{4} \\ x_2 &\geq 0 & \textcircled{5} \end{aligned}$$

Simplex can be started at the origin, which is specified by constraints ④ and ⑤. To move, we release the tight constraint $x_2 \geq 0$. As x_2 is gradually increased, the first constraint it runs into is $-x_1 + x_2 \leq 3$, and thus it has to stop at $x_2 = 3$, at which point this new inequality is tight. The new vertex is thus given by ③ and ④.

So we know what to do if we are at the origin. But what if our current vertex $\mathbf{u}$ is elsewhere? The trick is to transform $\mathbf{u}$ into the origin, by shifting the coordinate system from the usual $(x_1, \ldots, x_n)$ to the "local view" from $\mathbf{u}$. These local coordinates consist of (appropriately scaled) distances $y_1, \ldots, y_n$ to the n hyperplanes (inequalities) that define and enclose $\mathbf{u}$:

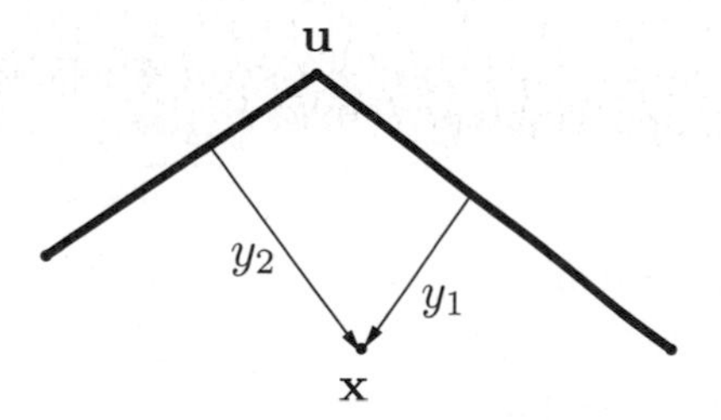

Specifically, if one of these enclosing inequalities is $\mathbf{a}_i \cdot \mathbf{x} \leq b_i$, then the distance from a point $\mathbf{x}$ to that particular "wall" is

$$y_i = b_i - \mathbf{a}_i \cdot \mathbf{x}.$$

The n equations of this type, one per wall, define the y_i's as linear functions of the x_i's, and this relationship can be inverted to express the x_i's as a linear function of the y_i's. Thus we can rewrite the entire LP in terms of the y's. This doesn't fundamentally change it (for instance, the optimal value stays the same), but expresses it in a different coordinate frame. The revised "local" LP has the following three properties:

1. It includes the inequalities $\mathbf{y} \geq 0$, which are simply the transformed versions of the inequalities defining $\mathbf{u}$.
2. $\mathbf{u}$ itself is the origin in y-space.
3. The cost function becomes $\max c_{\mathbf{u}} + \tilde{\mathbf{c}}^T \mathbf{y}$, where $c_{\mathbf{u}}$ is the value of the objective function at $\mathbf{u}$ and $\tilde{\mathbf{c}}$ is a transformed cost vector.

In short, we are back to the situation we know how to handle! Figure 7.13 shows this algorithm in action, continuing with our earlier example.

The simplex algorithm is now fully defined. It moves from vertex to neighboring vertex, stopping when the objective function is locally optimal, that is, when the coordinates of the local cost vector are all zero or negative. As we've just seen, a vertex with this property must also be globally optimal. On the other hand, if the current vertex is not locally optimal, then its local coordinate system includes some dimension along which the objective function can be improved, so we move along this direction—along this edge of the polyhedron—until we reach a neighboring vertex. By the nondegeneracy assumption (see footnote 3 in Section 7.6.1), this edge has nonzero length, and so we strictly improve the objective value. Thus the process must eventually halt.

7.6.3 Loose ends

There are several important issues in the simplex algorithm that we haven't yet mentioned.

The starting vertex

How do we find a vertex at which to start simplex? In our 2D and 3D examples we always started at the origin, which worked because the linear programs happened to have inequalities with positive right-hand sides. In a general LP we won't always be so fortunate. However, it turns out that finding a starting vertex *can be reduced to an LP* and solved by simplex!

To see how this is done, start with any linear program in standard form (recall Section 7.1.4), since we know LPs can always be rewritten this way.

$$\min \mathbf{c}^T\mathbf{x} \text{ such that } \mathbf{A}\mathbf{x} = \mathbf{b} \text{ and } \mathbf{x} \geq 0.$$

We first make sure that the right-hand sides of the equations are all nonnegative: if $b_i < 0$, just multiply both sides of the ith equation by -1.

Then we create a new LP as follows:

- Create m new *artificial variables* $z_1, \ldots, z_m \geq 0$, where m is the number of equations.
- Add z_i to the left-hand side of the ith equation.
- Let the objective, to be *minimized*, be $z_1 + z_2 + \cdots + z_m$.

For this new LP, it's easy to come up with a starting vertex, namely, the one with $z_i = b_i$ for all i and all other variables zero. Therefore we can solve it by simplex, to obtain the optimum solution.

Figure 7.13 Simplex in action.

Initial LP: $$\begin{array}{rcll} & \max 2x_1 + 5x_2 & & \\ 2x_1 - x_2 & \leq & 4 & ① \\ x_1 + 2x_2 & \leq & 9 & ② \\ -x_1 + x_2 & \leq & 3 & ③ \\ x_1 & \geq & 0 & ④ \\ x_2 & \geq & 0 & ⑤ \end{array}$$	*Current vertex:* {④, ⑤} (origin). *Objective value:* 0. *Move:* increase x_2. ⑤ is released, ③ becomes tight. Stop at $x_2 = 3$. New vertex {④, ③} has local coordinates (y_1, y_2): $$y_1 = x_1, \quad y_2 = 3 + x_1 - x_2$$
Rewritten LP: $$\begin{array}{rcll} & \max 15 + 7y_1 - 5y_2 & & \\ y_1 + y_2 & \leq & 7 & ① \\ 3y_1 - 2y_2 & \leq & 3 & ② \\ y_2 & \geq & 0 & ③ \\ y_1 & \geq & 0 & ④ \\ -y_1 + y_2 & \leq & 3 & ⑤ \end{array}$$	*Current vertex:* {④, ③}. *Objective value:* 15. *Move:* increase y_1. ④ is released, ② becomes tight. Stop at $y_1 = 1$. New vertex {②, ③} has local coordinates (z_1, z_2): $$z_1 = 3 - 3y_1 + 2y_2, \quad z_2 = y_2$$
Rewritten LP: $$\begin{array}{rcll} & \max 22 - \frac{7}{3}z_1 - \frac{1}{3}z_2 & & \\ -\frac{1}{3}z_1 + \frac{5}{3}z_2 & \leq & 6 & ① \\ z_1 & \geq & 0 & ② \\ z_2 & \geq & 0 & ③ \\ \frac{1}{3}z_1 - \frac{2}{3}z_2 & \leq & 1 & ④ \\ \frac{1}{3}z_1 + \frac{1}{3}z_2 & \leq & 4 & ⑤ \end{array}$$	*Current vertex:* {②, ③}. *Objective value:* 22. *Optimal:* all $c_i < 0$. Solve ②, ③ (in original LP) to get optimal solution $(x_1, x_2) = (1, 4)$.

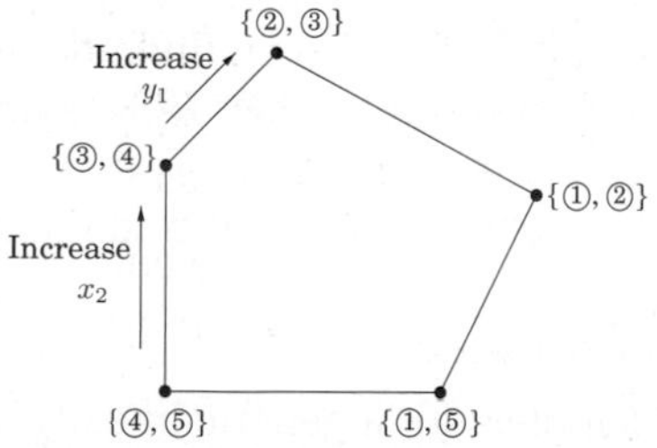

{②, ③}
Increase y_1
{③, ④}
{①, ②}
Increase x_2
{④, ⑤}
{①, ⑤}

There are two cases. If the optimum value of $z_1 + \cdots + z_m$ is zero, then all z_i's obtained by simplex are zero, and hence from the optimum vertex of the new LP we get a starting feasible vertex of the original LP, just by ignoring the z_i's. We can at last start simplex!

But what if the optimum objective turns out to be positive? Let us think. We tried to minimize the sum of the z_i's, but simplex decided that it cannot be zero. But this means that the original linear program is infeasible: it *needs* some nonzero z_i's to become feasible. This is how simplex discovers and reports that an LP is infeasible.

Degeneracy

In the polyhedron of Figure 7.12 vertex B is *degenerate*. Geometrically, this means that it is the intersection of more than $n = 3$ faces of the polyhedron (in this case, ②, ③, ④, ⑤). Algebraically, it means that if we choose any one of four sets of three inequalities ({②, ③, ④}, {②, ③, ⑤}, {②, ④, ⑤}, and {③, ④, ⑤}) and solve the corresponding system of three linear equations in three unknowns, we'll get the same solution in all four cases: $(0, 300, 100)$. This is a serious problem: simplex may return a suboptimal degenerate vertex simply because all its neighbors are identical to it and thus have no better objective. And if we modify simplex so that it detects degeneracy and continues to hop from vertex to vertex despite lack of any improvement in the cost, it may end up looping forever.

One way to fix this is by a *perturbation*: change each b_i by a tiny random amount to $b_i \pm \epsilon_i$. This doesn't change the essence of the LP since the ϵ_i's are tiny, but it has the effect of differentiating between the solutions of the linear systems. To see why geometrically, imagine that the four planes ②, ③, ④, ⑤ were jolted a little. Wouldn't vertex B split into two vertices, very close to one another?

Unboundedness

In some cases an LP is unbounded, in that its objective function can be made arbitrarily large (or small, if it's a minimization problem). If this is the case, simplex will discover it: in exploring the neighborhood of a vertex, it will notice that taking out an inequality and adding another leads to an underdetermined system of equations that has an infinity of solutions. And in fact (this is an easy test) the space of solutions contains a whole line across which the objective can become larger and larger, all the way to ∞. In this case simplex halts and complains.

7.6.4 The running time of simplex

What is the running time of simplex, for a generic linear program

$$\max \mathbf{c}^T\mathbf{x} \text{ such that } \mathbf{A}\mathbf{x} \leq 0 \text{ and } \mathbf{x} \geq 0,$$

where there are n variables and $\mathbf{A}$ contains m inequality constraints? Since it is an iterative algorithm that proceeds from vertex to vertex, let's start by computing the time taken for a single iteration. Suppose the current vertex is $\mathbf{u}$. By definition, it is the unique point at which n inequality constraints are satisfied with equality. Each of its neighbors shares $n - 1$ of these inequalities, so $\mathbf{u}$ can have at most $n \cdot m$ neighbors: choose which inequality to drop and which new one to add.

A naive way to perform an iteration would be to check each potential neighbor to see whether it really is a vertex of the polyhedron and to determine its cost.

Gaussian elimination

Under our algebraic definition, merely writing down the coordinates of a vertex involves solving a system of linear equations. How is this done?

We are given a system of n linear equations in n unknowns, say $n = 4$ and

$$\begin{aligned} x_1 \quad\quad - 2x_3 \quad\quad &= 2 \\ x_2 + x_3 \quad\quad &= 3 \\ x_1 + x_2 \quad\quad - x_4 &= 4 \\ x_2 + 3x_3 + x_4 &= 5 \end{aligned}$$

The high school method for solving such systems is to repeatedly apply the following rule: *if we add a multiple of one equation to another equation, the overall system of equations remains equivalent.* For example, adding -1 times the first equation to the third one, we get the equivalent system

$$\begin{aligned} x_1 \quad\quad - 2x_3 \quad\quad &= 2 \\ x_2 + x_3 \quad\quad &= 3 \\ x_2 + 2x_3 - x_4 &= 2 \\ x_2 + 3x_3 + x_4 &= 5 \end{aligned}$$

This transformation is clever in the following sense: it *eliminates* the variable x_1 from the third equation, leaving just one equation with x_1. In other words, ignoring the first equation, we have a system of *three* equations in *three* unknowns: we decreased n by 1! We can solve this smaller system to get x_2, x_3, x_4, and then plug these into the first equation to get x_1.

This suggests an algorithm—once more due to Gauss.

```
procedure gauss (E, X)
Input:  A system E = {e_1,...,e_n} of equations in n unknowns X = {x_1,...,x_n}:
        e_1 : a_11 x_1 + a_12 x_2 + ··· + a_1n x_n = b_1; ···; e_n : a_n1 x_1 + a_n2 x_2 + ··· + a_nn x_n = b_n
Output: A solution of the system, if one exists

If all coefficients a_i1 are zero:
   halt with message "either infeasible or not linearly independent"
if n = 1: return b_1/a_11

choose the coefficient a_p1 of largest magnitude, and swap equations e_1, e_p
for i = 2 to n:
    e_i = e_i - (a_i1/a_11)·e_i
(x_2,...,x_n) = gauss(E - {e_1}, X - {x_1})
x_1 = (b_1 - Σ_{j>1} a_1j x_j)/a_11
return (x_1,...,x_n)
```

(When choosing the equation to swap into first place, we pick the one with largest $|a_{p1}|$ for reasons of *numerical accuracy*; after all, we will be dividing by a_{p1}.)

Gaussian elimination uses $O(n^2)$ *arithmetic operations* to reduce the problem size from n to $n-1$, and thus uses $O(n^3)$ operations overall. To show that this is also a good estimate of the total *running time*, we need to argue that the numbers involved remain polynomially bounded—for instance, that the solution $(x_1, \ldots, x_n)$ does not require too much more precision to write down than the original coefficients a_{ij} and b_i. Do you see why this is true?

Finding the cost is quick, just a dot product, but checking whether it is a true vertex involves solving a system of n equations in n unknowns (that is, satisfying the n chosen inequalities exactly) and checking whether the result is feasible. By Gaussian elimination (see the following box) this takes $O(n^3)$ time, giving an unappetizing running time of $O(mn^4)$ per iteration.

Fortunately, there is a much better way, and this mn^4 factor can be improved to mn, making simplex a practical algorithm. Recall our earlier discussion (Section 7.6.2) about the *local view* from vertex $\mathbf{u}$. It turns out that the per-iteration overhead of rewriting the LP in terms of the current local coordinates is just $O((m+n)n)$; this exploits the fact that the local view changes only slightly between iterations, in just one of its defining inequalities.

Next, to select the best neighbor, we recall that the (local view of) the objective function is of the form "max $c_{\mathbf{u}} + \tilde{\mathbf{c}} \cdot \mathbf{y}$" where $c_{\mathbf{u}}$ is the value of the objective function at $\mathbf{u}$. This immediately identifies a promising direction to move: we pick any $\tilde{c}_i > 0$ (if there is none, then the current vertex is optimal and simplex halts). Since the rest of the LP has now been rewritten in terms of the $\mathbf{y}$-coordinates, it is easy to determine how much y_i can be increased before some other inequality is violated. (And if we can increase y_i indefinitely, we know the LP is unbounded.)

It follows that the running time per iteration of simplex is just $O(mn)$. But how many iterations could there be? Naturally, there can't be more than $\binom{m+n}{n}$, which is an upper bound on the number of vertices. But this upper bound is exponential in n. And in fact, there are examples of LPs for which simplex does indeed take an exponential number of iterations. In other words, *simplex is an exponential-time algorithm*. However, such exponential examples do not occur in practice, and it is this fact that makes simplex so valuable and so widely used.

Linear programming in polynomial time

Simplex is not a polynomial time algorithm. Certain rare kinds of linear programs cause it to go from one corner of the feasible region to a better corner and then to a still better one, and so on for an exponential number of steps. For a long time, linear programming was considered a paradox, a problem that can be solved in practice, but not in theory!

Then, in 1979, a young Soviet mathematician called Leonid Khachiyan came up with the *ellipsoid algorithm*, one that is very different from simplex, extremely simple in its conception (but sophisticated in its proof) and yet one that solves any linear program in polynomial time. Instead of chasing the solution from one corner of the polyhedron to the next, Khachiyan's algorithm confines it to smaller and smaller ellipsoids (skewed high-dimensional balls). When this algorithm was announced, it became a kind of "mathematical Sputnik," a splashy achievement that had the U.S. establishment worried, in the height of the Cold War, about the possible scientific superiority of the Soviet Union. The ellipsoid algorithm turned out to be an important theoretical advance, but did not compete well with simplex in practice. The paradox of linear programming deepened: A problem with two algorithms, one that is efficient in theory, and one that is efficient in practice!

Linear programming in polynomial time (*Continued*)

A few years later Narendra Karmarkar, a graduate student at UC Berkeley, came up with a completely different idea, which led to another provably polynomial algorithm for linear programming. Karmarkar's algorithm is known as *the interior point method*, because it does just that: it dashes to the optimum corner not by hopping from corner to corner on the surface of the polyhedron like simplex does, but by cutting a clever path in the interior of the polyhedron. And it does perform well in practice.

But perhaps the greatest advance in linear programming algorithms was not Khachiyan's theoretical breakthrough or Karmarkar's novel approach, but an unexpected consequence of the latter: the fierce competition between the two approaches, simplex and interior point, resulted in the development of very fast code for linear programming.

7.7 Postscript: circuit evaluation

The importance of linear programming stems from the astounding variety of problems that reduce to it and thereby bear witness to its expressive power. In a sense, this next one is the *ultimate* application.

We are given a *Boolean circuit*, that is, a dag of gates of the following types.

- *Input gates* have indegree zero, with value `true` or `false`.
- AND gates and OR gates have indegree 2.
- NOT gates have indegree 1.

In addition, one of the gates is designated as the *output*. Here's an example.

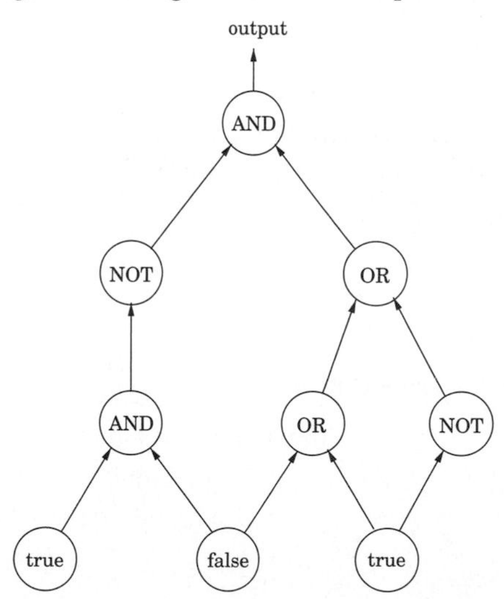

The CIRCUIT VALUE problem is the following: when the laws of Boolean logic are applied to the gates in topological order, does the output evaluate to `true`?

There is a simple, automatic way of translating this problem into a linear program. Create a variable x_g for each gate g, with constraints $0 \le x_g \le 1$. Add additional constraints for each type of gate:

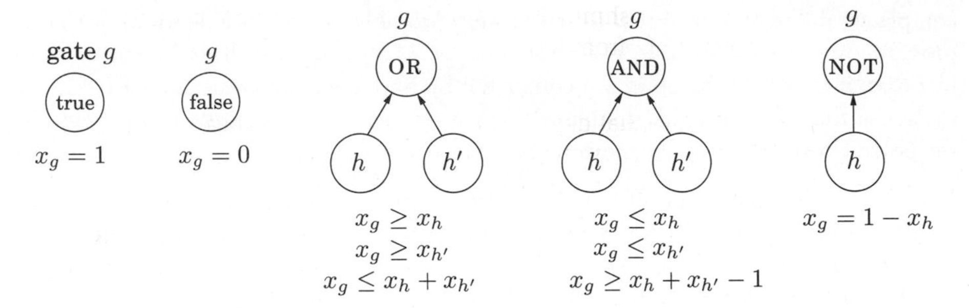

These constraints force all the gates to take on exactly the right values—0 for `false`, and 1 for `true`. We don't need to maximize or minimize anything, and we can read the answer off from the variable x_o corresponding to the output gate.

This is a straightforward reduction to linear programming, from a problem that may not seem very interesting at first. However, the CIRCUIT VALUE problem is in a sense *the most general problem solvable in polynomial time*! After all, any algorithm will eventually run on a computer, and the computer is ultimately a Boolean combinational circuit implemented on a chip. If the algorithm runs in polynomial time, it can be rendered as a Boolean circuit consisting of polynomially many copies of the computer's circuit, one per unit of time, with the values of the gates in one layer used to compute the values for the next. Hence, the fact that CIRCUIT VALUE reduces to linear programming means that *all problems that can be solved in polynomial time do!*

In our next topic, **NP**-*completeness,* we shall see that many *hard* problems reduce, much the same way, to *integer programming,* linear programming's difficult twin.

Another parting thought: by what other means can the circuit evaluation problem be solved? Let's think—a circuit is a dag. And what algorithmic technique is most appropriate for solving problems on dags? That's right: dynamic programming! Together with linear programming, the world's two most general algorithmic techniques.

Exercises

7.1. Consider the following linear program.

$$\begin{aligned} \text{maximize } & 5x + 3y \\ 5x - 2y &\geq 0 \\ x + y &\leq 7 \\ x &\leq 5 \\ x &\geq 0 \\ y &\geq 0 \end{aligned}$$

Plot the feasible region and identify the optimal solution.

7.2. Duckwheat is produced in Kansas and Mexico and consumed in New York and California. Kansas produces 15 shnupells of duckwheat and Mexico 8. Meanwhile, New York consumes 10 shnupells and California 13. The transportation costs per shnupell are $4 from Mexico to New York, $1 from Mexico to California, $2 from Kansas to New York, and $3 from Kansas to California.

Write a linear program that decides the amounts of duckwheat (in shnupells and fractions of a shnupell) to be transported from each producer to each consumer, so as to minimize the overall transportation cost.

7.3. A cargo plane can carry a maximum weight of 100 tons and a maximum volume of 60 cubic meters. There are three materials to be transported, and the cargo company may choose to carry any amount of each, up to the maximum available limits given below.

- Material 1 has density 2 tons/cubic meter, maximum available amount 40 cubic meters, and revenue $1,000 per cubic meter.
- Material 2 has density 1 ton/cubic meter, maximum available amount 30 cubic meters, and revenue $1,200 per cubic meter.
- Material 3 has density 3 tons/cubic meter, maximum available amount 20 cubic meters, and revenue $12,000 per cubic meter.

Write a linear program that optimizes revenue within the constraints.

7.4. Moe is deciding how much Regular Duff beer and how much Duff Strong beer to order each week. Regular Duff costs Moe $1 per pint and he sells it at $2 per pint; Duff Strong costs Moe $1.50 per pint and he sells it at $3 per pint. However, as part of a complicated marketing scam, the Duff company will only sell a pint of Duff Strong for each two pints or more of Regular Duff that Moe buys. Furthermore, due to past events that are better left untold, Duff will not sell Moe more than 3,000 pints per week. Moe knows that he can sell however much beer he has. Formulate a linear program for deciding how much Regular Duff and how much Duff Strong to buy, so as to maximize Moe's profit. Solve the program geometrically.

7.5. The Canine Products company offers two dog foods, Frisky Pup and Husky Hound, that are made from a blend of cereal and meat. A package of Frisky Pup requires 1 pound of cereal and 1.5 pounds of meat, and sells for $7. A package of Husky Hound uses 2 pounds of cereal and 1 pound of meat, and sells for $6. Raw cereal costs $1 per pound and raw meat costs $2 per pound. It also costs $1.40 to package the Frisky Pup and $0.60 to package the Husky Hound. A total of 240,000 pounds of cereal and 180,000 pounds of meat are available each month. The only production bottleneck is that the factory can only package 110,000 bags of Frisky Pup per month. Needless to say, management would like to maximize profit.

(a) Formulate the problem as a linear program in two variables.

(b) Graph the feasible region, give the coordinates of every vertex, and circle the vertex maximizing profit. What is the maximum profit possible?

7.6. Give an example of a linear program in two variables whose feasible region is infinite, but such that there is an optimum solution of bounded cost.

7.7. Find necessary and sufficient conditions on the reals a and b under which the linear program

$$\max\ x + y$$
$$ax + by \leq 1$$
$$x, y \geq 0$$

(a) Is infeasible.

(b) Is unbounded.

(c) Has a finite and unique optimal solution.

7.8. You are given the following points in the plane:

$$(1, 3),\ (2, 5),\ (3, 7),\ (5, 11),\ (7, 14),\ (8, 15),\ (10, 19).$$

You want to find a line $ax + by = c$ that approximately passes through these points (no line is a perfect fit). Write a linear program (you don't need to solve it) to find the line that minimizes the *maximum absolute error*,

$$\max_{1 \leq i \leq 7} |ax_i + by_i - c|.$$

7.9. A *quadratic programming* problem seeks to maximize a quadratic objective function (with terms like $3x_1^2$ or $5x_1x_2$) subject to a set of linear constraints. Give an example of a quadratic program in two variables x_1, x_2 such that the feasible region is nonempty and bounded, and yet none of the vertices of this region optimize the (quadratic) objective.

7.10. For the following network, with edge capacities as shown, find the maximum flow from S to T, along with a matching cut.

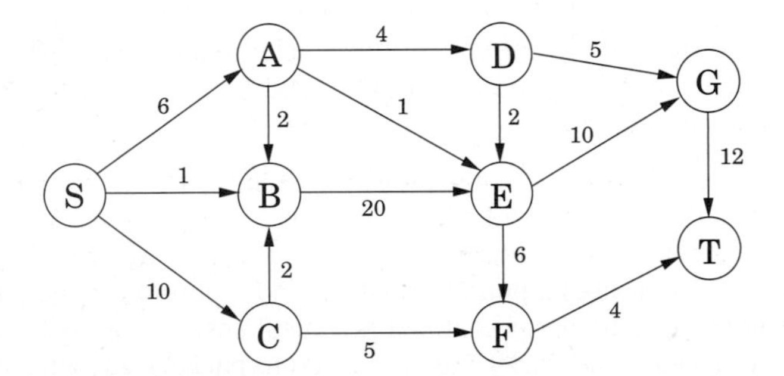

7.11. Write the dual to the following linear program.

$$\max\ x + y$$
$$2x + y \leq 3$$
$$x + 3y \leq 5$$
$$x, y \geq 0$$

Find the optimal solutions to both primal and dual LPs.

7.12. For the linear program

$$\begin{aligned} \max\ x_1 - 2x_3 \\ x_1 - x_2 \le 1 \\ 2x_2 - x_3 \le 1 \\ x_1, x_2, x_3 \ge 0 \end{aligned}$$

prove that the solution $(x_1, x_2, x_3) = (3/2, 1/2, 0)$ is optimal.

7.13. *Matching pennies.* In this simple two-player game, the players (call them R and C) each choose an outcome, *heads* or *tails*. If both outcomes are equal, C gives a dollar to R; if the outcomes are different, R gives a dollar to C.

(a) Represent the payoffs by a 2×2 matrix.

(b) What is the value of this game, and what are the optimal strategies for the two players?

7.14. The pizza business in Little Town is split between two rivals, Tony and Joey. They are each investigating strategies to steal business away from the other. Joey is considering either lowering prices or cutting bigger slices. Tony is looking into starting up a line of gourmet pizzas, or offering outdoor seating, or giving free sodas at lunchtime. The effects of these various strategies are summarized in the following payoff matrix (entries are dozens of pizzas, Joey's gain and Tony's loss).

		TONY		
		Gourmet	Seating	Free soda
JOEY	Lower price	+2	0	−3
	Bigger slices	−1	−2	+1

For instance, if Joey reduces prices and Tony goes with the gourmet option, then Tony will lose 2 dozen pizzas worth of business to Joey.

What is the value of this game, and what are the optimal strategies for Tony and Joey?

7.15. Find the value of the game specified by the following payoff matrix.

$$\begin{matrix} 0 & 0 & -1 & -1 \\ 0 & 1 & -2 & -1 \\ -1 & -1 & 1 & 1 \\ -1 & 0 & 0 & 1 \\ 1 & -2 & 0 & -3 \\ 0 & -3 & 2 & -1 \\ 0 & -2 & 1 & -1 \end{matrix}$$

(*Hint:* Consider the mixed strategies $(1/3, 0, 0, 1/2, 1/6, 0, 0)$ and $(2/3, 0, 0, 1/3)$.)

7.16. A salad is any combination of the following ingredients: (1) tomato, (2) lettuce, (3) spinach, (4) carrot, and (5) oil. Each salad must contain: (A) at least 15 grams of protein, (B) at least 2 and at most 6 grams of fat, (C) at least 4 grams of carbohydrates, (D) at most 100 milligrams of sodium. Furthermore, (E) you do not want your salad to be more than 50% greens by mass. The nutritional contents of these ingredients (per 100 grams) are

ingredient	energy (kcal)	protein (grams)	fat (grams)	carbohydrate (grams)	sodium (milligrams)
tomato	21	0.85	0.33	4.64	9.00
lettuce	16	1.62	0.20	2.37	8.00
spinach	371	12.78	1.58	74.69	7.00
carrot	346	8.39	1.39	80.70	508.20
oil	884	0.00	100.00	0.00	0.00

Find a linear programming applet on the Web and use it to make the salad with the fewest calories under the nutritional constraints. Describe your linear programming formulation and the optimal solution (the quantity of each ingredient and the value). Cite the Web resources that you used.

7.17. Consider the following network (the numbers are edge capacities).

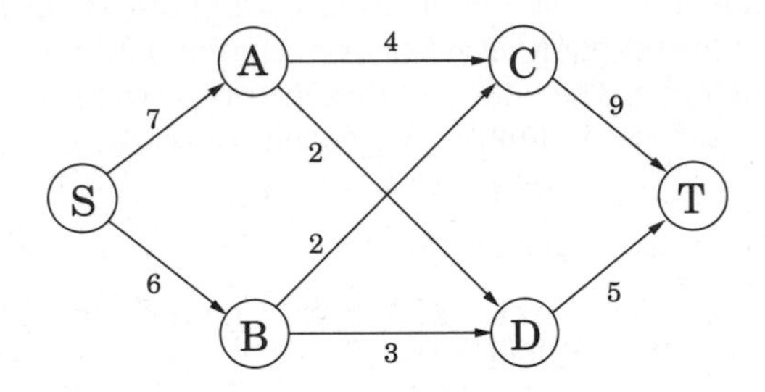

(a) Find the maximum flow f and a minimum cut.

(b) Draw the residual graph G_f (along with its edge capacities). In this residual network, mark the vertices reachable from S and the vertices from which T is reachable.

(c) An edge of a network is called a *bottleneck edge* if increasing its capacity results in an increase in the maximum flow. List all bottleneck edges in the above network.

(d) Give a very simple example (containing at most four nodes) of a network which has no bottleneck edges.

(e) Give an efficient algorithm to identify all bottleneck edges in a network. (*Hint:* Start by running the usual network flow algorithm, and then examine the residual graph.)

7.18. There are many common variations of the maximum flow problem. Here are four of them.

(a) There are many sources and many sinks, and we wish to maximize the total flow from all sources to all sinks.

(b) Each *vertex* also has a capacity on the maximum flow that can enter it.

(c) Each edge has not only a capacity, but also a *lower bound* on the flow it must carry.

(d) The outgoing flow from each node u is not the same as the incoming flow, but is smaller by a factor of $(1 - \epsilon_u)$, where ϵ_u is a loss coefficient associated with node u.

Each of these can be solved efficiently. Show this by reducing (a) and (b) to the original max-flow problem, and reducing (c) and (d) to linear programming.

7.19. Suppose someone presents you with a solution to a max-flow problem on some network. Give a *linear* time algorithm to determine whether the solution does indeed give a maximum flow.

7.20. Consider the following generalization of the maximum flow problem.

You are given a directed network $G = (V, E)$ with edge capacities $\{c_e\}$. Instead of a single (s, t) pair, you are given multiple pairs $(s_1, t_1), (s_2, t_2), \ldots, (s_k, t_k)$, where the s_i are sources of G and the t_i are sinks of G. You are also given k *demands* $d_1, \ldots, d_k$. The goal is to find k flows $f^{(1)}, \ldots, f^{(k)}$ with the following properties:

- $f^{(i)}$ is a valid flow from s_i to t_i.
- For each edge e, the total flow $f_e^{(1)} + f_e^{(2)} + \cdots + f_e^{(k)}$ does not exceed the capacity c_e.
- The size of each flow $f^{(i)}$ is at least the demand d_i.
- The size of the *total* flow (the sum of the flows) is as large as possible.

How would you solve this problem?

7.21. An edge of a flow network is called *critical* if decreasing the capacity of this edge results in a decrease in the maximum flow. Give an efficient algorithm that finds a critical edge in a network.

7.22. In a particular network $G = (V, E)$ whose edges have integer capacities c_e, we have already found the maximum flow f from node s to node t. However, we now find out that one of the capacity values we used was wrong: for edge (u, v) we used c_{uv} whereas it should have been $c_{uv} - 1$. This is unfortunate because the flow f uses that particular edge at full capacity: $f_{uv} = c_{uv}$.

We could redo the flow computation from scratch, but there's a faster way. Show how a new optimal flow can be computed in $O(|V| + |E|)$ time.

7.23. A *vertex cover* of an undirected graph $G = (V, E)$ is a subset of the vertices which touches every edge—that is, a subset $S \subset V$ such that for each edge $\{u, v\} \in E$, one or both of u, v are in S.

Show that the problem of finding the minimum vertex cover in a *bipartite* graph reduces to maximum flow. (*Hint:* Can you relate this problem to the minimum cut in an appropriate network?)

7.24. *Direct bipartite matching.* We've seen how to find a maximum matching in a bipartite graph via reduction to the maximum flow problem. We now develop a direct algorithm.

Let $G = (V_1 \cup V_2, E)$ be a bipartite graph (so each edge has one endpoint in V_1 and one endpoint in V_2), and let $M \in E$ be a matching in the graph (that is, a set of edges that don't touch). A vertex is said to be *covered* by M if it is the endpoint of one of the edges in M. An *alternating path* is a path of odd length that starts and ends with a non-covered vertex, and whose edges alternate between M and $E - M$.

(a) In the bipartite graph below, a matching M is shown in bold. Find an alternating path.

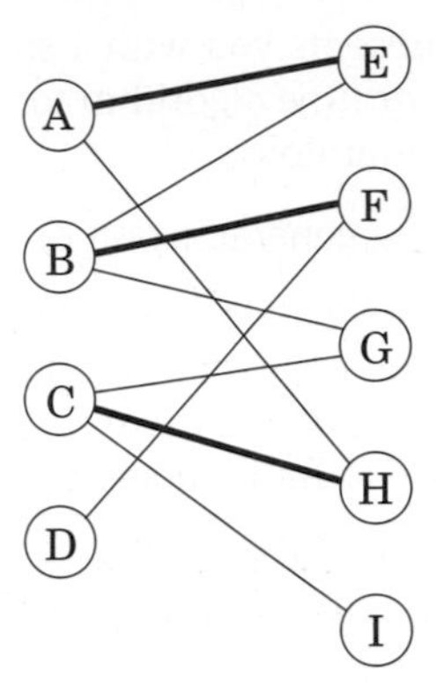

(b) Prove that a matching M is maximum if and only if there does not exist an alternating path with respect to it.

(c) Design an algorithm that finds an alternating path in $O(|V| + |E|)$ time using a variant of breadth-first search.

(d) Give a direct $O(|V| \cdot |E|)$ algorithm for finding a maximum matching in a bipartite graph.

7.25. *The dual of maximum flow.* Consider the following network with edge capacities.

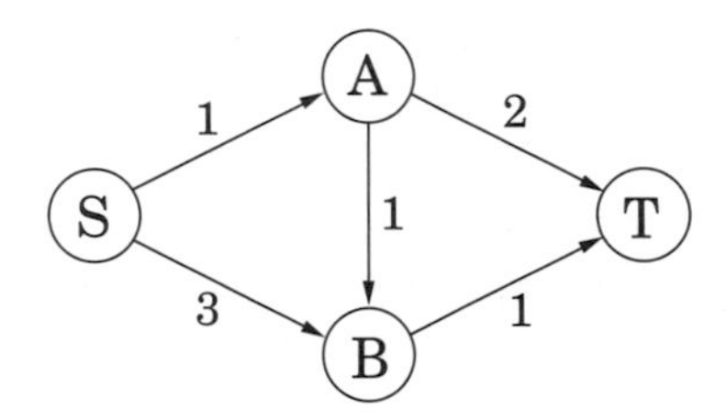

(a) Write the problem of finding the maximum flow from S to T as a linear program.

(b) Write down the dual of this linear program. There should be a dual variable for each edge of the network and for each vertex other than S, T.

Now we'll solve the same problem in full generality. Recall the linear program for a general maximum flow problem (Section 7.2).

(c) Write down the dual of this general flow LP, using a variable y_e for each edge and x_u for each vertex $u \neq s, t$.

(d) Show that any solution to the general dual LP must satisfy the following property: for any directed path from s to t in the network, the sum of the y_e values along the path must be at least 1.

(e) What are the intuitive meanings of the dual variables? Show that any $s-t$ cut in the network can be translated into a dual feasible solution whose cost is exactly the capacity of that cut.

7.26. In a satisfiable system of linear inequalities

$$a_{11}x_1 + \cdots + a_{1n}x_n \le b_1$$
$$\vdots$$
$$a_{m1}x_1 + \cdots + a_{mn}x_n \le b_m$$

we describe the jth inequality as *forced-equal* if it is satisfied with equality by *every* solution $\mathbf{x} = (x_1, \ldots, x_n)$ of the system. Equivalently, $\sum_i a_{ji}x_i \le b_j$ is *not* forced-equal if there exists an $\mathbf{x}$ that satisfies the whole system and such that $\sum_i a_{ji}x_i < b_j$.
For example, in

$$\begin{aligned} x_1 + x_2 &\le 2 \\ -x_1 - x_2 &\le -2 \\ x_1 &\le 1 \\ -x_2 &\le 0 \end{aligned}$$

the first two inequalities are forced-equal, while the third and fourth are not. A solution $\mathbf{x}$ to the system is called *characteristic* if, for every inequality I that is not forced-equal, $\mathbf{x}$ satisfies I without equality. In the instance above, such a solution is $(x_1, x_2) = (-1, 3)$, for which $x_1 < 1$ and $-x_2 < 0$ while $x_1 + x_2 = 2$ and $-x_1 - x_2 = -2$.

(a) Show that any satisfiable system has a characteristic solution.

(b) Given a satisfiable system of linear inequalities, show how to use linear programming to determine which inequalities are forced-equal, and to find a characteristic solution.

7.27. Show that the *change-making problem* (Exercise 6.17) can be formulated as an integer linear program. Can we solve this program as an LP, in the certainty that the solution will turn out to be integral (as in the case of bipartite matching)? Either prove it or give a counterexample.

7.28. *A linear program for shortest path.* Suppose we want to compute the shortest path from node s to node t in a directed graph with edge lengths $l_e > 0$.

(a) Show that this is equivalent to finding an $s-t$ flow f that minimizes $\sum_e l_e f_e$ subject to $\text{size}(f) = 1$. There are no capacity constraints.

(b) Write the shortest path problem as a linear program.

(c) Show that the dual LP can be written as

$$\max\ x_s - x_t$$
$$x_u - x_v \le l_{uv} \text{ for all } (u, v) \in E$$

(d) An interpretation for the dual is given in the box on page 209. Why isn't our dual LP identical to the one on that page?

7.29. *Hollywood.* A film producer is seeking actors and investors for his new movie. There are n available actors; actor i charges s_i dollars. For funding, there are m available investors. Investor j will provide p_j dollars, but only on the condition that certain actors $L_j \subseteq \{1, 2, \ldots, n\}$ are included in the cast (*all* of these actors L_j must be chosen in order to receive funding from investor j).

The producer's profit is the sum of the payments from investors minus the payments to actors. The goal is to maximize this profit.

(a) Express this problem as an integer linear program in which the variables take on values $\{0, 1\}$.

(b) Now relax this to a linear program, and show that there must in fact be an *integral* optimal solution (as is the case, for example, with maximum flow and bipartite matching).

7.30. *Hall's theorem.* Returning to the matchmaking scenario of Section 7.3, suppose we have a bipartite graph with boys on the left and an equal number of girls on the right. Hall's theorem says that there is a perfect matching if and only if the following condition holds: any subset S of boys is connected to at least $|S|$ girls.

Prove this theorem. (*Hint:* The max-flow min-cut theorem should be helpful.)

7.31. Consider the following simple network with edge capacities as shown.

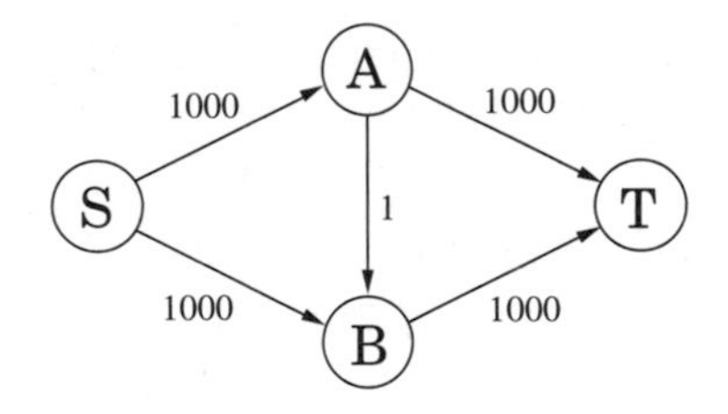

(a) Show that, if the Ford-Fulkerson algorithm is run on this graph, a careless choice of updates might cause it to take 1000 iterations. Imagine if the capacities were a million instead of 1000!

We will now find a strategy for choosing paths under which the algorithm is guaranteed to terminate in a reasonable number of iterations.

Consider an arbitrary directed network $(G = (V, E), s, t, \{c_e\})$ in which we want to find the maximum flow. Assume for simplicity that all edge capacities are at least 1, and define the capacity of an $s - t$ path to be the smallest capacity of its constituent edges. The *fattest path* from s to t is the path with the most capacity.

(b) Show that the fattest $s - t$ path in a graph can be computed by a variant of Dijkstra's algorithm.

(c) Show that the maximum flow in G is the sum of individual flows along at most $|E|$ paths from s to t.

(d) Now show that if we always increase flow along the fattest path in the residual graph, then the Ford-Fulkerson algorithm will terminate in at most $O(|E| \log F)$ iterations, where F is the size of the maximum flow. (*Hint:* It might help to recall the proof for the greedy set cover algorithm in Section 5.4.)

In fact, an even simpler rule—finding a path in the residual graph using breadth-first search—guarantees that at most $O(|V| \cdot |E|)$ iterations will be needed.

Chapter 8

NP-complete problems

8.1 Search problems

Over the past seven chapters we have developed algorithms for finding shortest paths and minimum spanning trees in graphs, matchings in bipartite graphs, maximum increasing subsequences, maximum flows in networks, and so on. All these algorithms are *efficient*, because in each case their time requirement grows as a polynomial function (such as n, n^2, or n^3) of the size of the input.

To better appreciate such efficient algorithms, consider the alternative: In all these problems we are searching for a solution (path, tree, matching, etc.) from among an *exponential* population of possibilities. Indeed, n boys can be matched with n girls in $n!$ different ways, a graph with n vertices has n^{n-2} spanning trees, and a typical graph has an exponential number of paths from s to t. All these problems could in principle be solved in exponential time by checking through all candidate solutions, one by one. But an algorithm whose running time is 2^n, or worse, is all but useless in practice (see the next box). The quest for efficient algorithms is about finding clever ways to bypass this process of exhaustive search, using clues from the input in order to dramatically narrow down the search space.

So far in this book we have seen the most brilliant successes of this quest, algorithmic techniques that defeat the specter of exponentiality: greedy algorithms, dynamic programming, linear programming (while divide-and-conquer typically yields faster algorithms for problems we can already solve in polynomial time). Now the time has come to meet the quest's most embarrassing and persistent failures. We shall see some other "search problems," in which again we are seeking a solution with particular properties among an exponential chaos of alternatives. But for these new problems no shortcut seems possible. The fastest algorithms we know for them are all exponential—not substantially better than an exhaustive search. We now introduce some important examples.

Satisfiability

SATISFIABILITY, or SAT (recall Exercise 3.28 and Section 5.3), is a problem of great practical importance, with applications ranging from chip testing and computer design

The story of Sissa and Moore

According to the legend, the game of chess was invented by the Brahmin Sissa to amuse and teach his king. Asked by the grateful monarch what he wanted in return, the wise man requested that the king place one grain of rice in the first square of the chessboard, two in the second, four in the third, and so on, doubling the amount of rice up to the 64th square. The king agreed on the spot, and as a result he was the first person to learn the valuable—albeit humbling—lesson of *exponential growth*. Sissa's request amounted to $2^{64} - 1 = 18{,}446{,}744{,}073{,}709{,}551{,}615$ grains of rice, enough rice to pave all of India several times over!

All over nature, from colonies of bacteria to cells in a fetus, we see systems that grow exponentially—for a while. In 1798, the British philosopher T. Robert Malthus published an essay in which he predicted that the exponential growth (he called it "geometric growth") of the human population would soon deplete linearly growing resources, an argument that influenced Charles Darwin deeply. Malthus knew the fundamental fact that an exponential sooner or later takes over any polynomial.

In 1965, computer chip pioneer Gordon E. Moore noticed that transistor density in chips had doubled every year in the early 1960s, and he predicted that this trend would continue. This prediction, moderated to a doubling every 18 months and extended to computer speed, is known as *Moore's law*. It has held remarkably well for 40 years. And these are the two root causes of the explosion of information technology in the past decades: *Moore's law and efficient algorithms*.

It would appear that Moore's law provides a disincentive for developing polynomial algorithms. After all, if an algorithm is exponential, why not wait it out until Moore's law makes it feasible? But in reality the exact opposite happens: Moore's law is a huge incentive for developing efficient algorithms, because such algorithms are needed in order to take advantage of the exponential increase in computer speed.

Here is why. If, for example, an $O(2^n)$ algorithm for Boolean satisfiability (SAT) were given an hour to run, it would have solved instances with 25 variables back in 1975, 31 variables on the faster computers available in 1985, 38 variables in 1995, and about 45 variables with today's machines. Quite a bit of progress—except that each extra variable requires a year and a half's wait, while the appetite of applications (many of which are, ironically, related to computer design) grows much faster. In contrast, the size of the instances solved by an $O(n)$ or $O(n \log n)$ algorithm would be *multiplied by a factor of about 100* each decade. In the case of an $O(n^2)$ algorithm, the instance size solvable in a fixed time would be multiplied by about 10 each decade. Even an $O(n^6)$ algorithm, polynomial yet unappetizing, would more than double the size of the instances solved each decade. When it comes to the growth of the size of problems we can attack with an algorithm, we have a reversal: exponential algorithms make polynomially slow progress, while polynomial algorithms advance exponentially fast! For Moore's law to be reflected in the world we *need* efficient algorithms.

The story of Sissa and Moore (*Continued*)

As Sissa and Malthus knew very well, exponential expansion cannot be sustained indefinitely in our finite world. Bacterial colonies run out of food; chips hit the atomic scale. Moore's law will stop doubling the speed of our computers within a decade or two. And then progress will depend on algorithmic ingenuity—or otherwise perhaps on novel ideas such as *quantum computation*, explored in Chapter 10.

to image analysis and software engineering. It is also a canonical hard problem. Here's what an instance of SAT looks like:

$$(x \vee y \vee z)\ (x \vee \overline{y})\ (y \vee \overline{z})\ (z \vee \overline{x})\ (\overline{x} \vee \overline{y} \vee \overline{z}).$$

This is a *Boolean formula in conjunctive normal form (CNF)*. It is a collection of *clauses* (the parentheses), each consisting of the disjunction (logical *or*, denoted $\vee$) of several *literals*, where a literal is either a Boolean variable (such as x) or the negation of one (such as $\overline{x}$). A *satisfying truth assignment* is an assignment of `false` or `true` to each variable so that every clause contains a literal whose value is `true`. The SAT problem is the following: given a Boolean formula in conjunctive normal form, either find a satisfying truth assignment or else report that none exists.

In the instance shown previously, setting all variables to `true`, for example, satisfies every clause except the last. Is there a truth assignment that satisfies *all* clauses?

With a little thought, it is not hard to argue that in this particular case no such truth assignment exists. (*Hint*: The three middle clauses constrain all three variables to have the same value.) But how do we decide this in general? Of course, we can always search through all truth assignments, one by one, but for formulas with n variables, the number of possible assignments is exponential, 2^n.

SAT is a typical *search problem*. We are given an *instance* I (that is, some input data specifying the problem at hand, in this case a Boolean formula in conjunctive normal form), and we are asked to find a *solution* S (an object that meets a particular specification, in this case an assignment that satisfies each clause). If no such solution exists, we must say so.

More specifically, a search problem must have the property that any proposed solution S to an instance I can be *quickly checked* for correctness. What does this entail? For one thing, S must at least be concise (quick to read), with length polynomially bounded by that of I. This is clearly true in the case of SAT, for which S is an assignment to the variables. To formalize the notion of quick checking, we will say that there is a polynomial-time algorithm that takes as input I and S and decides whether or not S is a solution of I. For SAT, this is easy as it just involves checking whether the assignment specified by S indeed satisfies every clause in I.

Later in this chapter it will be useful to shift our vantage point and to think of this efficient algorithm for checking proposed solutions as *defining* the search problem. Thus:

> *A* search problem *is specified by an algorithm C that takes two inputs, an instance I and a proposed solution S, and runs in time polynomial in $|I|$. We say S is a solution to I if and only if $C(I, S) =$* `true`.

Given the importance of the SAT search problem, researchers over the past 50 years have tried hard to find efficient ways to solve it, but without success. The fastest algorithms we have are still exponential on their worst-case inputs.

Yet, interestingly, there are two natural variants of SAT for which we do have good algorithms. If all clauses contain at most one positive literal, then the Boolean formula is called a *Horn formula*, and a satisfying truth assignment, if one exists, can be found by the greedy algorithm of Section 5.3. Alternatively, if all clauses have only *two* literals, then graph theory comes into play, and SAT can be solved in linear time by finding the strongly connected components of a particular graph constructed from the instance (recall Exercise 3.28). In fact, in Chapter 9 we'll see a different polynomial algorithm for this same special case, which is called 2SAT.

On the other hand, if we are just a little more permissive and allow clauses to contain *three* literals, then the resulting problem, known as 3SAT (an example of which we saw earlier), once again becomes hard to solve!

The traveling salesman problem

In the traveling salesman problem (TSP) we are given n vertices $1, \ldots, n$ and all $n(n-1)/2$ distances between them, as well as a *budget* b. We are asked to find a *tour*, a cycle that passes through every vertex exactly once, of total cost b or less—or to report that no such tour exists. That is, we seek a permutation $\tau(1), \ldots, \tau(n)$ of the vertices such that when they are toured in this order, the total distance covered is at most b:

$$d_{\tau(1),\tau(2)} + d_{\tau(2),\tau(3)} + \cdots + d_{\tau(n),\tau(1)} \leq b.$$

See Figure 8.1 for an example (only some of the distances are shown; assume the rest are very large).

Notice how we have defined the TSP as a *search problem*: given an instance, find a tour within the budget (or report that none exists). But why are we expressing the traveling salesman problem in this way, when in reality it is an *optimization problem*, in which the *shortest* possible tour is sought? Why dress it up as something else?

For a good reason. Our plan in this chapter is to compare and relate problems. The framework of search problems is helpful in this regard, because it encompasses optimization problems like the TSP in addition to true search problems like SAT.

Turning an optimization problem into a search problem does not change its difficulty at all, because the two versions *reduce to one another*. Any algorithm that solves the optimization TSP also readily solves the search problem: find the optimum tour and if it is within budget, return it; if not, there is no solution.

Figure 8.1 The optimal traveling salesman tour, shown in bold, has length 18.

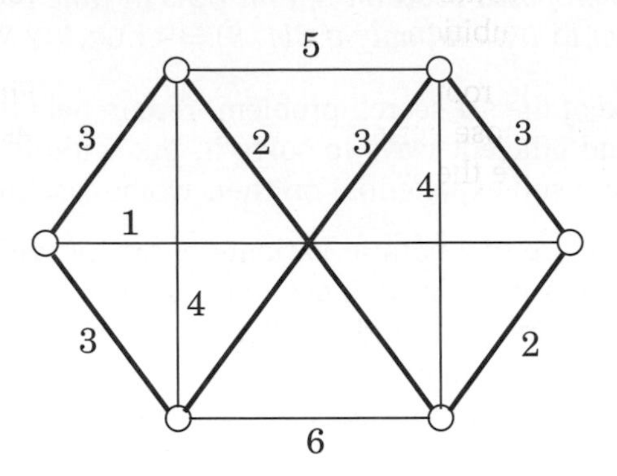

Conversely, an algorithm for the search problem can also be used to solve the optimization problem. To see why, first suppose that we somehow knew the *cost* of the optimum tour; then we could find this tour by calling the algorithm for the search problem, using the optimum cost as the budget. Fine, but how do we find the optimum cost? Easy: By binary search! (See Exercise 8.1.)

Incidentally, there is a subtlety here: Why do we have to introduce a budget? Isn't any optimization problem also a search problem in the sense that we are searching for a solution that has the property of being optimal? The catch is that the solution to a search problem should be easy to recognize, or as we put it earlier, polynomial-time checkable. Given a potential solution to the TSP, it is easy to check the properties "is a tour" (just check that each vertex is visited exactly once) and "has total length $\leq b$." But how could one check the property "is optimal"?

As with SAT, there are no known polynomial-time algorithms for the TSP, despite much effort by researchers over nearly a century. Of course, there is an exponential algorithm for solving it, by trying all $(n-1)!$ tours, and in Section 6.6 we saw a faster, yet still exponential, dynamic programming algorithm.

The minimum spanning tree (MST) problem, for which we *do* have efficient algorithms, provides a stark contrast here. To phrase it as a search problem, we are again given a distance matrix and a bound b, and are asked to find a tree T with total weight $\sum_{(i,j)\in T} d_{ij} \leq b$. The TSP can be thought of as a tough cousin of the MST problem, in which the tree is not allowed to branch and is therefore a path.[1] This extra restriction on the structure of the tree results in a much harder problem.

Euler and Rudrata

In the summer of 1735 Leonhard Euler (pronounced "Oiler"), the famous Swiss mathematician, was walking the bridges of the East Prussian town of Königsberg.

[1] Actually the TSP demands a cycle, but one can define an alternative version that seeks a path, and it is not hard to see that this is just as hard as the TSP itself.

After a while, he noticed in frustration that, no matter where he started his walk, no matter how cleverly he continued, it was impossible to cross each bridge exactly once. And from this silly ambition, the field of graph theory was born.

Euler identified at once the roots of the park's deficiency. First, you turn the map of the park into a graph whose vertices are the four land masses (two islands, two banks) and whose edges are the seven bridges:

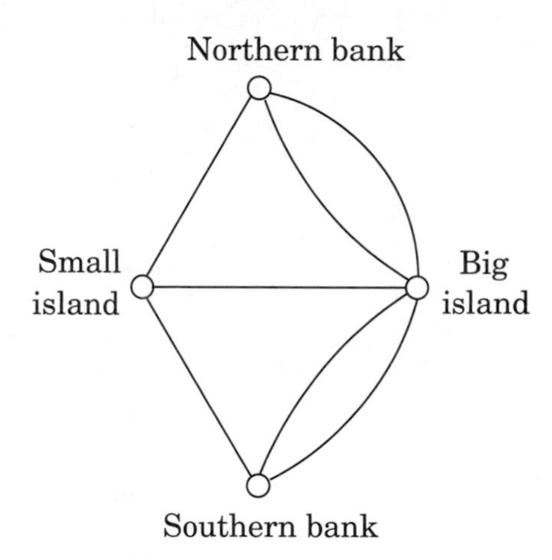

This graph has multiple edges between two vertices—a feature we have not been allowing so far in this book, but one that is meaningful for this particular problem, since each bridge must be accounted for separately. We are looking for a path that goes through each edge exactly once (the path is allowed to repeat vertices). In other words, we are asking this question: *When can a graph be drawn without lifting the pencil from the paper?*

The answer discovered by Euler is simple, elegant, and intuitive: *If and only if (a) the graph is connected and (b) every vertex, with the possible exception of two vertices (the start and final vertices of the walk), has even degree* (Exercise 3.26). This is why Königsberg's park was impossible to traverse: all four vertices have odd degree.

To put it in terms of our present concerns, let us define a search problem called EULER PATH: Given a graph, find a path that contains each edge exactly once. It follows from Euler's observation, and a little more thinking, that this search problem can be solved in polynomial time.

Almost a millennium before Euler's fateful summer in East Prussia, a Kashmiri poet named Rudrata had asked this question: Can one visit all the squares of the chessboard, without repeating any square, in one long walk that ends at the starting square and at each step makes a legal knight move? This is again a graph problem: the graph now has 64 vertices, and two squares are joined by an edge if a knight can go from one to the other in a single move (that is, if their coordinates differ by 2 in one dimension and by 1 in the other). See Figure 8.2 for the portion of the graph

Figure 8.2 Knight's moves on a corner of a chessboard.

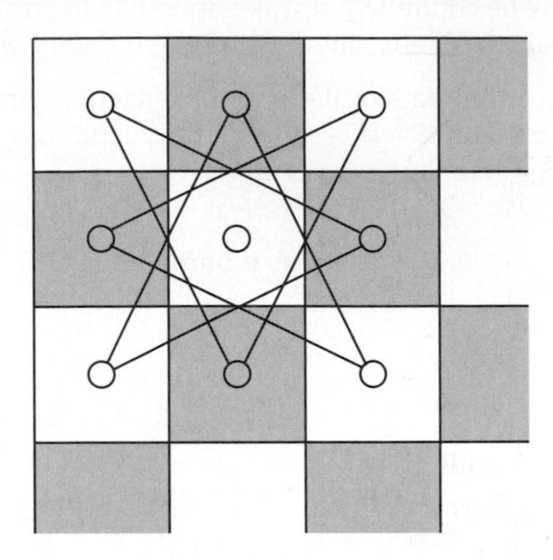

corresponding to the upper left corner of the board. Can you find a knight's tour on your chessboard?

This is a different kind of search problem in graphs: we want a cycle that goes through all *vertices* (as opposed to all edges in Euler's problem), without repeating any vertex. And there is no reason to stick to chessboards; this question can be asked of any graph. Let us define the RUDRATA CYCLE search problem to be the following: given a graph, find a cycle that visits each vertex exactly once—or report that no such cycle exists.[2] This problem is ominously reminiscent of the TSP, and indeed no polynomial algorithm is known for it.

There are two differences between the definitions of the Euler and Rudrata problems. The first is that Euler's problem visits all *edges* while Rudrata's visits all *vertices*. But there is also the issue that one of them demands a path while the other requires a cycle. Which of these differences accounts for the huge disparity in computational complexity between the two problems? It must be the first, because the second difference can be shown to be purely cosmetic. Indeed, define the RUDRATA PATH problem to be just like RUDRATA CYCLE, except that the goal is now to find a *path* that goes through each vertex exactly once. As we will soon see, there is a precise equivalence between the two versions of the Rudrata problem.

Cuts and bisections

A *cut* is a set of edges whose removal leaves a graph disconnected. It is often of interest to find small cuts, and the MINIMUM CUT problem is, given a graph and a

[2]In the literature this problem is known as the *Hamilton cycle* problem, after the great Irish mathematician who rediscovered it in the 19th century.

Figure 8.3 What is the smallest cut in this graph?

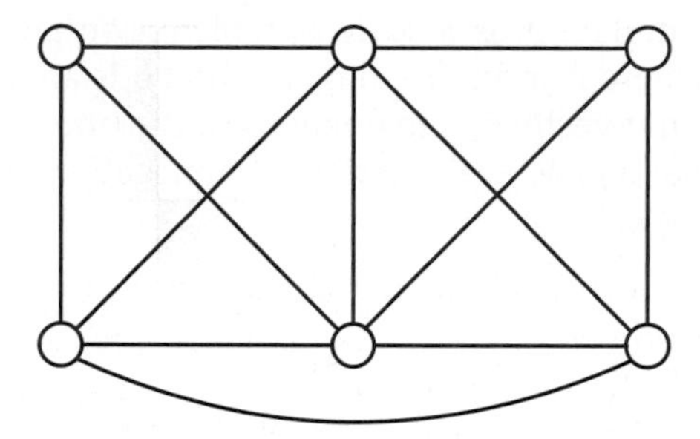

budget b, to find a cut with at most b edges. For example, the smallest cut in Figure 8.3 is of size 3. This problem can be solved in polynomial time by $n - 1$ max-flow computations: give each edge a capacity of 1, and find the maximum flow between some fixed node and every single other node. The smallest such flow will correspond (via the max-flow min-cut theorem) to the smallest cut. Can you see why? We've also seen a very different, randomized algorithm for this problem (page 140).

In many graphs, such as the one in Figure 8.3, the smallest cut leaves just a singleton vertex on one side—it consists of all edges adjacent to this vertex. Far more interesting are small cuts that partition the vertices of the graph into nearly equal-sized sets. More precisely, the BALANCED CUT problem is this: given a graph with n vertices and a budget b, partition the vertices into two sets S and T such that $|S|, |T| \geq n/3$ and such that there are at most b edges between S and T. Another hard problem.

Balanced cuts arise in a variety of important applications, such as *clustering*. Consider for example the problem of segmenting an image into its constituent components (say, an elephant standing in a grassy plain with a clear blue sky above). A good way of doing this is to create a graph with a node for each pixel of the image and to put an edge between nodes whose corresponding pixels are spatially close together and are also similar in color. A single object in the image (like the elephant, say) then corresponds to a set of highly connected vertices in the graph. A balanced cut is therefore likely to divide the pixels into two clusters without breaking apart any of the primary constituents of the image. The first cut might, for instance, separate the elephant on the one hand from the sky and from grass on the other. A further cut would then be needed to separate the sky from the grass.

Integer linear programming

Even though the simplex algorithm is not polynomial time, we mentioned in Chapter 7 that there *is* a different, polynomial algorithm for linear programming. Therefore, linear programming is efficiently solvable both in practice and in theory. But the situation changes completely if, in addition to specifying a linear objective function and linear inequalities, we also constrain the solution (the values for the variables) to be *integer*. This latter problem is called INTEGER LINEAR PROGRAMMING (ILP). Let's see how we might formulate it as a search problem. We are given a set of linear inequalities $\mathbf{Ax} \leq \mathbf{b}$, where $\mathbf{A}$ is an $m \times n$ matrix and $\mathbf{b}$ is an m-vector; an objective function specified by an n-vector $\mathbf{c}$; and finally, a *goal* g (the counterpart of a budget in maximization problems). We want to find a nonnegative *integer* n-vector $\mathbf{x}$ such that $\mathbf{Ax} \leq \mathbf{b}$ and $\mathbf{c} \cdot \mathbf{x} \geq g$.

But there is a redundancy here: the last constraint $\mathbf{c} \cdot \mathbf{x} \geq g$ is itself a linear inequality and can be absorbed into $\mathbf{Ax} \leq \mathbf{b}$. So, we define ILP to be following search problem: given $\mathbf{A}$ and $\mathbf{b}$, find a nonnegative integer vector $\mathbf{x}$ satisfying the inequalities $\mathbf{Ax} \leq \mathbf{b}$, or report that none exists. Despite the many crucial applications of this problem, and intense interest by researchers, no efficient algorithm is known for it.

There is a particularly clean special case of ILP that is very hard in and of itself: the goal is to find a vector $\mathbf{x}$ of 0's and 1's satisfying $\mathbf{Ax} = \mathbf{1}$, where $\mathbf{A}$ is an $m \times n$ matrix with $0 - 1$ entries and $\mathbf{1}$ is the m-vector of all 1's. It should be apparent from the reductions in Section 7.1.4 that this is indeed a special case of ILP. We call it ZERO-ONE EQUATIONS (ZOE).

We have now introduced a number of important search problems, some of which are familiar from earlier chapters and for which there are efficient algorithms, and others which are different in small but crucial ways that make them very hard computational problems. To complete our story we will introduce a few more hard problems, which will play a role later in the chapter, when we relate the computational difficulty of all these problems. The reader is invited to skip ahead to Section 8.2 and then return to the definitions of these problems as required.

Three-dimensional matching

Recall the BIPARTITE MATCHING problem: given a bipartite graph with n nodes on each side (the *boys* and the *girls*), find a set of n disjoint edges, or decide that no such set exists. In Section 7.3, we saw how to efficiently solve this problem by a reduction to maximum flow. However, there is an interesting generalization, called 3D MATCHING, for which no polynomial algorithm is known. In this new setting, there are n boys and n girls, but also n *pets*, and the compatibilities among them are specified by a set of *triples*, each containing a boy, a girl, and a pet. Intuitively, a triple (b, g, p) means that boy b, girl g, and pet p get along well together. We want to find n disjoint triples and thereby create n harmonious households.

Can you spot a solution in Figure 8.4?

Independent set, vertex cover, and clique

In the INDEPENDENT SET problem (recall Section 6.7) we are given a graph and an integer g, and the aim is to find g vertices that are independent, that is, no two of which have an edge between them. Can you find an independent set of three vertices in Figure 8.5? How about four vertices? We saw in Section 6.7 that this

Figure 8.4 A more elaborate matchmaking scenario. Each triple is shown as a triangular-shaped node joining boy, girl, and pet.

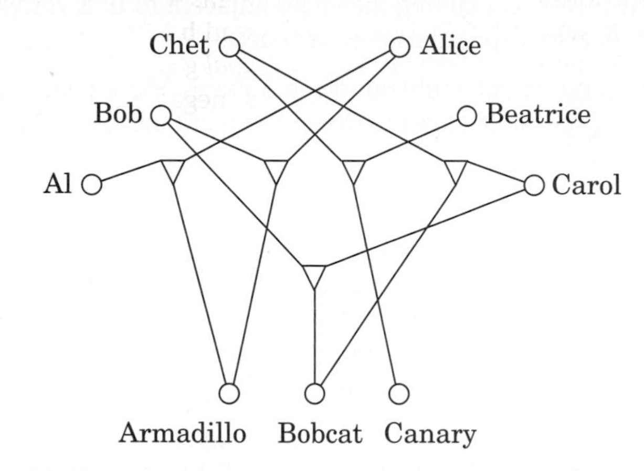

problem can be solved efficiently on trees, but for general graphs no polynomial algorithm is known.

There are many other search problems about graphs. In VERTEX COVER, for example, the input is a graph and a budget b, and the idea is to find b vertices that cover (touch) every edge. Can you cover all edges of Figure 8.5 with seven vertices? With six? (And do you see the intimate connection to the INDEPENDENT SET problem?)

VERTEX COVER is a special case of SET COVER, which we encountered in Chapter 5. In that problem, we are given a set E and several subsets of it, $S_1, \ldots, S_m$, along with

Figure 8.5 What is the size of the largest independent set in this graph?

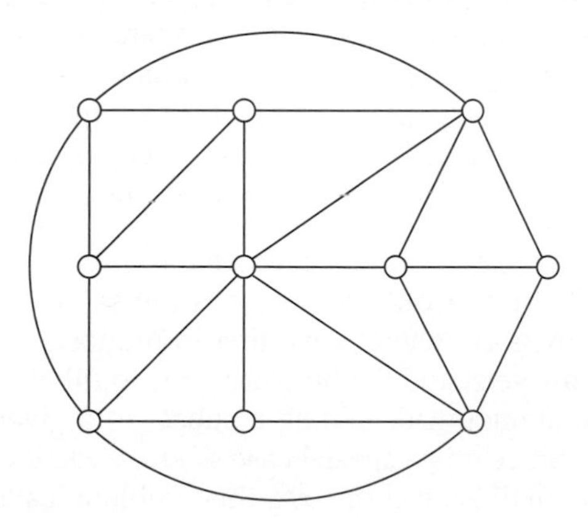

a budget b. We are asked to select b of these subsets so that their union is E. VERTEX COVER is the special case in which E consists of the edges of a graph, and there is a subset S_i for each vertex, containing the edges adjacent to that vertex. Can you see why 3D MATCHING is also a special case of SET COVER?

And finally there is the CLIQUE problem: given a graph and a goal g, find a set of g vertices such that all possible edges between them are present. What is the largest clique in Figure 8.5?

Longest path

We know the shortest-path problem can be solved very efficiently, but how about the LONGEST PATH problem? Here we are given a graph G with nonnegative edge weights and two distinguished vertices s and t, along with a goal g. We are asked to find a path from s to t with total weight at least g. Naturally, to avoid trivial solutions we require that the path be *simple*, containing no repeated vertices.

No efficient algorithm is known for this problem (which sometimes also goes by the name of TAXICAB RIP-OFF).

Knapsack and subset sum

Recall the KNAPSACK problem (Section 6.4): we are given integer weights $w_1, \ldots, w_n$ and integer values $v_1, \ldots, v_n$ for n items. We are also given a weight capacity W and a goal g (the former is present in the original optimization problem, the latter is added to make it a search problem). We seek a set of items whose total weight is at most W and whose total value is at least g. As always, if no such set exists, we should say so.

In Section 6.4, we developed a dynamic programming scheme for KNAPSACK with running time $O(nW)$, which we noted is exponential in the input size, since it involves W rather than $\log W$. And we have the usual exhaustive algorithm as well, which looks at all subsets of items—all 2^n of them. Is there a polynomial algorithm for KNAPSACK? Nobody knows of one.

But suppose that we are interested in the variant of the knapsack problem in which the integers are coded in *unary*—for instance, by writing $IIIIIIIIIIII$ for 12. This is admittedly an exponentially wasteful way to represent integers, but it does define a legitimate problem, which we could call UNARY KNAPSACK. It follows from our discussion that this somewhat artificial problem does have a polynomial algorithm.

A different variation: suppose now that each item's value is equal to its weight (all given in binary), and to top it off, the goal g is the same as the capacity W. (To adapt the silly break-in story whereby we first introduced the knapsack problem, the items are all gold nuggets, and the burglar wants to fill his knapsack to the hilt.) This special case is tantamount to finding a subset of a given set of integers that adds up to exactly W. Since it is a special case of KNAPSACK, it cannot be any harder. But could it be polynomial? As it turns out, this problem, called SUBSET SUM, is also very hard.

At this point one could ask: If SUBSET SUM is a special case that happens to be as hard as the general KNAPSACK problem, why are we interested in it? The reason is *simplicity*. In the complicated calculus of reductions between search problems that we shall develop in this chapter, conceptually simple problems like SUBSET SUM and 3SAT are invaluable.

8.2 NP-complete problems

Hard problems, easy problems:
In short, the world is full of search problems, some of which can be solved efficiently, while others seem to be very hard. This is depicted in the following table.

Hard problems (**NP**-complete)	Easy problems (in **P**)
3SAT	2SAT, HORN SAT
TRAVELING SALESMAN PROBLEM	MINIMUM SPANNING TREE
LONGEST PATH	SHORTEST PATH
3D MATCHING	BIPARTITE MATCHING
KNAPSACK	UNARY KNAPSACK
INDEPENDENT SET	INDEPENDENT SET on trees
INTEGER LINEAR PROGRAMMING	LINEAR PROGRAMMING
RUDRATA PATH	EULER PATH
BALANCED CUT	MINIMUM CUT

This table is worth contemplating. On the right we have problems that can be solved efficiently. On the left, we have a bunch of hard nuts that have escaped efficient solution over many decades or centuries.

The various problems on the right can be solved by algorithms that are specialized and diverse: dynamic programming, network flow, graph search, greedy. These problems are easy for a variety of different reasons.

In stark contrast, the problems on the left *are all difficult for the same reason*! At their core, they are all the same problem, just in different disguises! They are all *equivalent*: as we shall see in Section 8.3, each of them can be reduced to any of the others—and back.

P and NP
It's time to introduce some important concepts. We know what a search problem is: its defining characteristic is that any proposed solution can be quickly checked for correctness, in the sense that there is an efficient checking algorithm C that takes as input the given instance I (the data specifying the problem to be solved), as well as the proposed solution S, and outputs `true` if and only if S really is a solution

Why P and NP?

Okay, **P** must stand for "polynomial." But why use the initials **NP** (the common chatroom abbreviation for "no problem") to describe the class of search problems, some of which are terribly hard?

NP stands for "nondeterministic polynomial time," a term going back to the roots of complexity theory. Intuitively, it means that a solution to any search problem can be found and verified in polynomial time by a special (and quite unrealistic) sort of algorithm, called a *nondeterministic algorithm*. Such an algorithm has the power of *guessing* correctly at every step.

Incidentally, the original definition of **NP** (and its most common usage to this day) was not as a class of search problems, but as a class of *decision problems*: algorithmic questions that can be answered by yes or no. Example: "Is there a truth assignment that satisfies this Boolean formula?" But this too reflects a historical reality: At the time the theory of **NP**-completeness was being developed, researchers in the theory of computation were interested in formal languages, a domain in which such decision problems are of central importance.

to instance I. Moreover the running time of $\mathcal{C}(I, S)$ is bounded by a polynomial in $|I|$, the length of the instance. *We denote the class of all search problems by* **NP.**

We've seen many examples of **NP** search problems that are solvable in polynomial time. In such cases, there is an algorithm that takes as input an instance I and has a running time polynomial in $|I|$. If I has a solution, the algorithm returns such a solution; and if I has no solution, the algorithm correctly reports so. *The class of all search problems that can be solved in polynomial time is denoted* **P.** Hence, all the search problems on the right-hand side of the table are in **P.**

Are there search problems that cannot be solved in polynomial time? In other words, is **P** $\neq$ **NP?** Most algorithms researchers think so. It is hard to believe that exponential search can always be avoided, that a simple trick will crack all these hard problems, famously unsolved for decades and centuries. And there is a good reason for mathematicians to believe that **P** $\neq$ **NP**—the task of finding a proof for a given mathematical assertion is a search problem and is therefore in **NP** (after all, when a formal proof of a mathematical statement is written out in excruciating detail, it can be checked mechanically, line by line, by an efficient algorithm). So if **P** $=$ **NP,** there would be an efficient way to prove theorems, thus eliminating the need for mathematicians! All in all, there are a variety of reasons why it is widely believed that **P** $\neq$ **NP.** However, proving this has turned out to be extremely difficult, one of the deepest and most important unsolved puzzles of mathematics.

Reductions, again

Even if we accept that **P** $\neq$ **NP,** what about the specific problems on the left side of the table? On the basis of what evidence do we believe that these particular problems have no efficient algorithm (besides, of course, the historical fact that many clever mathematicians and computer scientists have tried hard and failed to find any)? Such evidence is provided by *reductions*, which translate one search problem into another. What they demonstrate is that the problems on the left side of the table are all, in some sense, *exactly the same problem*, except that they are stated in different languages. What's more, we will also use reductions to show that these problems are the *hardest* search problems in **NP**—if even one of them has a polynomial time algorithm, then *every* problem in **NP** has a polynomial time algorithm. Thus if we believe that **P** $\neq$ **NP**, then all these search problems are hard.

We defined reductions in Chapter 7 and saw many examples of them. Let's now specialize this definition to search problems. A *reduction* from search problem A to search problem B is a polynomial-time algorithm f that transforms any instance I of A into an instance $f(I)$ of B, together with another polynomial-time algorithm h that maps any solution S of $f(I)$ back into a solution $h(S)$ of I; see the following diagram. If $f(I)$ has no solution, then neither does I. These two translation procedures f and h imply that any algorithm for B can be converted into an algorithm for A by bracketing it between f and h.

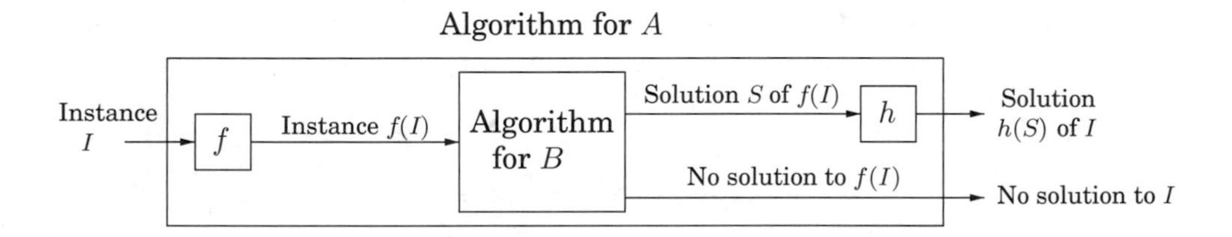

And now we can finally define the class of the hardest search problems.

A search problem is **NP***-complete if all other search problems reduce to it.*

This is a very strong requirement indeed. For a problem to be **NP**-complete, it must be useful in solving every search problem in the world! It is remarkable that such problems exist. But they do, and the first column of the table we saw earlier is filled with the most famous examples. In Section 8.3 we shall see how all these problems reduce to one another, and also why all other search problems reduce to them.

Factoring

One last point: we started off this book by introducing another famously hard search problem: FACTORING, the task of finding all prime factors of a given integer. But the difficulty of FACTORING is of a different nature than that of the other hard search problems we have just seen. For example, nobody believes that FACTORING is **NP**-complete. One major difference is that, in the case of FACTORING, the definition does not contain the now familiar clause "or report that none exists." A number can *always* be factored into primes.

The two ways to use reductions

So far in this book the purpose of a reduction from a problem A to a problem B has been straightforward and honorable: We know how to solve B efficiently, and we want to use this knowledge to solve A. In this chapter, however, reductions from A to B serve a somewhat perverse goal: we know A is hard, and we use the reduction to prove that B is hard as well!

If we denote a reduction from A to B by

$$A \longrightarrow B$$

then we can say that *difficulty* flows in the direction of the arrow, while *efficient algorithms* move in the opposite direction. It is through this propagation of difficulty that we know **NP**-complete problems are hard: all other search problems reduce to them, and thus each **NP**-complete problem contains the complexity of all search problems. If even one **NP**-complete problem is in **P**, then **P** = **NP**.

Reductions also have the convenient property that they *compose*.

$$\text{If } A \longrightarrow B \text{ and } B \longrightarrow C, \text{ then } A \longrightarrow C\,.$$

To see this, observe first of all that any reduction is completely specified by the pre- and postprocessing functions f and h (see the reduction diagram). If (f_{AB}, h_{AB}) and (f_{BC}, h_{BC}) define the reductions from A to B and from B to C, respectively, then a reduction from A to C is given by compositions of these functions: $f_{BC} \circ f_{AB}$ maps an instance of A to an instance of C and $h_{AB} \circ h_{BC}$ sends a solution of C back to a solution of A.

This means that once we know a problem A is **NP**-complete, we can use it to prove that a new search problem B is also **NP**-complete, simply by reducing A to B. Such a reduction establishes that all problems in **NP** reduce to B, via A.

Figure 8.6 The space **NP** of all search problems, assuming **P** $\neq$ **NP**.

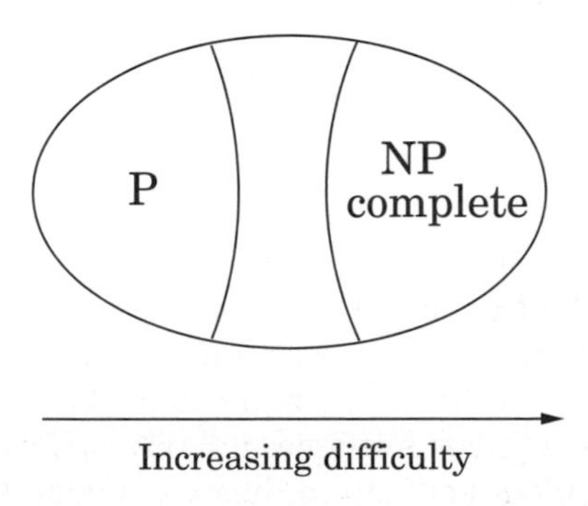

Figure 8.7 Reductions between search problems.

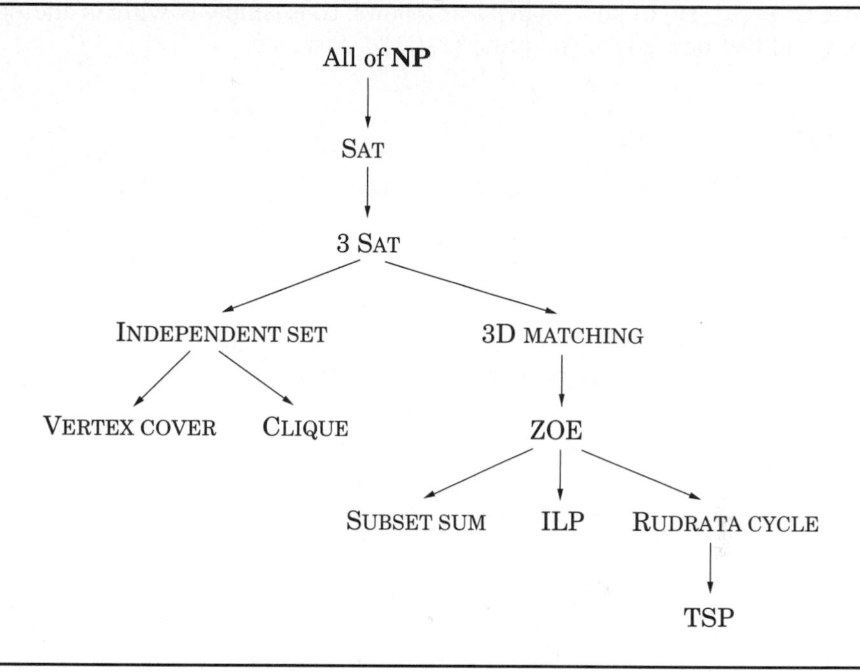

Another difference (possibly not completely unrelated) is this: as we shall see in Chapter 10, FACTORING succumbs to the power of *quantum computation*—while SAT, TSP, and the other **NP**-complete problems do not seem to.

8.3 The reductions

We shall now see that the search problems of Section 8.1 can be reduced to one another as depicted in Figure 8.7. As a consequence, they are all **NP**-complete.

Before we tackle the specific reductions in the tree, let's warm up by relating two versions of the Rudrata problem.

RUDRATA (s,t)-PATH$\longrightarrow$RUDRATA CYCLE

Recall the RUDRATA CYCLE problem: given a graph, is there a cycle that passes through each vertex exactly once? We can also formulate the closely related RUDRATA (s,t)-PATH problem, in which two vertices s and t are specified, and we want a path starting at s and ending at t that goes through each vertex exactly once. Is it possible that RUDRATA CYCLE is easier than RUDRATA (s,t)-PATH? We will show by a reduction that the answer is no.

The reduction maps an instance $(G = (V, E), s, t)$ of RUDRATA (s, t)-PATH into an instance $G' = (V', E')$ of RUDRATA CYCLE as follows: G' is simply G with an additional vertex x and two new edges $\{s, x\}$ and $\{x, t\}$. For instance:

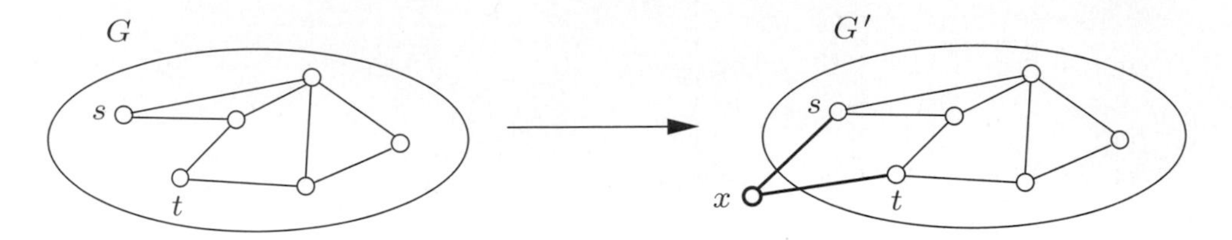

So $V' = V \cup \{x\}$, and $E' = E \cup \{\{s, x\}, \{x, t\}\}$. How do we recover a Rudrata (s, t)-*path* in G given any Rudrata *cycle* in G'? Easy, we just delete the edges $\{s, x\}$ and $\{x, t\}$ from the cycle.

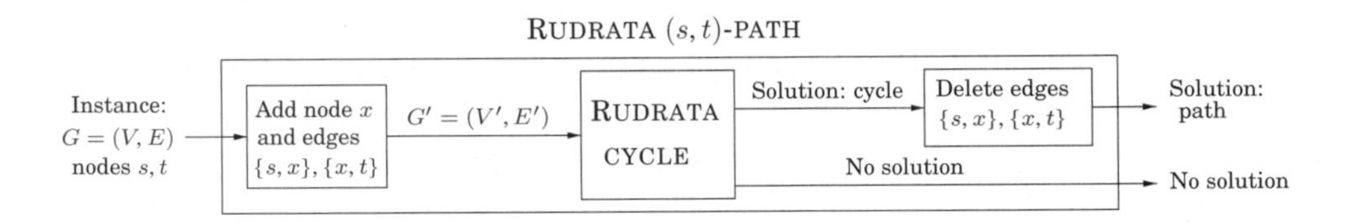

To confirm the validity of this reduction, we have to show that it works in the case of either outcome depicted.

1. When the instance of RUDRATA CYCLE has a solution.

Since the new vertex x has only two neighbors, s and t, any Rudrata cycle in G' must consecutively traverse the edges $\{t, x\}$ and $\{x, s\}$. The rest of the cycle then traverses every other vertex en route from s to t. Thus deleting the two edges $\{t, x\}$ and $\{x, s\}$ from the Rudrata cycle gives a Rudrata path from s to t in the original graph G.

2. When the instance of RUDRATA CYCLE does not have a solution.

In this case we must show that the original instance of RUDRATA (s, t)-PATH cannot have a solution either. It is usually easier to prove the contrapositive, that is, to show that if there is a Rudrata (s, t)-path in G, then there is also a Rudrata cycle in G'. But this is easy: just add the two edges $\{t, x\}$ and $\{x, s\}$ to the Rudrata path to close the cycle.

One last detail, crucial but typically easy to check, is that the pre- and postprocessing functions take time polynomial in the size of the instance (G, s, t).

It is also possible to go in the other direction and reduce RUDRATA CYCLE to RUDRATA (s, t)-PATH. Together, these reductions demonstrate that the two Rudrata variants are in essence the same problem—which is not too surprising, given that their descriptions are almost the same. But most of the other reductions we will see are between pairs of problems that, on the face of it, look quite different. To show that they are essentially the same, our reductions will have to cleverly translate between them.

3SAT$\longrightarrow$INDEPENDENT SET

One can hardly think of two more different problems. In 3SAT the input is a set of clauses, each with three or fewer literals, for example

$$(\overline{x} \vee y \vee \overline{z})\ (x \vee \overline{y} \vee z)\ (x \vee y \vee z)\ (\overline{x} \vee \overline{y}),$$

and the aim is to find a satisfying truth assignment. In INDEPENDENT SET the input is a graph and a number g, and the problem is to find a set of g pairwise non-adjacent vertices. We must somehow relate Boolean logic with graphs!

Let us think. To form a satisfying truth assignment we must pick one literal from each clause and give it the value `true`. But our choices must be consistent: if we choose $\overline{x}$ in one clause, we cannot choose x in another. Any consistent choice of literals, one from each clause, specifies a truth assignment (variables for which neither literal has been chosen can take on either value).

So, let us represent a clause, say $(x \vee \overline{y} \vee z)$, by a triangle, with vertices labeled $x, \overline{y}, z$. Why triangle? Because a triangle has its three vertices maximally connected, and thus forces us to pick only one of them for the independent set. Repeat this construction for all clauses—a clause with two literals will be represented simply by an edge joining the literals. (A clause with one literal is silly and can be removed in a preprocessing step, since the value of the variable is determined.) In the resulting graph, an independent set has to pick at most one literal from each group (clause). To force exactly one choice from each clause, take the goal g to be the number of clauses; in our example, $g = 4$.

All that is missing now is a way to prevent us from choosing opposite literals (that is, both x and $\overline{x}$) in different clauses. But this is easy: put an edge between any two vertices that correspond to opposite literals. The resulting graph for our example is shown in Figure 8.8.

Let's recap the construction. Given an instance I of 3SAT, we create an instance (G, g) of INDEPENDENT SET as follows.

- Graph G has a triangle for each clause (or just an edge, if the clause has two literals), with vertices labeled by the clause's literals, and has additional edges between any two vertices that represent opposite literals.
- The goal g is set to the number of clauses.

Clearly, this construction takes polynomial time. However, recall that for a reduction we do not just need an efficient way to map instances of the first problem to instances of the second (the function f in the diagram on page 245), but also a way to

Figure 8.8 The graph corresponding to $(\overline{x} \vee y \vee \overline{z})\ (x \vee \overline{y} \vee z)\ (x \vee y \vee z)\ (\overline{x} \vee \overline{y})$.

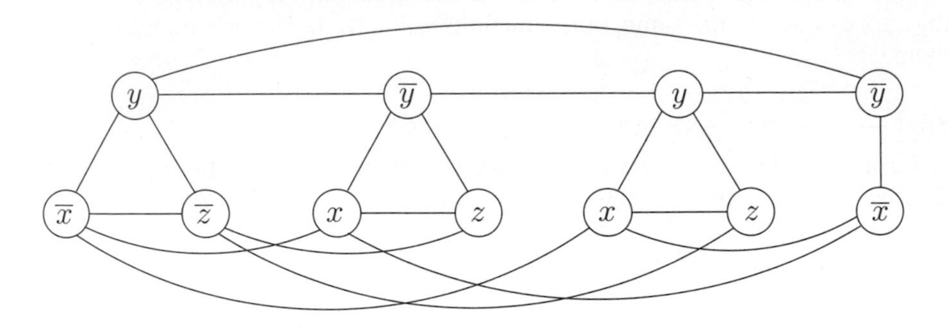

reconstruct a solution to the first instance from any solution of the second (the function h). As always, there are two things to show.

1. Given an independent set S of g vertices in G, it is possible to efficiently recover a satisfying truth assignment to I.

For any variable x, the set S cannot contain vertices labeled both x and $\overline{x}$, because any such pair of vertices is connected by an edge. So assign x a value of `true` if S contains a vertex labeled x, and a value of `false` if S contains a vertex labeled $\overline{x}$ (if S contains neither, then assign either value to x). Since S has g vertices, it must have one vertex per clause; this truth assignment satisfies those particular literals, and thus satisfies all clauses.

2. If graph G has no independent set of size g, then the Boolean formula I is unsatisfiable.

It is usually cleaner to prove the contrapositive, that if I has a satisfying assignment then G has an independent set of size g. This is easy: for each clause, pick any literal whose value under the satisfying assignment is `true` (there must be at least one such literal), and add the corresponding vertex to S. Do you see why set S must be independent?

SAT⟶3SAT

This is an interesting and common kind of reduction, from a problem to a *special case* of itself. We want to show that the problem remains hard even if its inputs are restricted somehow—in the present case, even if all clauses are restricted to have ≤ 3 literals. Such reductions modify the given instance so as to get rid of the forbidden feature (clauses with ≥ 4 literals) while keeping the instance essentially the same, in that we can read off a solution to the original instance from any solution of the modified one.

Here's the trick for reducing SAT to 3SAT: given an instance I of SAT, use exactly the same instance for 3SAT, except that any clause with more than three literals, $(a_1 \vee a_2 \vee \cdots \vee a_k)$ (where the a_i's are literals and $k > 3$), is replaced by a set of clauses,

$$(a_1 \vee a_2 \vee y_1)\ (\overline{y}_1 \vee a_3 \vee y_2)\ (\overline{y}_2 \vee a_4 \vee y_3)\ \cdots\ (\overline{y}_{k-3} \vee a_{k-1} \vee a_k),$$

where the y_i's are new variables. Call the resulting 3SAT instance I'. The conversion from I to I' is clearly polynomial time.

Why does this reduction work? I' is equivalent to I in terms of satisfiability, because for any assignment to the a_i's,

$$\left\{ \begin{array}{c} (a_1 \vee a_2 \vee \cdots \vee a_k) \\ \text{is satisfied} \end{array} \right\} \iff \left\{ \begin{array}{c} \text{there is a setting of the } y_i\text{'s for which} \\ (a_1 \vee a_2 \vee y_1)\ (\overline{y}_1 \vee a_3 \vee y_2)\ \cdots\ (\overline{y}_{k-3} \vee a_{k-1} \vee a_k) \\ \text{are all satisfied} \end{array} \right\}.$$

To see this, first suppose that the clauses on the right are all satisfied. Then at least one of the literals $a_1, \ldots, a_k$ must be true—otherwise y_1 would have to be true, which would in turn force y_2 to be true, and so on, eventually falsifying the last clause. But this means $(a_1 \vee a_2 \vee \cdots \vee a_k)$ is also satisfied.

Conversely, if $(a_1 \vee a_2 \vee \cdots \vee a_k)$ is satisfied, then some a_i must be true. Set $y_1, \ldots, y_{i-2}$ to `true` and the rest to `false`. This ensures that the clauses on the right are all satisfied.

Thus, any instance of SAT can be transformed into an equivalent instance of 3SAT. In fact, 3SAT remains hard even under the further restriction that no variable appears in more than three clauses. To show this, we must somehow get rid of any variable that appears too many times.

Here's the reduction from 3SAT to its constrained version. Suppose that in the 3SAT instance, variable x appears in $k > 3$ clauses. Then replace its first appearance by x_1, its second appearance by x_2, and so on, replacing each of its k appearances by a different new variable. Finally, add the clauses

$$(\overline{x}_1 \vee x_2)\ (\overline{x}_2 \vee x_3)\ \cdots\ (\overline{x}_k \vee x_1).$$

And repeat for every variable that appears more than three times.

It is easy to see that in the new formula no variable appears more than three times (and in fact, no literal appears more than twice). Furthermore, the extra clauses involving $x_1, x_2, \ldots, x_k$ constrain these variables to have the same value; do you see why? Hence the original instance of 3SAT is satisfiable if and only if the constrained instance is satisfiable.

Figure 8.9 S is a vertex cover if and only if $V - S$ is an independent set.

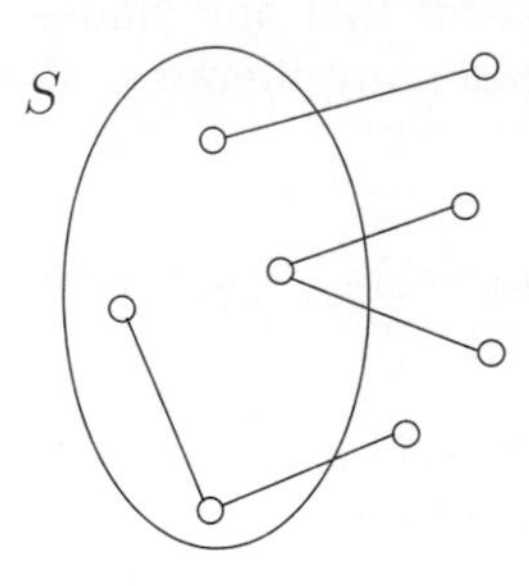

INDEPENDENT SET⟶VERTEX COVER

Some reductions rely on ingenuity to relate two very different problems. Others simply record the fact that one problem is a thin disguise of another. To reduce INDEPENDENT SET to VERTEX COVER we just need to notice that a set of nodes S is a vertex cover of graph $G = (V, E)$ (that is, S touches every edge in E) if and only if the remaining nodes, $V - S$, are an independent set of G (Figure 8.9).

Therefore, to solve an instance (G, g) of INDEPENDENT SET, simply look for a vertex cover of G with $|V| - g$ nodes. If such a vertex cover exists, then take all nodes *not* in it. If no such vertex cover exists, then G cannot possibly have an independent set of size g.

INDEPENDENT SET⟶CLIQUE

INDEPENDENT SET and CLIQUE are also easy to reduce to one another. Define the *complement* of a graph $G = (V, E)$ to be $\overline{G} = (V, \overline{E})$, where $\overline{E}$ contains precisely those unordered pairs of vertices that are not in E. Then a set of nodes S is an independent set of G if and only if S is a clique of $\overline{G}$. To paraphrase, these nodes have no edges between them in G if and only if they have all possible edges between them in $\overline{G}$.

Therefore, we can reduce INDEPENDENT SET to CLIQUE by mapping an instance (G, g) of INDEPENDENT SET to the corresponding instance $(\overline{G}, g)$ of CLIQUE; the solution to both is identical.

3SAT⟶3D MATCHING

Again, two very different problems. We must reduce 3SAT to the problem of finding, among a set of boy-girl-pet triples, a subset that contains each boy, each girl, and each pet exactly once. In short, we must design sets of boy-girl-pet triples that somehow behave like Boolean variables and gates!

Consider the following set of four triples, each represented by a triangular node joining a boy, girl, and pet:

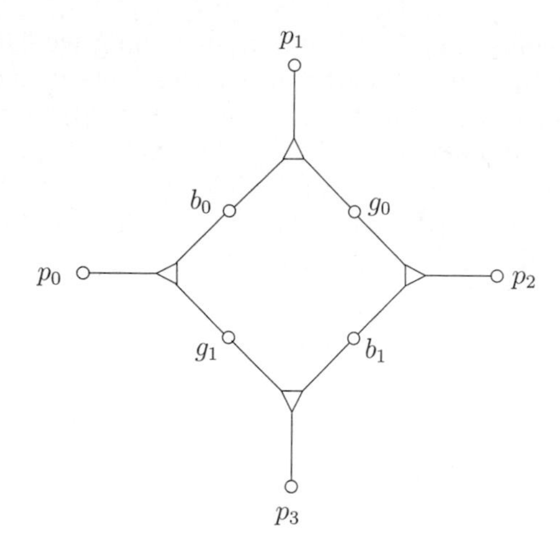

Suppose that the two boys b_0 and b_1 and the two girls g_0 and g_1 are not involved in any other triples. (The four pets $p_0, \ldots, p_3$ will of course belong to other triples as well; for otherwise the instance would trivially have no solution.) Then any matching must contain either the two triples $(b_0, g_1, p_0), (b_1, g_0, p_2)$ or the two triples $(b_0, g_0, p_1), (b_1, g_1, p_3)$, because these are the only ways in which these two boys and girls can find any match. Therefore, this "gadget" has two possible states: it behaves like a Boolean variable!

To then transform an instance of 3SAT to one of 3D MATCHING, we start by creating a copy of the preceding gadget for *each* variable x. Call the resulting nodes p_{x1}, b_{x0}, g_{x1}, and so on. The intended interpretation is that boy b_{x0} is matched with girl g_{x1} if $x = \texttt{true}$, and with girl g_{x0} if $x = \texttt{false}$.

Next we must create triples that somehow mimic clauses. For each clause, say $c = (x \vee \overline{y} \vee z)$, introduce a new boy b_c and a new girl g_c. They will be involved in three triples, one for each literal in the clause. And the pets in these triples must reflect the three ways whereby the clause can be satisfied: (1) $x = \texttt{true}$, (2) $y = \texttt{false}$, (3) $z = \texttt{true}$. For (1), we have the triple (b_c, g_c, p_{x1}), where p_{x1} is the pet p_1 in the gadget for x. Here is why we chose p_1: if $x = \texttt{true}$, then b_{x0} is matched with g_{x1} and b_{x1} with g_{x0}, and so pets p_{x0} and p_{x2} are taken. In which case b_c and g_c can be matched with p_{x1}. But if $x = \texttt{false}$, then p_{x1} and p_{x3} are taken, and so g_c and b_c cannot be accommodated this way. We do the same thing for the other two literals of the clause, which yield triples involving b_c and g_c with either p_{y0} or p_{y2} (for the negated variable y) and with either p_{z1} or p_{z3} (for variable z).

We have to make sure that for every occurrence of a literal in a clause c there is a different pet to match with b_c and g_c. But this is easy: by an earlier reduction we can assume that no literal appears more than twice, and so each variable gadget has enough pets, two for negated occurrences and two for unnegated.

The reduction now seems complete: from any matching we can recover a satisfying truth assignment by simply looking at each variable gadget and seeing with which girl b_{x0} was matched. And from any satisfying truth assignment we can match the gadget corresponding to each variable x so that triples (b_{x0}, g_{x1}, p_{x0}) and (b_{x1}, g_{x0}, p_{x2}) are chosen if $x = \texttt{true}$ and triples (b_{x0}, g_{x0}, p_{x1}) and (b_{x1}, g_{x1}, p_{x3}) are chosen if $x = \texttt{false}$; and for each clause c match b_c and g_c with the pet that corresponds to one of its satisfying literals.

But one last problem remains: in the matching defined at the end of the last paragraph, *some pets may be left unmatched*. In fact, if there are n variables and m clauses, then exactly $2n - m$ pets *will* be left unmatched (you can check that this number is sure to be positive, because we have at most three occurrences of every variable, and at least two literals in every clause). But this is easy to fix: Add $2n - m$ new boy-girl couples that are "generic animal-lovers," and match them by triples with all the pets!

3D MATCHING⟶ZOE

Recall that in ZOE we are given an $m \times n$ matrix $\mathbf{A}$ with $0-1$ entries, and we must find a $0-1$ vector $\mathbf{x} = (x_1, \ldots, x_n)$ such that the m equations

$$\mathbf{A}\mathbf{x} = \mathbf{1}$$

are satisfied, where by $\mathbf{1}$ we denote the column vector of all 1's. How can we express the 3D MATCHING problem in this framework?

ZOE and ILP are very useful problems precisely because they provide a format in which many combinatorial problems can be expressed. In such a formulation we think of the $0-1$ variables as describing a solution, and we write equations expressing the constraints of the problem.

For example, here is how we express an instance of 3D MATCHING (m boys, m girls, m pets, and n boy-girl-pet triples) in the language of ZOE. We have $0-1$ variables $x_1, \ldots, x_n$, one per triple, where $x_i = 1$ means that the ith triple is chosen for the matching, and $x_i = 0$ means that it is not chosen.

Now all we have to do is write equations stating that the solution described by the x_i's is a legitimate matching. For each boy (or girl, or pet), suppose that the triples containing him (or her, or it) are those numbered $j_1, j_2, \ldots, j_k$; the appropriate equation is then

$$x_{j_1} + x_{j_2} + \cdots + x_{j_k} = 1,$$

which states that exactly one of these triples must be included in the matching. For example, here is the $\mathbf{A}$ matrix for an instance of 3D MATCHING we saw earlier.

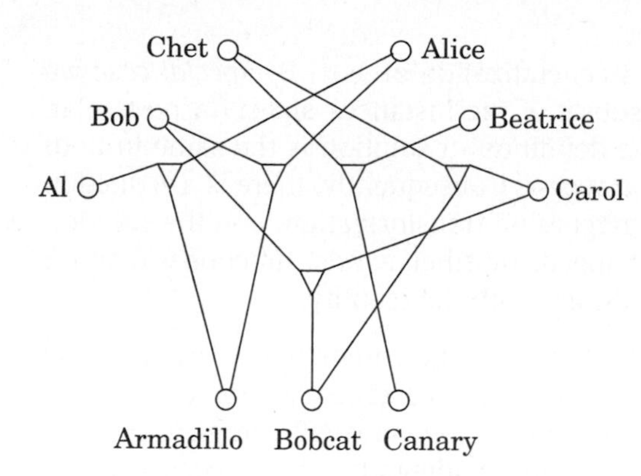

$$\mathbf{A} = \begin{pmatrix} 1 & 0 & 0 & 0 & 0 \\ 0 & 0 & 0 & 1 & 1 \\ 0 & 1 & 1 & 0 & 0 \\ 1 & 0 & 0 & 0 & 1 \\ 0 & 1 & 0 & 0 & 0 \\ 0 & 0 & 1 & 1 & 0 \\ 1 & 0 & 0 & 0 & 1 \\ 0 & 0 & 1 & 1 & 0 \\ 0 & 1 & 0 & 0 & 0 \end{pmatrix}$$

The five columns of **A** correspond to the five triples, while the nine rows are for Al, Bob, Chet, Alice, Beatrice, Carol, Armadillo, Bobcat, and Canary, respectively.

It is straightforward to argue that solutions to the two instances translate back and forth.

ZOE$\longrightarrow$SUBSET SUM

This is a reduction between two special cases of ILP: one with many equations but only $0-1$ coefficients, and the other with a single equation but arbitrary integer coefficients. The reduction is based on a simple and time-honored idea: $0-1$ vectors can encode numbers!

For example, given this instance of ZOE:

$$\mathbf{A} = \begin{pmatrix} 1 & 0 & 0 & 0 \\ 0 & 0 & 0 & 1 \\ 0 & 1 & 1 & 0 \\ 1 & 0 & 0 & 0 \\ 0 & 1 & 0 & 0 \end{pmatrix},$$

we are looking for a set of columns of **A** that, added together, make up the all-1's vector. But if we think of the columns as binary integers (read from top to bottom), we are looking for a subset of the integers 18, 5, 4, 8 that add up to the binary integer $11111_2 = 31$. And this is an instance of SUBSET SUM. The reduction is complete!

Except for one detail, the one that usually spoils the close connection between $0-1$ vectors and binary integers: *carry*. Because of carry, 5-bit binary integers can add up to 31 (for example, $5+6+20=31$ or, in binary, $00101_2 + 00110_2 + 10100_2 = 11111_2$) even when the sum of the corresponding vectors is not $(1,1,1,1,1)$. But this is easy to fix: Think of the column vectors not as integers in base 2, but as integers in base $n+1$—one more than the number of columns. This way, since at most n integers are added, and all their digits are 0 and 1, there can be no carry, and our reduction works.

ZOE⟶ILP

3SAT is a special case of SAT—or, SAT is a generalization of 3SAT. By *special case* we mean that the instances of 3SAT are a subset of the instances of SAT (in particular, the ones with no long clauses), and the definition of solution is the same in both problems (an assignment satisfying all clauses). Consequently, there is a reduction from 3SAT to SAT, in which the input undergoes no transformation, and the solution to the target instance is also kept unchanged. In other words, functions f and h from the reduction diagram (on page 245) are both the identity.

This sounds trivial enough, but it is a very useful and common way of establishing that a problem is **NP**-complete: Simply notice that it is a generalization of a known **NP**-complete problem. For example, the SET COVER problem is **NP**-complete because it is a generalization of VERTEX COVER (and also, incidentally, of 3D MATCHING). See Exercise 8.10 for more examples.

Often it takes a little work to establish that one problem is a special case of another. The reduction from ZOE to ILP is a case in point. In ILP we are looking for an integer vector $\mathbf{x}$ that satisfies $\mathbf{Ax} \leq \mathbf{b}$, for given matrix $\mathbf{A}$ and vector $\mathbf{b}$. To write an instance of ZOE in this precise form, we need to rewrite each equation of the ZOE instance as two inequalities (recall the transformations of Section 7.1.4), and to add for each variable x_i the inequalities $x_i \leq 1$ and $-x_i \leq 0$.

ZOE⟶RUDRATA CYCLE

In the RUDRATA CYCLE problem we seek a cycle in a graph that visits every vertex exactly once. We shall prove it **NP**-complete in two stages: first we will reduce ZOE to a generalization of RUDRATA CYCLE, called RUDRATA CYCLE WITH PAIRED EDGES, and then

Figure 8.10 Rudrata cycle with paired edges: $C = \{(e_1, e_3), (e_5, e_6), (e_4, e_5), (e_3, e_7), (e_3, e_8)\}$.

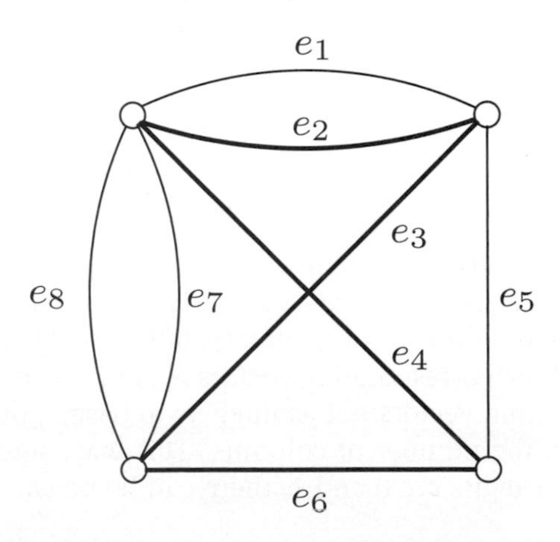

Figure 8.11 Reducing ZOE to RUDRATA CYCLE WITH PAIRED EDGES.

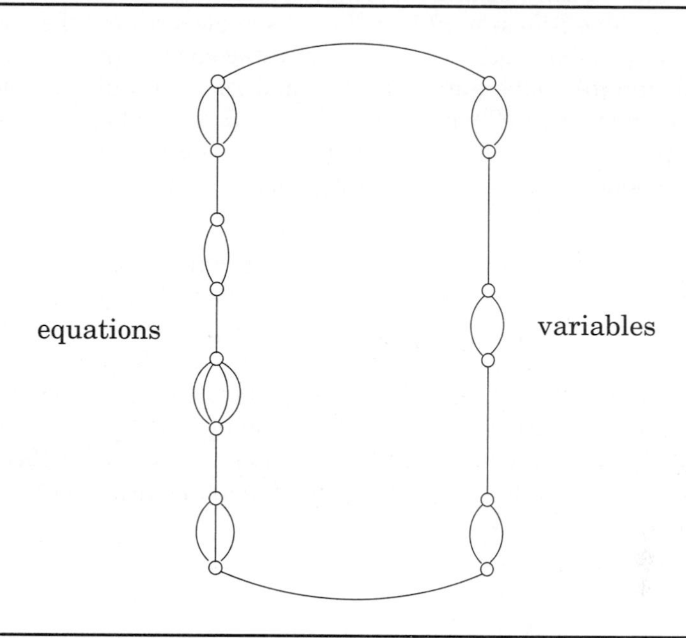

we shall see how to get rid of the extra features of that problem and reduce it to the plain RUDRATA CYCLE problem.

In an instance of RUDRATA CYCLE WITH PAIRED EDGES we are given a graph $G = (V, E)$ and a set $C \subseteq E \times E$ of pairs of edges. We seek a cycle that (1) visits all vertices once, like a Rudrata cycle should, and (2) for every pair of edges (e, e') in C, traverses either edge e or edge e'—*exactly one* of them. In the simple example of Figure 8.10 a solution is shown in bold. Notice that we allow two or more parallel edges between two nodes—a feature that doesn't make sense in most graph problems—since now the different copies of an edge can be paired with other copies of edges in ways that do make a difference.

Now for the reduction of ZOE to RUDRATA CYCLE WITH PAIRED EDGES. Given an instance of ZOE, $\mathbf{Ax} = \mathbf{1}$ (where $\mathbf{A}$ is an $m \times n$ matrix with $0-1$ entries, and thus describes m equations in n variables), the graph we construct has the very simple structure shown in Figure 8.11: a cycle that connects $m+n$ collections of parallel edges. For each variable x_i we have two parallel edges (corresponding to $x_i = 1$ and $x_i = 0$). And for each equation $x_{j_1} + \cdots + x_{j_k} = 1$ involving k variables we have k parallel edges, one for every variable appearing in the equation. This is the whole graph. Evidently, any Rudrata cycle in this graph must traverse the $m+n$ collections of parallel edges one by one, choosing one edge from each collection. This way, the cycle "chooses" for each variable a value—0 or 1—and, for each equation, a variable appearing in it.

The whole reduction can't be this simple, of course. The structure of the matrix **A** (and not just its dimensions) must be reflected somewhere, and there is one place left: the set C of pairs of edges such that exactly one edge in each pair is traversed. For every equation (recall there are m in total), and for every variable x_i appearing in it, we add to C the pair (e, e') where e is the edge corresponding to the appearance of x_i in that particular equation (on the left-hand side of Figure 8.11), and e' is the edge corresponding to the variable assignment $x_i = 0$ (on the right side of the figure). This completes the construction.

Take any solution of this instance of RUDRATA CYCLE WITH PAIRED EDGES. As discussed before, it picks a value for each variable and a variable for every equation. We claim that the values thus chosen are a solution to the original instance of ZOE. If a variable x_i has value 1, then the edge $x_i = 0$ is not traversed, and thus all edges associated with x_i on the equation side must be traversed (since they are paired in C with the $x_i = 0$ edge). So, in each equation exactly one of the variables appearing in it has value 1—which is the same as saying that all equations are satisfied. The other direction is straightforward as well: from a solution to the instance of ZOE one easily obtains an appropriate Rudrata cycle.

Getting Rid of the Edge Pairs

So far we have a reduction from ZOE to RUDRATA CYCLE WITH PAIRED EDGES; but we are really interested in RUDRATA CYCLE, which is a special case of the problem with paired edges: the one in which the set of pairs C is empty. To accomplish our goal, we need, as usual, to find a way of getting rid of the unwanted feature—in this case the edge pairs.

Consider the graph shown in Figure 8.12, and suppose that it is a part of a larger graph G in such a way that only the four endpoints a, b, c, d touch the rest of the graph. We claim that this graph has the following important property: *in any Rudrata cycle of G the subgraph shown must be traversed in one of the two ways shown in bold in Figure 8.12(b) and (c).* Here is why. Suppose that the cycle first enters the subgraph from vertex a continuing to f. Then it must continue to vertex g, because g has degree 2 and so it must be visited immediately after one of its adjacent nodes is visited—otherwise there is no way to include it in the cycle. Hence we must go on to node h, and here we seem to have a choice. We could continue on to j, or return to c. But if we take the second option, how are we going to visit the rest of the subgraph? (A Rudrata cycle must leave no vertex unvisited.) It is easy to see that this would be impossible, and so from h we have no choice but to continue to j and from there to visit the rest of the graph as shown in Figure 8.12(b). By symmetry, if the Rudrata cycle enters this subgraph at c, it must traverse it as in Figure 8.12(c). And these are the only two ways.

But this property tells us something important: this gadget behaves just like two edges $\{a, b\}$ and $\{c, d\}$ that are paired up in the RUDRATA CYCLE WITH PAIRED EDGES problem (see Figure 8.12(d)).

The rest of the reduction is now clear: to reduce RUDRATA CYCLE WITH PAIRED EDGES to RUDRATA CYCLE we go through the pairs in C one by one. To get rid of each

Figure 8.12 A gadget for enforcing paired behavior.

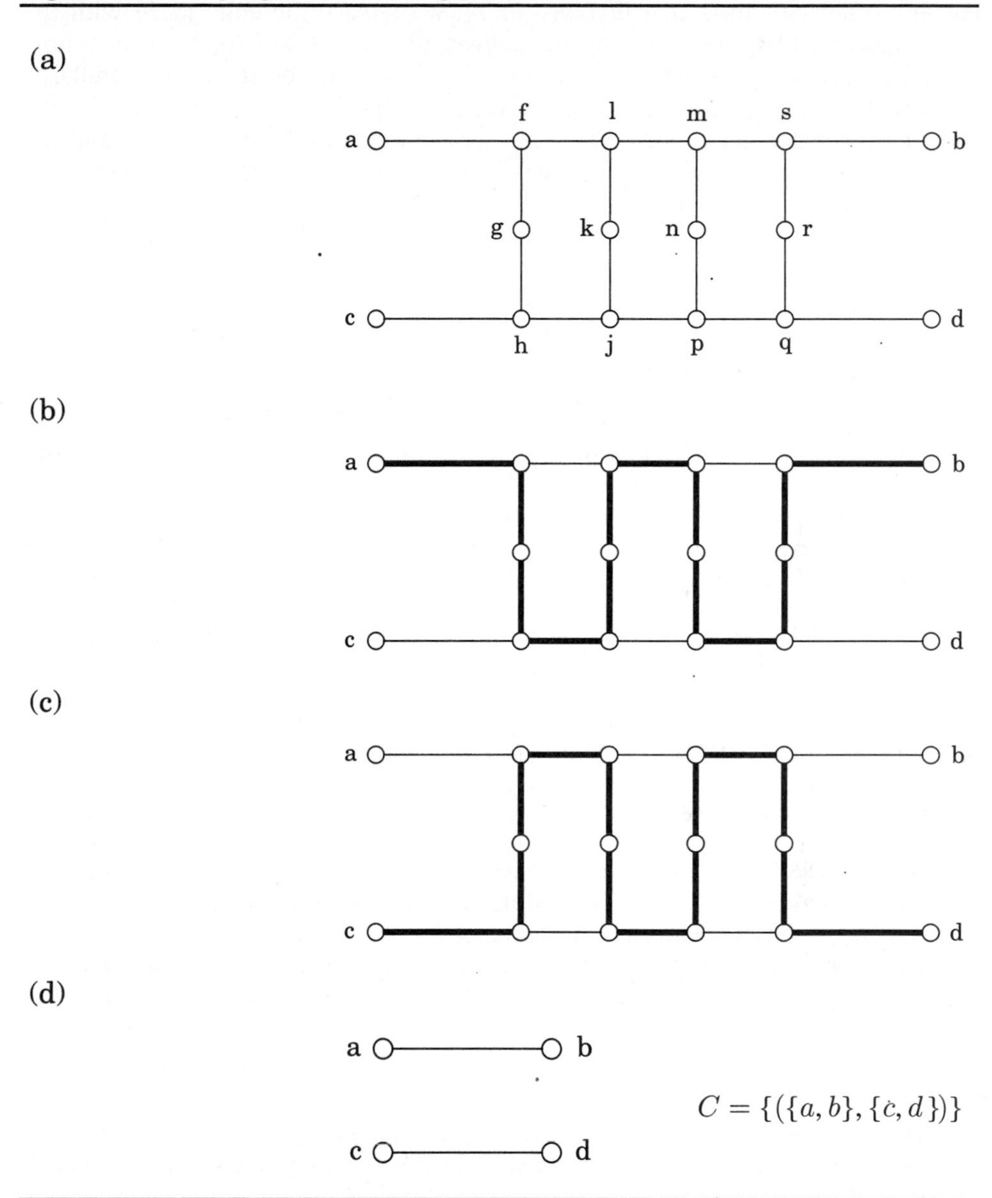

pair $(\{a, b\}, \{c, d\})$ we replace the two edges with the gadget in Figure 8.12(a). For any other pair in C that involves $\{a, b\}$, we replace the edge $\{a, b\}$ with the new edge $\{a, f\}$, where f is from the gadget: the traversal of $\{a, f\}$ is from now on an indication that edge $\{a, b\}$ in the old graph would be traversed. Similarly, $\{c, h\}$ replaces $\{c, d\}$. After $|C|$ such replacements (performed in polynomial time, since each replacement adds only 12 vertices to the graph) we are done, and the Rudrata cycles in the resulting graph will be in one-to-one correspondence with the Rudrata cycles in the original graph that conform to the constraints in C.

RUDRATA CYCLE⟶TSP

Given a graph $G = (V, E)$, construct the following instance of the TSP: the set of cities is the same as V, and the distance between cities u and v is 1 if $\{u, v\}$ is an edge of G and $1 + \alpha$ otherwise, for some $\alpha > 1$ to be determined. The budget of the TSP instance is equal to the number of nodes, $|V|$.

It is easy to see that if G has a Rudrata cycle, then the same cycle is also a tour within the budget of the TSP instance; and that conversely, if G has no Rudrata cycle, then there is no solution: the cheapest possible TSP tour has cost at least $n + \alpha$ (it must use at least one edge of length $1 + \alpha$, and the total length of all $n - 1$ others is at least $n - 1$). Thus RUDRATA CYCLE reduces to TSP.

In this reduction, we introduced the parameter α because by varying it, we can obtain two interesting results. If $\alpha = 1$, then all distances are either 1 or 2, and so this instance of the TSP satisfies the triangle inequality: if i, j, k are cities, then $d_{ij} + d_{jk} \geq d_{ik}$ (proof: $a + b \geq c$ holds for any numbers $1 \leq a, b, c \leq 2$). This is a special case of the TSP which is of practical importance and which, as we shall see in Chapter 9, is in a certain sense easier, because it can be efficiently *approximated.*

If on the other hand α is large, then the resulting instance of the TSP may not satisfy the triangle inequality, but has another important property: either it has a solution of cost n or less, or all its solutions have cost at least $n + \alpha$ (which now can be arbitrarily larger than n). There can be nothing in between! As we shall see in Chapter 9, this important *gap* property implies that, unless $\mathbf{P} = \mathbf{NP}$, no approximation algorithm is possible.

ANY PROBLEM IN **NP**⟶SAT

We have reduced SAT to the various search problems in Figure 8.7. Now we come full circle and argue that all these problems—and in fact all problems in **NP**—reduce to SAT.

In particular, we shall show that all problems in **NP** can be reduced to a generalization of SAT which we call CIRCUIT SAT. In CIRCUIT SAT we are given a (*Boolean*) *circuit* (see Figure 8.13, and recall Section 7.7), a dag whose vertices are *gates* of five different types:

- AND gates and OR gates have indegree 2.
- NOT gates have indegree 1.

Figure 8.13 An instance of CIRCUIT SAT.

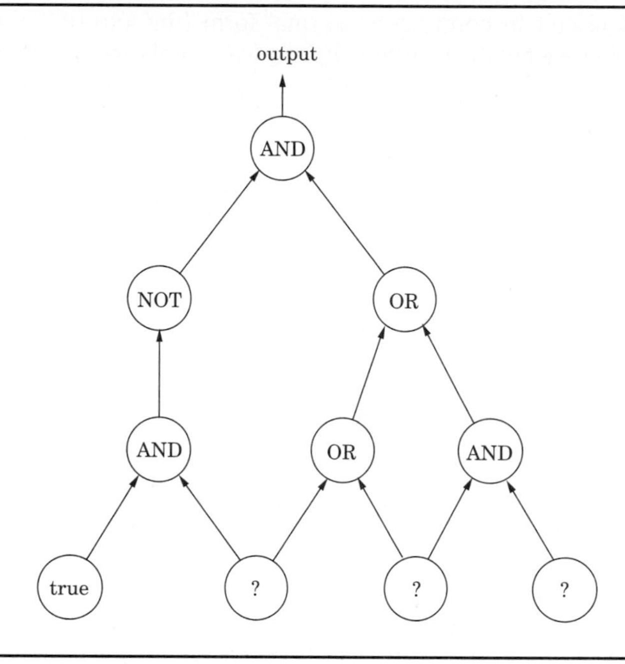

- *Known input* gates have no incoming edges and are labeled `false` or `true`.
- *Unknown input* gates have no incoming edges and are labeled "?".

One of the sinks of the dag is designated as the *output* gate.

Given an assignment of values to the unknown inputs, we can evaluate the gates of the circuit in topological order, using the rules of Boolean logic (such as `false` $\vee$ `true` = `true`), until we obtain the value at the output gate. This is the value of the circuit for the particular assignment to the inputs. For instance, the circuit in Figure 8.13 evaluates to `false` under the assignment `true`, `false`, `true` (from left to right).

CIRCUIT SAT is then the following search problem: Given a circuit, find a truth assignment for the unknown inputs such that the output gate evaluates to `true`, or report that no such assignment exists. For example, if presented with the circuit in Figure 8.13 we could have returned the assignment (`false`, `true`, `true`) because, if we substitute these values to the unknown inputs (from left to right), the output becomes `true`.

CIRCUIT SAT is a generalization of SAT. To see why, notice that SAT asks for a satisfying truth assignment for a circuit that has this simple structure: a bunch of AND gates at the top join the clauses, and the result of this big AND is the output. Each clause is the OR of its literals. And each literal is either an unknown input gate or the NOT of one. There are no known input gates.

Going in the other direction, CIRCUIT SAT can also be reduced to SAT. Here is how we can rewrite any circuit in conjunctive normal form (the AND of clauses): for each gate g in the circuit we create a variable g, and we model the effect of the gate using a few clauses:

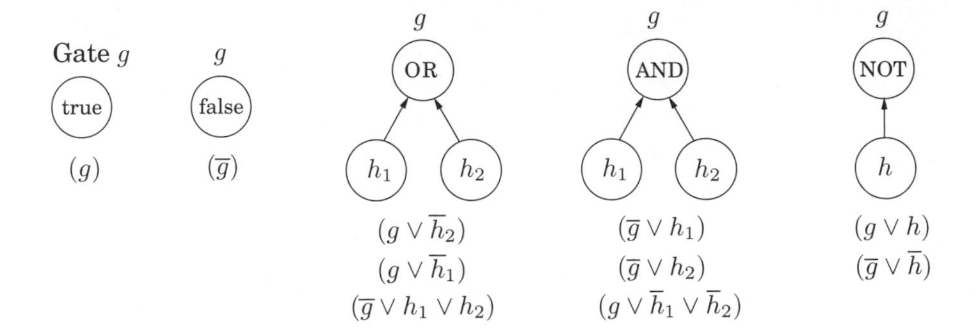

(Do you see that these clauses do, in fact, force exactly the desired effect?) And to finish up, if g is the output gate, we force it to be `true` by adding the clause (g). The resulting instance of SAT is equivalent to the given instance of CIRCUIT SAT: the satisfying truth assignments of this conjunctive normal form are in one-to-one correspondence with those of the circuit.

Now that we know CIRCUIT SAT reduces to SAT, we turn to our main job, showing that *all* search problems reduce to CIRCUIT SAT. So, suppose that A is a problem in **NP.** We must discover a reduction from A to CIRCUIT SAT. This sounds very difficult, *because we know almost nothing about* A!

All we know about A is that it is a search problem, so we must put this knowledge to work. The main feature of a search problem is that any solution to it can quickly be checked: there is an algorithm C that checks, given an instance I and a proposed solution S, whether or not S is a solution of I. Moreover, C makes this decision in time polynomial in the length of I (we can assume that S is itself encoded as a binary string, and we know that the length of this string is polynomial in the length of I).

Recall now our argument in Section 7.7 that any polynomial algorithm can be rendered as a circuit, whose input gates encode the input to the algorithm. Naturally, for any input length (number of input bits) the circuit will be scaled to the appropriate number of inputs, but the total number of gates of the circuit will be polynomial in the number of inputs. If the polynomial algorithm in question solves a problem that requires a yes or no answer (as is the situation with C: "Does S encode a solution to the instance encoded by I?"), then this answer is given at the output gate.

We conclude that, given any instance I of problem A, we can construct in polynomial time a circuit whose known inputs are the bits of I, and whose unknown inputs are the bits of S, such that the output is `true` if and only if the unknown inputs spell a solution S of I. In other words, *the satisfying truth assignments to the unknown inputs of the circuit are in one-to-one correspondence with the solutions of instance I of A.* The reduction is complete.

Unsolvable problems

At least an **NP**-complete problem can be solved by *some* algorithm—the trouble is that this algorithm will be exponential. But it turns out there are perfectly decent computational problems for which *no algorithms exist at all*!

One famous problem of this sort is an arithmetical version of SAT. Given a polynomial equation in many variables, perhaps

$$x^3yz + 2y^4z^2 - 7xy^5z = 6,$$

are there integer values of x, y, z that satisfy it? There is no algorithm that solves this problem. No algorithm at all, polynomial, exponential, doubly exponential, or worse! Such problems are called *unsolvable*.

The first unsolvable problem was discovered in 1936 by Alan M. Turing, then a student of mathematics at Cambridge, England. When Turing came up with it, there were no computers or programming languages (in fact, it can be argued that these things came about later *exactly because* this brilliant thought occurred to Turing). But today we can state it in familiar terms.

Suppose that you are given a program in your favorite programming language, along with a particular input. Will the program ever terminate, once started on this input? This is a very reasonable question. Many of us would be ecstatic if we had an algorithm, call it `terminates(p,x)`, that took as input a file containing a program `p`, and a file of data `x`, and after grinding away, finally told us whether or not `p` would ever stop if started on `x`.

But how would you go about writing the program `terminates`? (If you haven't seen this before, it's worth thinking about it for a while, to appreciate the difficulty of writing such a "universal infinite-loop detector.")

Well, you can't. *Such an algorithm does not exist*!

And here is the proof: Suppose we actually had such a program `terminates(p,x)`. Then we could use it as a subroutine of the following evil program:

```
function paradox(z:file)
1: if terminates(z,z) goto 1
```

Unsolvable problems (*Continued*)

Notice what `paradox` does: it terminates if and only if program `z` does *not* terminate when given its own code as input.

You should smell trouble. What if we put this program in a file named `paradox` and we executed `paradox(paradox)`? Would this execution ever stop? Or not? Neither answer is possible. Since we arrived at this contradiction by assuming that there is an algorithm for telling whether programs terminate, we must conclude that this problem cannot be solved by any algorithm.

By the way, all this tells us something important about programming: It will never be automated, it will forever depend on discipline, ingenuity, and hackery. We now know that you can't tell whether a program has an infinite loop. But can you tell if it has a buffer overrun? Do you see how to use the unsolvability of the "halting problem" to show that this, too, is unsolvable?

Exercises

8.1. *Optimization versus search.* Recall the traveling salesman problem:

> TSP
> *Input:* A matrix of distances; a budget b
> *Output:* A tour which passes through all the cities and has length $\leq b$, if such a tour exists.

The optimization version of this problem asks directly for the shortest tour.

> TSP-OPT
> *Input:* A matrix of distances
> *Output:* The shortest tour which passes through all the cities.

Show that if TSP can be solved in polynomial time, then so can TSP-OPT.

8.2. *Search versus decision.* Suppose you have a procedure which runs in polynomial time and tells you whether or not a graph has a Rudrata path. Show that you can use it to develop a polynomial-time algorithm for RUDRATA PATH (which returns the actual path, if it exists).

8.3. STINGY SAT is the following problem: given a set of clauses (each a disjunction of literals) and an integer k, find a satisfying assignment in which at most k variables are `true`, if such an assignment exists. Prove that STINGY SAT is **NP**-complete.

8.4. Consider the CLIQUE problem restricted to graphs in which every vertex has degree at most 3. Call this problem CLIQUE-3.

(a) Prove that CLIQUE-3 is in **NP.**

(b) What is wrong with the following proof of **NP**-completeness for CLIQUE-3? We know that the CLIQUE problem in general graphs is **NP**-complete, so it is enough to present a reduction from CLIQUE-3 to CLIQUE. Given a graph G with vertices of degree ≤ 3, and a parameter g, the reduction leaves the graph and the parameter unchanged: clearly the output of the reduction is a possible input for the CLIQUE problem. Furthermore, the answer to both problems is identical. This proves the correctness of the reduction and, therefore, the **NP**-completeness of CLIQUE-3.

(c) It is true that the VERTEX COVER problem remains **NP**-complete even when restricted to graphs in which every vertex has degree at most 3. Call this problem VC-3. What is wrong with the following proof of **NP**-completeness for CLIQUE-3?

We present a reduction from VC-3 to CLIQUE-3. Given a graph $G = (V, E)$ with node degrees bounded by 3, and a parameter b, we create an instance of CLIQUE-3 by leaving the graph unchanged and switching the parameter to $|V| - b$. Now, a subset $C \subseteq V$ is a vertex cover in G if and only if the complementary set $V - C$ is a clique in G. Therefore G has a vertex cover of size $\leq b$ if and only if it has a clique of size $\geq |V| - b$. This proves the correctness of the reduction and, consequently, the **NP**-completeness of CLIQUE-3.

(d) Describe an $O(|V|)$ algorithm for CLIQUE-3.

8.5. Give a simple reduction from 3D MATCHING to SAT, and another from RUDRATA CYCLE to SAT. (*Hint:* In the latter case you may use variables x_{ij} whose intuitive meaning is "vertex i is the jth vertex of the Rudrata cycle"; you then need to write clauses that express the constraints of the problem.)

8.6. On page 251 we saw that 3SAT remains **NP**-complete even when restricted to formulas in which each literal appears at most twice.

(a) Show that if each literal appears at most *once*, then the problem is solvable in polynomial time.

(b) Show that INDEPENDENT SET remains **NP**-complete even in the special case when all the nodes in the graph have degree at most 4.

8.7. Consider a special case of 3SAT in which all clauses have exactly three literals, and each variable appears exactly three times. Show that this problem can be solved in polynomial time. (*Hint:* Create a bipartite graph with clauses on the left, variables on the right, and edges whenever a variable appears in a clause. Use Exercise 7.30 to show that this graph has a matching.)

8.8. In the EXACT 4SAT problem, the input is a set of clauses, each of which is a disjunction of exactly four literals, and such that each variable occurs at most once in each clause. The goal is to find a satisfying assignment, if one exists. Prove that EXACT 4SAT is NP-complete.

8.9. In the HITTING SET problem, we are given a family of sets $\{S_1, S_2, \ldots, S_n\}$ and a budget b, and we wish to find a set H of size $\leq b$ which intersects every S_i, if such an H exists. In other words, we want $H \cap S_i \neq \emptyset$ for all i.

Show that HITTING SET is **NP**-complete.

8.10. *Proving* **NP**-*completeness by generalization.* For each of the problems below, prove that it is **NP**-complete by showing that it is a *generalization* of some **NP**-complete problem we have seen in this chapter.

(a) SUBGRAPH ISOMORPHISM: Given as input two undirected graphs G and H, determine whether G is a subgraph of H (that is, whether by deleting certain vertices and edges of H we obtain a graph that is, up to renaming of vertices, identical to G), and if so, return the corresponding mapping of $V(G)$ into $V(H)$.

(b) LONGEST PATH: Given a graph G and an integer g, find in G a simple path of length g.

(c) MAX SAT: Given a CNF formula and an integer g, find a truth assignment that satisfies at least g clauses.

(d) DENSE SUBGRAPH: Given a graph and two integers a and b, find a set of a vertices of G such that there are at least b edges between them.

(e) SPARSE SUBGRAPH: Given a graph and two integers a and b, find a set of a vertices of G such that there are at most b edges between them.

(f) SET COVER. (This problem generalizes *two* known **NP**-complete problems.)

(g) RELIABLE NETWORK: We are given two $n \times n$ matrices, a distance matrix d_{ij} and a *connectivity requirement* matrix r_{ij}, as well as a budget b; we must find a graph $G = (\{1, 2, \ldots, n\}, E)$ such that (1) the total cost of all edges is b or less and (2) between any two distinct vertices i and j there are r_{ij} vertex-disjoint paths. (*Hint:* Suppose that all d_{ij}'s are 1 or 2, $b = n$, and all r_{ij}'s are 2. Which well known **NP**-complete problem is this?)

8.11. There are many variants of Rudrata's problem, depending on whether the graph is undirected or directed, and whether a cycle or path is sought. Reduce the DIRECTED RUDRATA PATH problem to each of the following.

(a) The (undirected) RUDRATA PATH problem.

(b) The undirected RUDRATA (s, t)-PATH problem, which is just like RUDRATA PATH except that the endpoints of the path are specified in the input.

8.12. The k-SPANNING TREE problem is the following.

Input: An undirected graph $G = (V, E)$
Output: A spanning tree of G in which each node has degree $\leq k$, if such a tree exists.

Show that for any $k \geq 2$:

(a) k-SPANNING TREE is a search problem.

(b) k-SPANNING TREE is **NP**-complete. (*Hint:* Start with $k = 2$ and consider the relation between this problem and RUDRATA PATH.)

8.13. Determine which of the following problems are **NP**-complete and which are solvable in polynomial time. In each problem you are given an undirected graph $G = (V, E)$, along with:

(a) A set of nodes $L \subseteq V$, and you must find a spanning tree such that its set of leaves includes the set L.

(b) A set of nodes $L \subseteq V$, and you must find a spanning tree such that its set of leaves is precisely the set L.

(c) A set of nodes $L \subseteq V$, and you must find a spanning tree such that its set of leaves is included in the set L.

(d) An integer k, and you must find a spanning tree with k or fewer leaves.

(e) An integer k, and you must find a spanning tree with k or more leaves.

(f) An integer k, and you must find a spanning tree with exactly k leaves.

(*Hint:* All the **NP**-completeness proofs are by generalization, except for one.)

8.14. Prove that the following problem is **NP**-complete: given an undirected graph $G = (V, E)$ and an integer k, return a clique of size k *as well as* an independent set of size k, provided both exist.

8.15. Show that the following problem is **NP**-complete.

MAXIMUM COMMON SUBGRAPH
Input: Two graphs $G_1 = (V_1, E_1)$ and $G_2 = (V_2, E_2)$; a budget b.
Output: Two set of nodes $V_1' \subseteq V_1$ and $V_2' \subseteq V_2$ whose deletion leaves at least b nodes in each graph, and makes the two graphs identical.

8.16. We are feeling experimental and want to create a new dish. There are various ingredients we can choose from and we'd like to use as many of them as possible, but some ingredients don't go well with others. If there are n possible ingredients (numbered 1 to n), we write down an $n \times n$ matrix giving the *discord* between any pair of ingredients. This *discord* is a real number between 0.0 and 1.0, where 0.0 means "they go together perfectly" and 1.0 means "they really don't go together." Here's an example matrix when there are five possible ingredients.

	1	2	3	4	5
1	0.0	0.4	0.2	0.9	1.0
2	0.4	0.0	0.1	1.0	0.2
3	0.2	0.1	0.0	0.8	0.5
4	0.9	1.0	0.8	0.0	0.2
5	1.0	0.2	0.5	0.2	0.0

In this case, ingredients 2 and 3 go together pretty well whereas 1 and 5 clash badly. Notice that this matrix is necessarily symmetric; and that the diagonal entries are always 0.0. Any set of ingredients incurs a *penalty* which is *the sum of all discord values between pairs of ingredients*. For instance, the set of ingredients $\{1, 3, 5\}$ incurs a penalty of $0.2 + 1.0 + 0.5 = 1.7$. We want this penalty to be small.

> EXPERIMENTAL CUISINE
> *Input:* n, the number of ingredients to choose from; D, the $n \times n$ "discord" matrix; some number $p \geq 0$
> *Output:* The maximum number of ingredients we can choose with penalty $\leq p$.

Show that if EXPERIMENTAL CUISINE is solvable in polynomial time, then so is 3SAT.

8.17. Show that for any problem Π in **NP**, there is an algorithm which solves Π in time $O(2^{p(n)})$, where n is the size of the input instance and $p(n)$ is a polynomial (which may depend on Π).

8.18. Show that if $\mathbf{P} = \mathbf{NP}$ then the RSA cryptosystem (Section 1.4.2) can be broken in polynomial time.

8.19. A *kite* is a graph on an even number of vertices, say $2n$, in which n of the vertices form a clique and the remaining n vertices are connected in a "tail" that consists of a path joined to one of the vertices of the clique. Given a graph and a goal g, the KITE problem asks for a subgraph which is a kite and which contains $2g$ nodes. Prove that KITE is **NP**-complete.

8.20. In an undirected graph $G = (V, E)$, we say $D \subseteq V$ is a *dominating set* if every $v \in V$ is either in D or adjacent to at least one member of D. In the DOMINATING SET problem, the input is a graph and a budget b, and the aim is to find a dominating set in the graph of size at most b, if one exists. Prove that this problem is **NP**-complete.

8.21. *Sequencing by hybridization.* One experimental procedure for identifying a new DNA sequence repeatedly probes it to determine which k-mers (substrings of length k) it contains. Based on these, the full sequence must then be reconstructed.

Let's now formulate this as a combinatorial problem. For any string x (the DNA sequence), let $\Gamma(x)$ denote the multiset of all of its k-mers. In particular, $\Gamma(x)$ contains exactly $|x| - k + 1$ elements.

The reconstruction problem is now easy to state: given a multiset of k-length strings, find a string x such that $\Gamma(x)$ is exactly this multiset.

(a) Show that the reconstruction problem reduces to RUDRATA PATH. (*Hint:* Construct a directed graph with one node for each k-mer, and with an edge from a to b if the last $k-1$ characters of a match the first $k-1$ characters of b.)

(b) But in fact, there is much better news. Show that the same problem also reduces to EULER PATH. (*Hint:* This time, use one directed edge for each k-mer.)

8.22. In task scheduling, it is common to use a graph representation with a node for each task and a directed edge from task i to task j if i is a precondition for j. This directed graph depicts the precedence constraints in the scheduling problem. Clearly, a schedule is possible if and only if the graph is acyclic; if it isn't, we'd like to identify the smallest number of constraints that must be dropped so as to make it acyclic.

Given a directed graph $G = (V, E)$, a subset $E' \subseteq E$ is called a *feedback arc set* if the removal of edges E' renders G acyclic.

FEEDBACK ARC SET (FAS): Given a directed graph $G = (V, E)$ and a budget b, find a feedback arc set of $\leq b$ edges, if one exists.

(a) Show that FAS is in **NP**.

FAS can be shown to be **NP**-complete by a reduction from VERTEX COVER. Given an instance (G, b) of VERTEX COVER, where G is an undirected graph and we want a vertex cover of size $\leq b$, we construct a instance (G', b) of FAS as follows. If $G = (V, E)$ has n vertices $v_1, \ldots, v_n$, then make $G' = (V', E')$ a directed graph with $2n$ vertices $w_1, w_1', \ldots, w_n, w_n'$, and $n + 2|E|$ (directed) edges:

- (w_i, w_i') for all $i = 1, 2, \ldots, n$.
- (w_i', w_j) and (w_j', w_i) for every $(v_i, v_j) \in E$.

(b) Show that if G contains a vertex cover of size b, then G' contains a feedback arc set of size b.

(c) Show that if G' contains a feedback arc set of size b, then G contains a vertex cover of size (at most) b. (*Hint:* Given a feedback arc set of size b in G', you may need to first modify it slightly to obtain another one which is of a more convenient form, but is of the same size or smaller. Then, argue that G must contain a vertex cover of the same size as the modified feedback arc set.)

8.23. In the NODE-DISJOINT PATHS problem, the input is an undirected graph in which some vertices have been specially marked: a certain number of "sources" $s_1, s_2, \ldots s_k$ and an equal number of "destinations" $t_1, t_2, \ldots t_k$. The goal is to find k node-disjoint paths (that is, paths which have no nodes in common) where the ith path goes from s_i to t_i. Show that this problem is **NP**-complete.

Here is a sequence of progressively stronger hints.

(i) Reduce from 3SAT.

(ii) For a 3SAT formula with m clauses and n variables, use $k = m + n$ sources and destinations. Introduce one source/destination pair (s_x, t_x) for each variable x, and one source/destination pair (s_c, t_c) for each clause c.

(iii) For each 3SAT clause, introduce 6 new intermediate vertices, one for each literal occurring in that clause and one for its complement.

(iv) Notice that if the path from s_c to t_c goes through some intermediate vertex representing, say, an occurrence of variable x, then no other path can go through that vertex. What vertex would you like the other path to be forced to go through instead?

Chapter 9

Coping with NP-completeness

You are the junior member of a seasoned project team. Your current task is to write code for solving a simple-looking problem involving graphs and numbers. What are you supposed to do?

If you are very lucky, your problem will be among the half-dozen problems concerning graphs with weights (shortest path, minimum spanning tree, maximum flow, etc.), that we have solved in this book. Even if this is the case, recognizing such a problem in its natural habitat—grungy and obscured by reality and context—requires practice and skill. It is more likely that you will need to reduce your problem to one of these lucky ones—or to solve it using dynamic programming or linear programming.

But chances are that nothing like this will happen. The world of search problems is a bleak landscape. There are a few spots of light—brilliant algorithmic ideas—each illuminating a small area around it (the problems that reduce to it; two of these areas, linear and dynamic programming, are in fact decently large). But the remaining vast expanse is pitch dark: **NP**-complete. What are you to do?

You can start by proving that your problem is actually **NP**-complete. Often a proof by generalization (recall the discussion on page 256 and Exercise 8.10) is all that you need; and sometimes a simple reduction from 3SAT or ZOE is not too difficult to find. This sounds like a theoretical exercise, but, if carried out successfully, it does bring some tangible rewards: now your status in the team has been elevated, you are no longer the kid who can't do, and you have become the noble knight with the impossible quest.

But, unfortunately, a problem does not go away when proved **NP**-complete. The real question is, *What do you do next?*

This is the subject of the present chapter and also the inspiration for some of the most important modern research on algorithms and complexity. **NP**-completeness is not a death certificate—it is only the beginning of a fascinating adventure.

Your problem's **NP**-completeness proof probably constructs graphs that are complicated and weird, very much unlike those that come up in your application. For example, even though SAT is **NP**-complete, satisfying assignments for HORN SAT (the instances of SAT that come up in logic programming) can be found efficiently (recall Section 5.3). Or, suppose the graphs that arise in your application are trees. In this case, many **NP**-complete problems, such as INDEPENDENT SET, can be solved in linear time by dynamic programming (recall Section 6.7).

Unfortunately, this approach does not always work. For example, we know that 3SAT is **NP**-complete. And the INDEPENDENT SET problem, along with many other **NP**-complete problems, remains so even for planar graphs (graphs that can be drawn in the plane without crossing edges). Moreover, often you cannot neatly characterize the instances that come up in your application. Instead, you will have to rely on some form of *intelligent exponential search*—procedures such as *backtracking* and *branch and bound* which are exponential time in the worst-case, but, with the right design, could be very efficient on typical instances that come up in your application. We discuss these methods in Section 9.1.

Or you can develop an algorithm for your **NP**-complete optimization problem that falls short of the optimum *but never by too much*. For example, in Section 5.4 we saw that the greedy algorithm always produces a set cover that is no more than $\log n$ times the optimal set cover. An algorithm that achieves such a guarantee is called an *approximation algorithm*. As we will see in Section 9.2, such algorithms are known for many **NP**-complete optimization problems, and they are some of the most clever and sophisticated algorithms around. And the theory of **NP**-completeness can again be used as a guide in this endeavor, by showing that, for some problems, there are even severe limits to how well they can be approximated—unless of course $\mathbf{P} = \mathbf{NP}$.

Finally, there are *heuristics*, algorithms with no guarantees on either the running time or the degree of approximation. Heuristics rely on ingenuity, intuition, a good understanding of the application, meticulous experimentation, and often insights from physics or biology, to attack a problem. We see some common kinds in Section 9.3.

9.1 Intelligent exhaustive search

9.1.1 Backtracking

Backtracking is based on the observation that it is often possible to reject a solution by looking at just a small portion of it. For example, if an instance of SAT contains the clause $(x_1 \vee x_2)$, then all assignments with $x_1 = x_2 = 0$ (i.e., `false`) can be instantly eliminated. To put it differently, by quickly checking and discrediting this *partial assignment*, we are able to prune a quarter of the entire search space. A promising direction, but can it be systematically exploited?

Here's how it is done. Consider the Boolean formula $\phi(w, x, y, z)$ specified by the set of clauses

$$(w \vee x \vee y \vee z),\ (w \vee \overline{x}),\ (x \vee \overline{y}),\ (y \vee \overline{z}),\ (z \vee \overline{w}),\ (\overline{w} \vee \overline{z}).$$

We will incrementally grow a tree of partial solutions. We start by branching on any one variable, say w:

Initial formula ϕ

$w = 0$ $w = 1$

Plugging $w = 0$ and $w = 1$ into ϕ, we find that no clause is immediately violated and thus neither of these two partial assignments can be eliminated outright. So we need to keep branching. We can expand either of the two available nodes, and on any variable of our choice. Let's try this one:

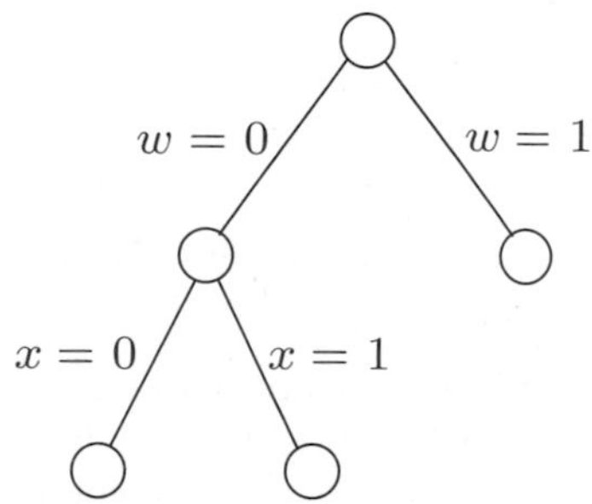

This time, we are in luck. The partial assignment $w = 0, x = 1$ violates the clause $(w \vee \overline{x})$ and can be terminated, thereby pruning a good chunk of the search space. We backtrack out of this cul-de-sac and continue our explorations at one of the two remaining active nodes.

In this manner, backtracking explores the space of assignments, growing the tree only at nodes where there is uncertainty about the outcome, and stopping if at any stage a satisfying assignment is encountered.

In the case of Boolean satisfiability, each node of the search tree can be described either by a partial assignment or by the clauses that remain when those values are plugged into the original formula. For instance, if $w = 0$ and $x = 0$ then any clause with $\overline{w}$ or $\overline{x}$ is instantly satisfied and any literal w or x is not satisfied and can be removed. What's left is

$$(y \vee z), (\overline{y}), (y \vee \overline{z}).$$

Figure 9.1 Backtracking reveals that ϕ is not satisfiable.

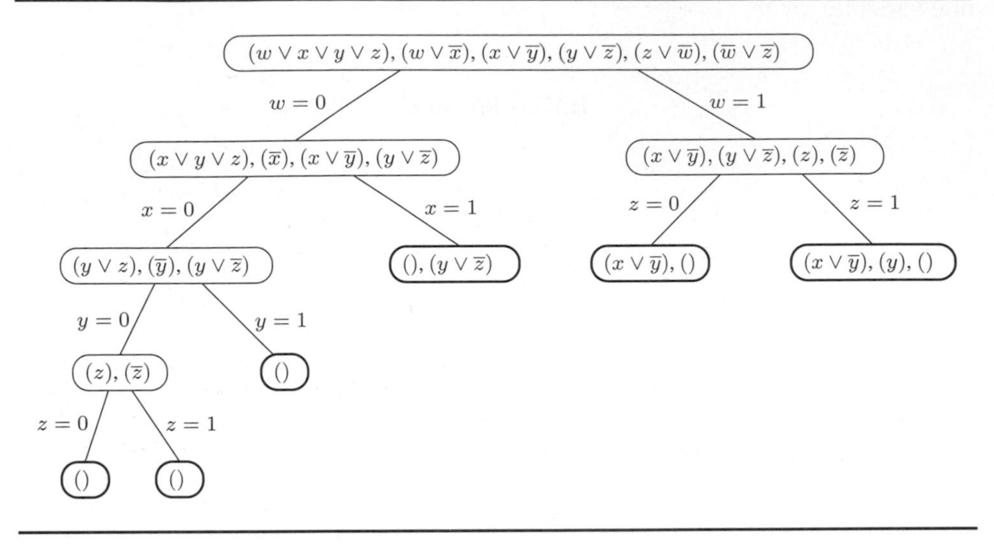

Likewise, $w = 0$ and $x = 1$ leaves

$$(), (y \vee \overline{z}),$$

with the "empty clause" $()$ ruling out satisfiability. Thus the nodes of the search tree, representing partial assignments, are themselves SAT *subproblems*.

This alternative representation is helpful for making the two decisions that repeatedly arise: which subproblem to expand next, and which branching variable to use. Since the benefit of backtracking lies in its ability to eliminate portions of the search space, and since this happens only when an empty clause is encountered, it makes sense to choose the subproblem that contains the *smallest* clause and to then branch on a variable in that clause. If this clause happens to be a singleton, then at least one of the resulting branches will be terminated. (If there is a tie in choosing subproblems, one reasonable policy is to pick the one lowest in the tree, in the hope that it is close to a satisfying assignment.) See Figure 9.1 for the conclusion of our earlier example.

More abstractly, a backtracking algorithm requires a *test* that looks at a subproblem and quickly declares one of three outcomes:

1. Failure: the subproblem has no solution.
2. Success: a solution to the subproblem is found.
3. Uncertainty.

In the case of SAT, this test declares failure if there is an empty clause, success if there are no clauses, and uncertainty otherwise. The backtracking procedure then has the following format.

```
Start with some problem P_0
Let S = {P_0}, the set of active subproblems
Repeat while S is nonempty:
   choose a subproblem P ∈ S and remove it from S
   expand it into smaller subproblems P_1, P_2, ..., P_k
   For each P_i:
      If test(P_i) succeeds: halt and announce this solution
      If test(P_i) fails: discard P_i
      Otherwise: add P_i to S
Announce that there is no solution
```

For SAT, the `choose` procedure picks a clause, and `expand` picks a variable within that clause. We have already discussed some reasonable ways of making such choices.

With the right `test`, `expand`, and `choose` routines, backtracking can be remarkably effective in practice. The backtracking algorithm we showed for SAT is the basis of many successful satisfiability programs. Another sign of quality is this: if presented with a 2SAT instance, it will always find a satisfying assignment, if one exists, in polynomial time (Exercise 9.1)!

9.1.2 Branch-and-bound

The same principle can be generalized from search problems such as SAT to optimization problems. For concreteness, let's say we have a minimization problem; maximization will follow the same pattern.

As before, we will deal with partial solutions, each of which represents a *subproblem*, namely, what is the (cost of the) best way to complete this solution? And as before, we need a basis for eliminating partial solutions, since there is no other source of efficiency in our method. To reject a subproblem, we must be certain that its cost exceeds that of some other solution we have already encountered. But its exact cost is unknown to us and is generally not efficiently computable. So instead we use a quick *lower bound* on this cost.

```
Start with some problem P_0
Let S = {P_0}, the set of active subproblems
bestsofar = ∞
Repeat while S is nonempty:
  choose a subproblem P ∈ S and remove
  it from S
  expand it into smaller subproblems P_1, P_2, ..., P_k
  For each P_i:
    If P_i is a complete solution: update bestsofar
    else if lowerbound(P_i) < bestsofar: add P_i to S
return bestsofar
```

Let's see how this works for the traveling salesman problem on a graph $G = (V, E)$ with edge lengths $d_e > 0$. A partial solution is a simple path $a \rightsquigarrow b$ passing through some vertices $S \subseteq V$, where S includes the endpoints a and b. We can denote such a partial solution by the tuple $[a, S, b]$—in fact, a will be fixed throughout the algorithm. The corresponding subproblem is to find the best completion of the tour, that is, the cheapest complementary path $b \rightsquigarrow a$ with intermediate nodes $V - S$. Notice that the initial problem is of the form $[a, \{a\}, a]$ for any $a \in V$ of our choosing.

At each step of the branch-and-bound algorithm, we extend a particular partial solution $[a, S, b]$ by a single edge (b, x), where $x \in V - S$. There can be up to $|V - S|$ ways to do this, and each of these branches leads to a subproblem of the form $[a, S \cup \{x\}, x]$.

How can we lower-bound the cost of completing a partial tour $[a, S, b]$? Many sophisticated methods have been developed for this, but let's look at a rather simple one. The remainder of the tour consists of a path through $V - S$, plus edges from a and b to $V - S$. Therefore, its cost is at least the sum of the following:

1. The lightest edge from a to $V - S$.
2. The lightest edge from b to $V - S$.
3. The minimum spanning tree of $V - S$.

(Do you see why?) And this lower bound can be computed quickly by a minimum spanning tree algorithm. Figure 9.2 runs through an example: each node of the tree represents a partial tour (specifically, the path from the root to that node) that at some stage is considered by the branch-and-bound procedure. Notice how just 28 partial solutions are considered, instead of the $7! = 5{,}040$ that would arise in a brute-force search.

9.2 Approximation algorithms

In an optimization problem we are given an instance I and are asked to find the optimum solution—the one with the maximum gain if we have a maximization problem like INDEPENDENT SET, or the minimum cost if we are dealing with a minimization problem such as the TSP. For every instance I, let us denote by $\text{OPT}(I)$ the value (benefit or cost) of the optimum solution. It makes the math a little simpler (and is not too far from the truth) to *assume that* $\text{OPT}(I)$ *is always a positive integer.*

We have already seen an example of a (famous) approximation algorithm in Section 5.4: the greedy scheme for SET COVER. For any instance I of size n, we showed that this greedy algorithm is guaranteed to quickly find a set cover of cardinality at most $\text{OPT}(I) \log n$. This $\log n$ factor is known as the approximation guarantee of the algorithm.

More generally, consider any minimization problem. Suppose now that we have an algorithm $\mathcal{A}$ for our problem which, given an instance I, returns a solution with value $\mathcal{A}(I)$. The *approximation ratio* of algorithm $\mathcal{A}$ is defined to be

$$\alpha_{\mathcal{A}} = \max_I \frac{\mathcal{A}(I)}{\text{OPT}(I)}.$$

Figure 9.2 (a) A graph and its optimal traveling salesman tour. (b) The branch-and-bound search tree, explored left to right. Boxed numbers indicate lower bounds on cost.

(a)

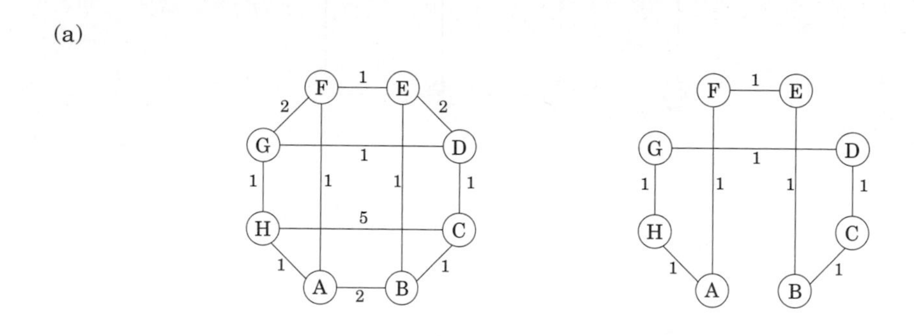

(b)

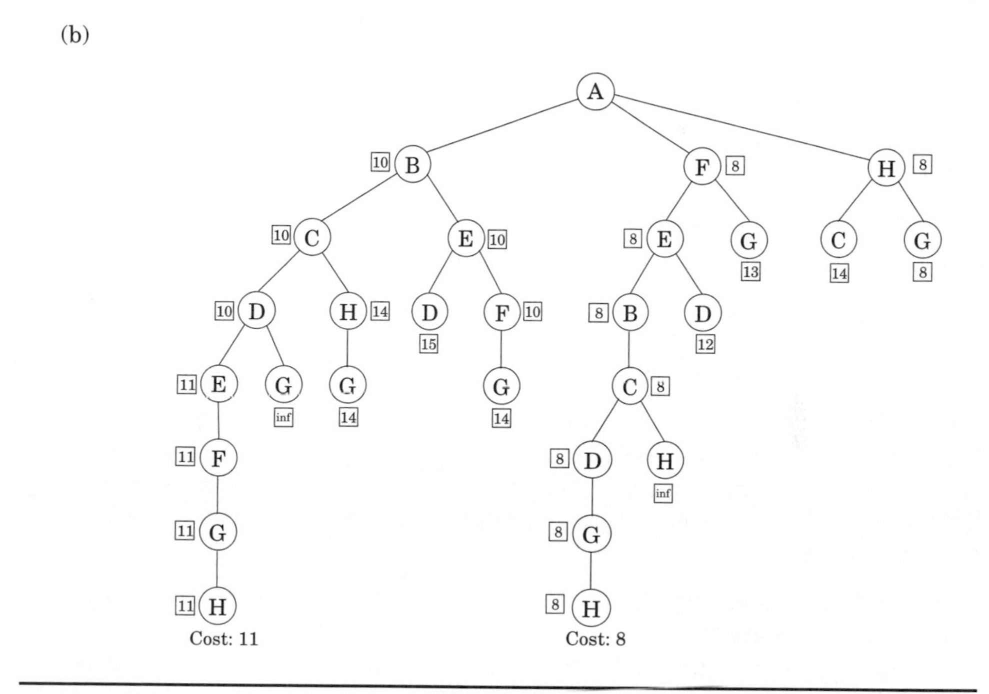

In other words, $\alpha_{\mathcal{A}}$ measures by the factor by which the output of algorithm $\mathcal{A}$ exceeds the optimal solution, on the worst-case input. The approximation ratio can also be defined for maximization problems, such as INDEPENDENT SET, in the same way—except that to get a number larger than 1 we take the reciprocal.

So, when faced with an **NP**-complete optimization problem, a reasonable goal is to look for an approximation algorithm $\mathcal{A}$ whose $\alpha_{\mathcal{A}}$ is as small as possible. But this kind of guarantee might seem a little puzzling: How can we come close to the optimum if we cannot determine the optimum? Let's look at a simple example.

Figure 9.3 A graph whose optimal vertex cover, shown shaded, is of size 8.

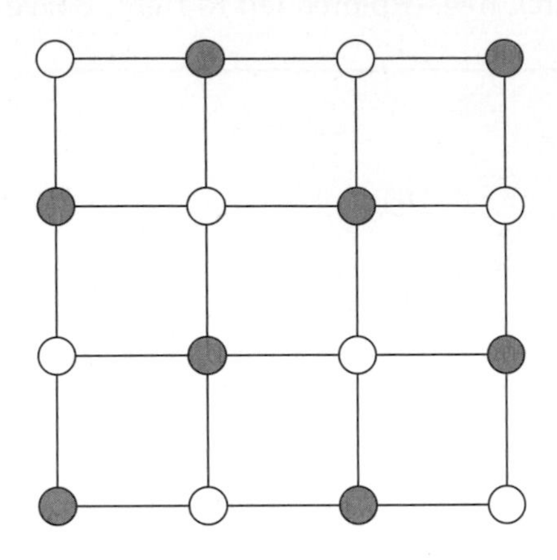

9.2.1 Vertex cover

We already know the VERTEX COVER problem is **NP**-hard.

> **Vertex Cover**
>
> *Input:* An undirected graph $G = (V, E)$.
>
> *Output:* A subset of the vertices $S \subseteq V$ that touches every edge.
>
> *Goal:* Minimize $|S|$.

See Figure 9.3 for an example.

Since VERTEX COVER is a special case of SET COVER, we know from Chapter 5 that it can be approximated within a factor of $O(\log n)$ by the greedy algorithm: repeatedly delete the vertex of highest degree and include it in the vertex cover. And there are graphs on which the greedy algorithm returns a vertex cover that is indeed $\log n$ times the optimum.

A better approximation algorithm for VERTEX COVER is based on the notion of a *matching*, a subset of edges that have no vertices in common (Figure 9.4). A matching is *maximal* if no more edges can be added to it. Maximal matchings will help us find good vertex covers, and moreover, they are easy to generate: repeatedly pick edges that are disjoint from the ones chosen already, until this is no longer possible.

What is the relationship between matchings and vertex covers? Here is the crucial fact: any vertex cover of a graph G must be at least as large as the number of edges in any matching in G; that is, *any matching provides a lower bound on* OPT. This is simply because each edge of the matching must be covered by one of its endpoints in any vertex cover! Finding such a lower bound is a key step in designing an approximation algorithm, because we must compare the quality of the solution found by our algorithm to OPT, which is **NP**-complete to compute.

Figure 9.4 (a) A matching, (b) its completion to a maximal matching, and (c) the resulting vertex cover.

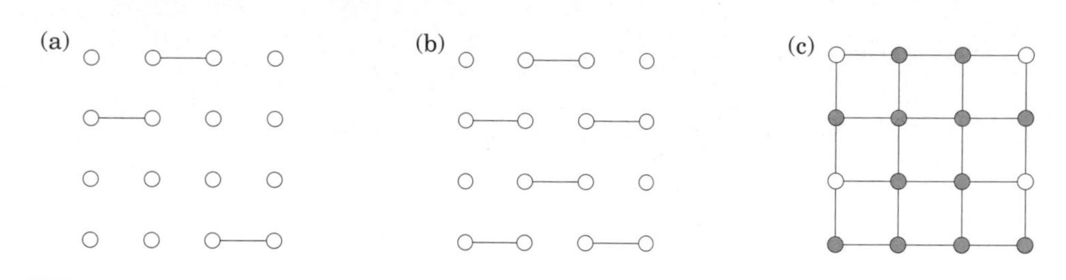

One more observation completes the design of our approximation algorithm: let S be a set that contains both endpoints of each edge in a maximal matching M. Then S must be a vertex cover—if it isn't, that is, if it doesn't touch some edge $e \in E$, then M could not possibly be maximal since we could still add e to it. But our cover S has $2|M|$ vertices. And from the previous paragraph we know that *any* vertex cover must have size at least $|M|$. So we're done.

Here's the algorithm for VERTEX COVER.

```
Find a maximal matching M ⊆ E
Return S = {all endpoints of edges in M}
```

This simple procedure always returns a vertex cover whose size is at most twice optimal!

In summary, even though we have no way of finding the best vertex cover, we can easily find another structure, a maximal matching, with two key properties:

1. Its size gives us a lower bound on the optimal vertex cover.
2. It can be used to build a vertex cover, whose size can be related to that of the optimal cover using property 1.

Thus, this simple algorithm has an approximation ratio of $\alpha_{\mathcal{A}} \leq 2$. In fact, it is not hard to find examples on which it does make a 100% error; hence $\alpha_{\mathcal{A}} = 2$.

9.2.2 Clustering

We turn next to a *clustering* problem, in which we have some data (text documents, say, or images, or speech samples) that we want to divide into groups. It is often useful to define "distances" between these data points, numbers that capture how close or far they are from one another. Often the data are true points in some high-dimensional space and the distances are the usual Euclidean distance; in other cases, the distances are the result of some "similarity tests" to which we have subjected the data points. Assume that we have such distances and that they satisfy the usual *metric* properties:

1. $d(x, y) \geq 0$ for all x, y.

Figure 9.5 Some data points and the optimal $k = 4$ clusters.

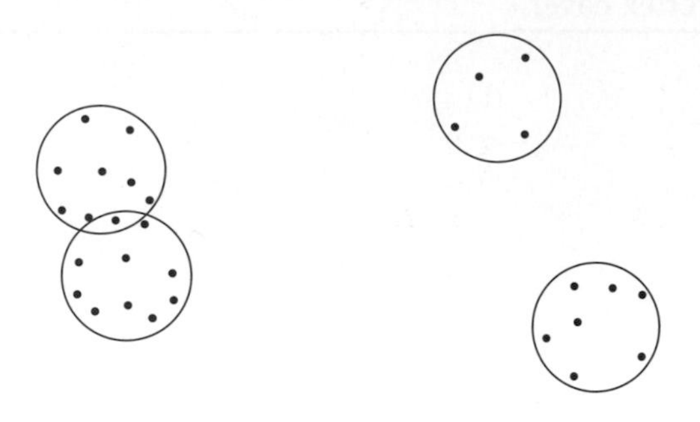

2. $d(x, y) = 0$ if and only if $x = y$.
3. $d(x, y) = d(y, x)$.
4. (Triangle inequality) $d(x, y) \leq d(x, z) + d(z, y)$.

We would like to partition the data points into groups that are compact in the sense of having small diameter.

k-Cluster

Input: Points $X = \{x_1, \ldots, x_n\}$ with underlying distance metric $d(\cdot, \cdot)$; integer k.

Output: A partition of the points into k clusters $C_1, \ldots, C_k$.

Goal: Minimize the diameter of the clusters,

$$\max_j \max_{x_a, x_b \in C_j} d(x_a, x_b).$$

One way to visualize this task is to imagine n points in space, which are to be covered by k spheres of equal size. What is the smallest possible diameter of the spheres? Figure 9.5 shows an example.

This problem is **NP**-hard, but has a very simple approximation algorithm. The idea is to pick k of the data points as *cluster centers* and to then assign each of the remaining points to the center closest to it, thus creating k clusters. The centers are picked one at a time, using an intuitive rule: always pick the next center to be as far as possible from the centers chosen so far (see Figure 9.6).

```
Pick any point μ_1 ∈ X as the first cluster center
for i = 2 to k:
  Let μ_i be the point in X farthest from μ_1, ..., μ_{i-1}
  (i.e., that maximizes min_{j<i} d(·, μ_j))
Create k clusters: C_i = {all x ∈ X whose closest center is μ_i}
```

It's clear that this algorithm returns a valid partition. What's more, the resulting diameter is guaranteed to be at most twice optimal.

Figure 9.6 (a) Four centers chosen by farthest-first traversal. (b) The resulting clusters.

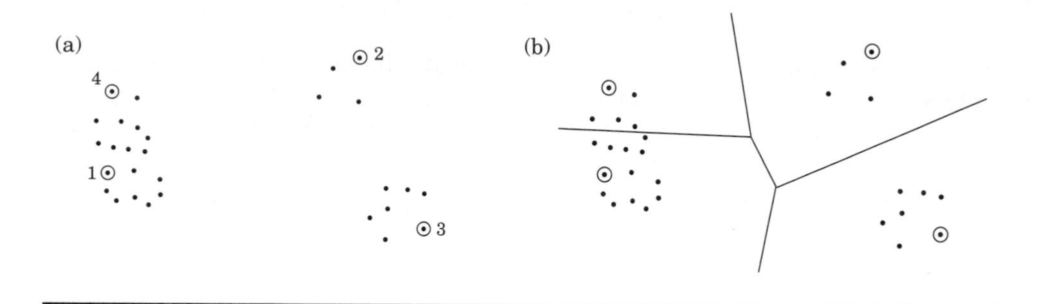

Here's the argument. Let $x \in X$ be the point farthest from $\mu_1, \ldots, \mu_k$ (in other words the next center we would have chosen, if we wanted $k+1$ of them), and let r be its distance to its closest center. Then every point in X must be within distance r of its cluster center. By the triangle inequality, this means that every cluster has diameter at most $2r$.

But how does r relate to the diameter of the optimal clustering? Well, we have identified $k+1$ points $\{\mu_1, \mu_2, \ldots, \mu_k, x\}$ that are all at a distance at least r from each other (why?). *Any* partition into k clusters must put two of these points in the same cluster and must therefore have diameter at least r.

This algorithm has a certain high-level similarity to our scheme for VERTEX COVER. Instead of a maximal matching, we use a different easily computable structure—a set of k points that cover all of X within some radius r, while at the same time being mutually separated by a distance of at least r. This structure is used both to generate a clustering and to give a lower bound on the optimal clustering.

We know of no better approximation algorithm for this problem.

9.2.3 TSP

The triangle inequality played a crucial role in making the k-CLUSTER problem approximable. It also helps with the TRAVELING SALESMAN PROBLEM: if the distances between cities satisfy the metric properties, then there is an algorithm that outputs a tour of length at most 1.5 times optimal. We'll now look at a slightly weaker result that achieves a factor of 2.

Continuing with the thought processes of our previous two approximation algorithms, we can ask whether there is some structure that is easy to compute and that is plausibly related to the best traveling salesman tour (as well as providing a good lower bound on OPT). A little thought and experimentation reveals the answer to be the *minimum spanning tree*.

Let's understand this relation. Removing any edge from a traveling salesman tour leaves a path through all the vertices, which is a spanning tree. Therefore,

$$\text{TSP cost} \geq \text{cost of this path} \geq \text{MST cost}.$$

Now, we somehow need to use the MST to build a traveling salesman tour. If we can use each edge *twice*, then by following the shape of the MST we end up with a tour that visits all the cities, some of them more than once. Here's an example, with the MST on the left and the resulting tour on the right (the numbers show the order in which the edges are taken).

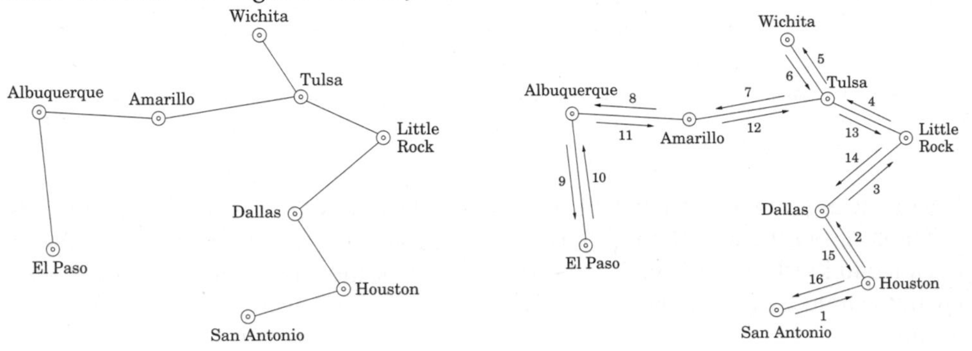

Therefore, this tour has a length at most twice the MST cost, which as we've already seen is at most twice the TSP cost.

This is the result we wanted, but we aren't quite done because our tour visits some cities multiple times and is therefore not legal. To fix the problem, the tour should simply skip any city it is about to revisit, and instead move directly to the next *new* city in its list:

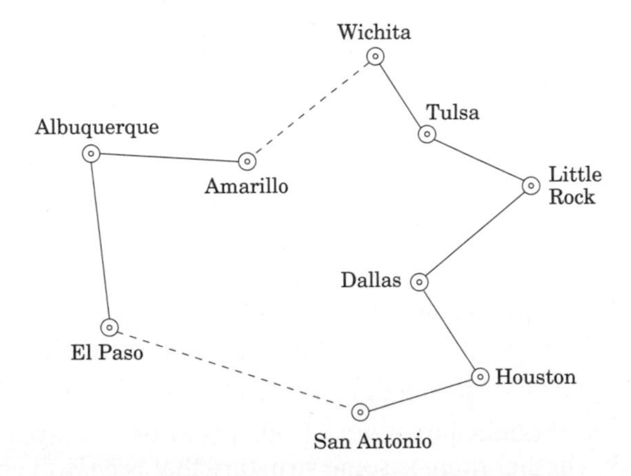

By the triangle inequality, these bypasses can only make the overall tour shorter.

General TSP

General TSP

But what if we are interested in instances of TSP that do not satisfy the triangle inequality? It turns out that this is a *much* harder problem to approximate.

Here is why: Recall that on page 260 we gave a polynomial-time reduction which given any graph G and any integer $C > 0$ produces an instance $I(G, C)$ of the TSP such that:

(i) If G has a Rudrata path, then $\text{OPT}(I(G, C)) = n$, the number of vertices in G.
(ii) If G has no Rudrata path, then $\text{OPT}(I(G, C)) \geq n + C$.

This means that even an approximate solution to TSP would enable us to solve RUDRATA PATH! Let's work out the details.

Consider an approximation algorithm $\mathcal{A}$ for TSP and let $\alpha_{\mathcal{A}}$ denote its approximation ratio. From any instance G of RUDRATA PATH, we will create an instance $I(G, C)$ of TSP using the specific constant $C = n\alpha_{\mathcal{A}}$. What happens when algorithm $\mathcal{A}$ is run on this TSP instance? In case (i), it must output a tour of length at most $\alpha_{\mathcal{A}}\text{OPT}(I(G, C)) = n\alpha_{\mathcal{A}}$, whereas in case (ii) it must output a tour of length at least $\text{OPT}(I(G, C)) > n\alpha_{\mathcal{A}}$. Thus we can figure out whether G has a Rudrata path! Here is the resulting procedure:

```
Given any graph G:
  compute I(G,C) (with C = n·α_A) and run algorithm A on it
  if the resulting tour has length ≤ nα_A:
    conclude that G has a Rudrata path
  else: conclude that G has no Rudrata path
```

This tells us whether or not G has a Rudrata path; by calling the procedure a polynomial number of times, we can find the actual path (Exercise 8.2).

We've shown that if TSP has a polynomial-time approximation algorithm, then there is a polynomial algorithm for the **NP**-complete RUDRATA PATH problem. So, unless **P** = **NP**, there cannot exist an efficient approximation algorithm for the TSP.

9.2.4 Knapsack

Our last approximation algorithm is for a maximization problem and has a very impressive guarantee: given any $\epsilon > 0$, it will return a solution of value at least $(1 - \epsilon)$ times the optimal value, in time that scales only polynomially in the input size and in $1/\epsilon$.

The problem is KNAPSACK, which we first encountered in Chapter 6. There are n items, with weights $w_1, \ldots, w_n$ and values $v_1, \ldots, v_n$ (all positive integers), and the goal is to pick the most valuable combination of items subject to the constraint that their total weight is at most W.

Earlier we saw a dynamic programming solution to this problem with running time $O(nW)$. Using a similar technique, a running time of $O(nV)$ can also be achieved, where V is the sum of the values. Neither of these running times is polynomial, because W and V can be very large, exponential in the size of the input.

Let's consider the $O(nV)$ algorithm. In the bad case when V is large, what if we simply scale down all the values in some way? For instance, if

$$v_1 = 117{,}586{,}003,\ v_2 = 738{,}493{,}291,\ v_3 = 238{,}827{,}453,$$

we could simply knock off some precision and instead use 117, 738, and 238. This doesn't change the problem all that much and will make the algorithm much, much faster!

Now for the details. Along with the input, the user is assumed to have specified some approximation factor $\epsilon > 0$.

```
Discard any item with weight > W
Let v_max = max_i v_i
Rescale values v̂_i = ⌊v_i · n/(ε v_max)⌋
Run the dynamic programming algorithm with values {v̂_i}
Output the resulting choice of items
```

Let's see why this works. First of all, since the rescaled values $\widehat{v}_i$ are all at most n/ϵ, the dynamic program is efficient, running in time $O(n^3/\epsilon)$.

Now suppose the optimal solution to the original problem is to pick some subset of items S, with total value K^*. The rescaled value of this same assignment is

$$\sum_{i \in S} \widehat{v}_i = \sum_{i \in S} \left\lfloor v_i \cdot \frac{n}{\epsilon v_{\max}} \right\rfloor \geq \sum_{i \in S} \left(v_i \cdot \frac{n}{\epsilon v_{\max}} - 1 \right) \geq K^* \cdot \frac{n}{\epsilon v_{\max}} - n.$$

Therefore, the optimal assignment for the shrunken problem, call it $\widehat{S}$, has a rescaled value of at least this much. In terms of the original values, assignment $\widehat{S}$ has a value of at least

$$\sum_{i \in \widehat{S}} v_i \geq \sum_{i \in \widehat{S}} \widehat{v}_i \cdot \frac{\epsilon v_{\max}}{n} \geq \left(K^* \cdot \frac{n}{\epsilon v_{\max}} - n \right) \cdot \frac{\epsilon v_{\max}}{n} = K^* - \epsilon v_{\max} \geq K^*(1 - \epsilon).$$

9.2.5 The approximability hierarchy

Given any **NP**-complete optimization problem, we seek the best approximation algorithm possible. Failing this, we try to prove *lower bounds* on the approximation ratios that are achievable in polynomial time (we just carried out such a proof for the general TSP). All told, **NP**-complete optimization problems are classified as follows:

- Those for which, like the TSP, no finite approximation ratio is possible.
- Those for which an approximation ratio is possible, but there are limits to how small this can be. VERTEX COVER, k-CLUSTER, and the TSP with triangle inequality belong here. (For these problems we have not established limits to their approximability, but these limits do exist, and their proofs constitute some of the most sophisticated results in this field.)
- Down below we have a more fortunate class of **NP**-complete problems for which approximability has no limits, and polynomial approximation algorithms with error ratios arbitrarily close to zero exist. KNAPSACK resides here.

- Finally, there is another class of problems, between the first two given here, for which the approximation ratio is about $\log n$. SET COVER is an example.

(A humbling reminder: All this is contingent upon the assumption **P** $\neq$ **NP.** Failing this, this hierarchy collapses down to **P**, and all **NP**-complete optimization problems can be solved exactly in polynomial time.)

A final point on approximation algorithms: often these algorithms, or their variants, perform much better on typical instances than their worst-case approximation ratio would have you believe.

9.3 Local search heuristics

Our next strategy for coping with **NP**-completeness is inspired by evolution (which is, after all, the world's best-tested optimization procedure)—by its incremental process of introducing small mutations, trying them out, and keeping them if they work well. This paradigm is called *local search* and can be applied to any optimization task. Here's how it looks for a minimization problem.

```
let s be any initial solution
while there is some solution s' in the neighborhood of s
   for which cost(s') < cost(s): replace s by s'
return s
```

On each iteration, the current solution is replaced by a better one close to it, in its *neighborhood*. This neighborhood structure is something we impose upon the problem and is the central design decision in local search. As an illustration, let's revisit the traveling salesman problem.

9.3.1 Traveling salesman, once more

Assume we have all interpoint distances between n cities, giving a search space of $(n-1)!$ different tours. What is a good notion of neighborhood?

The most obvious notion is to consider two tours as being close if they differ in just a few edges. They can't differ in just one edge (do you see why?), so we will consider differences of two edges. We define the *2-change* neighborhood of tour s as being the set of tours that can be obtained by removing two edges of s and then putting in two other edges. Here's an example of a local move:

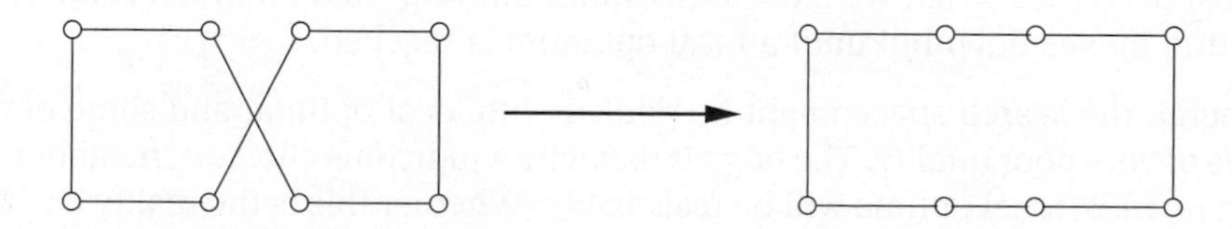

We now have a well-defined local search procedure. How does it measure up under our two standard criteria for algorithms—what is its overall running time, and does it always return the best solution?

Embarrassingly, neither of these questions has a satisfactory answer. Each iteration is certainly fast, because a tour has only $O(n^2)$ neighbors. However, it is not clear how many iterations will be needed: whether for instance, there might be an exponential number of them. Likewise, all we can easily say about the final tour is that it is *locally optimal*—that is, it is superior to the tours in its immediate neighborhood. There might be better solutions further away. For instance, the following picture shows a possible final answer that is clearly suboptimal; the range of local moves is simply too limited to improve upon it.

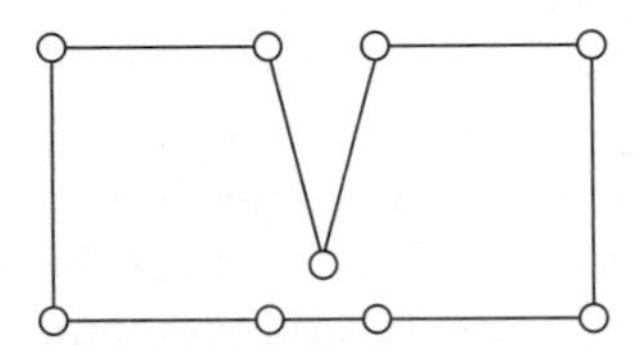

To overcome this, we may try a more generous neighborhood, for instance *3-change*, consisting of tours that differ on up to three edges. And indeed, the preceding bad case gets fixed:

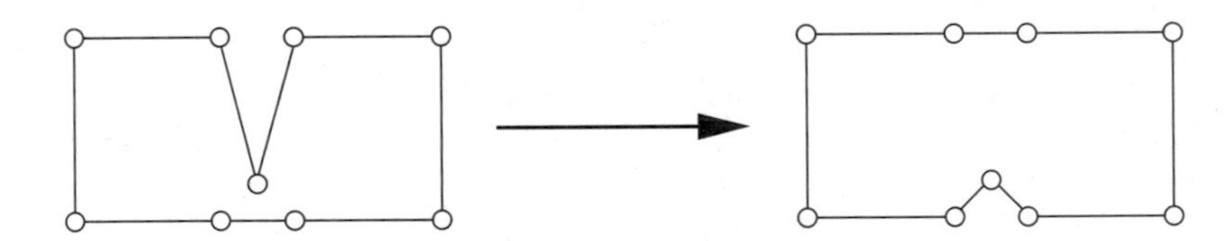

But there is a downside, in that the size of a neighborhood becomes $O(n^3)$, making each iteration more expensive. Moreover, there may still be suboptimal local minima, although fewer than before. To avoid these, we would have to go up to *4-change*, or higher. In this manner, efficiency and quality often turn out to be competing considerations in a local search. Efficiency demands neighborhoods that can be searched quickly, but smaller neighborhoods can increase the abundance of low-quality local optima. The appropriate compromise is typically determined by experimentation.

Figure 9.7 shows a specific example of local search at work. Figure 9.8 is a more abstract, stylized depiction of local search. The solutions crowd the unshaded area, and cost decreases when we move downward. Starting from an initial solution, the algorithm moves downhill until a local optimum is reached.

In general, the search space might be riddled with local optima, and some of them may be of very poor quality. The hope is that with a judicious choice of neighborhood structure, most local optima will be reasonable. Whether this is the reality or merely

Figure 9.7 (a) Nine American cities. (b) Local search, starting at a random tour, and using 3-change. The traveling salesman tour is found after three moves.

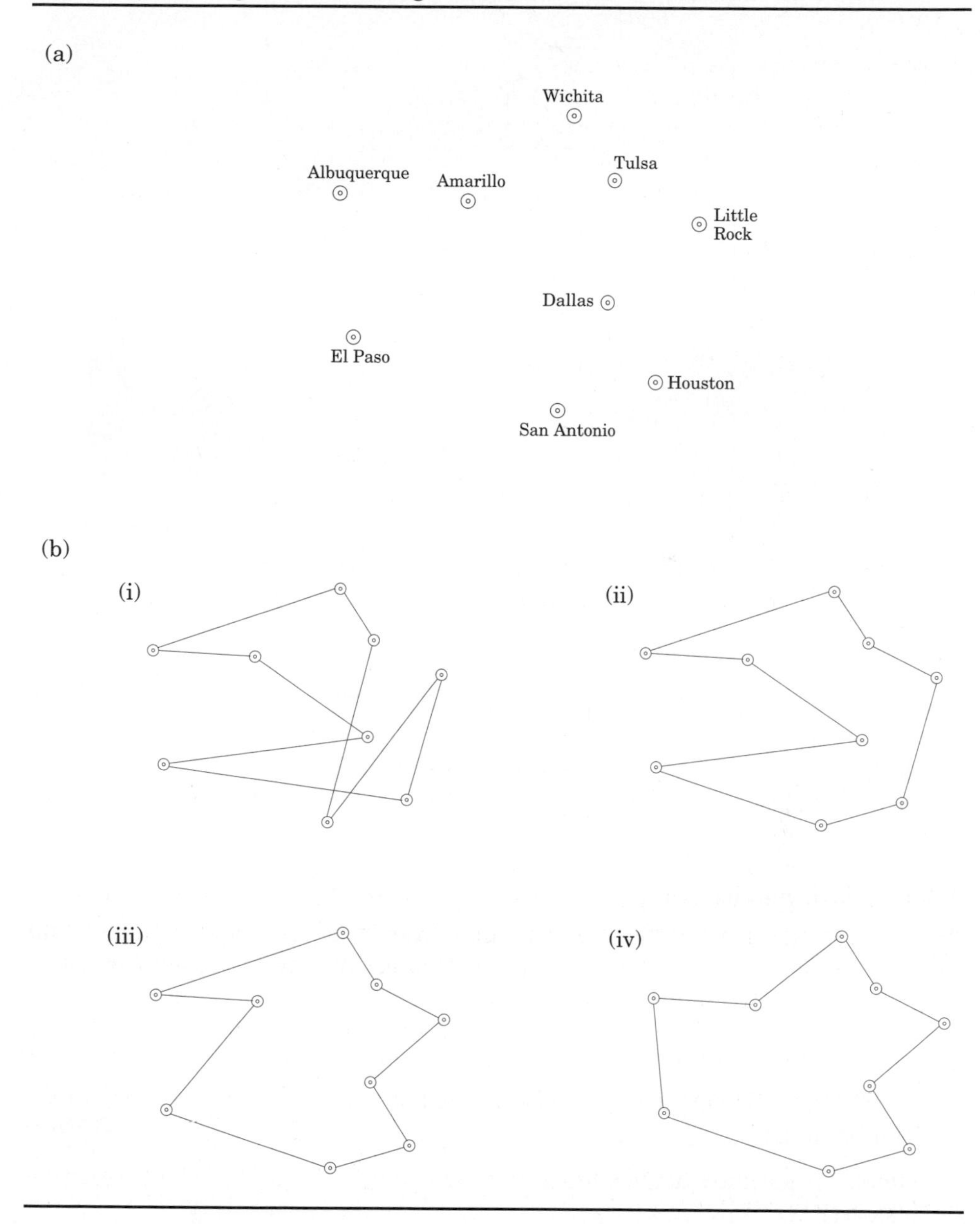

Figure 9.8 Local search.

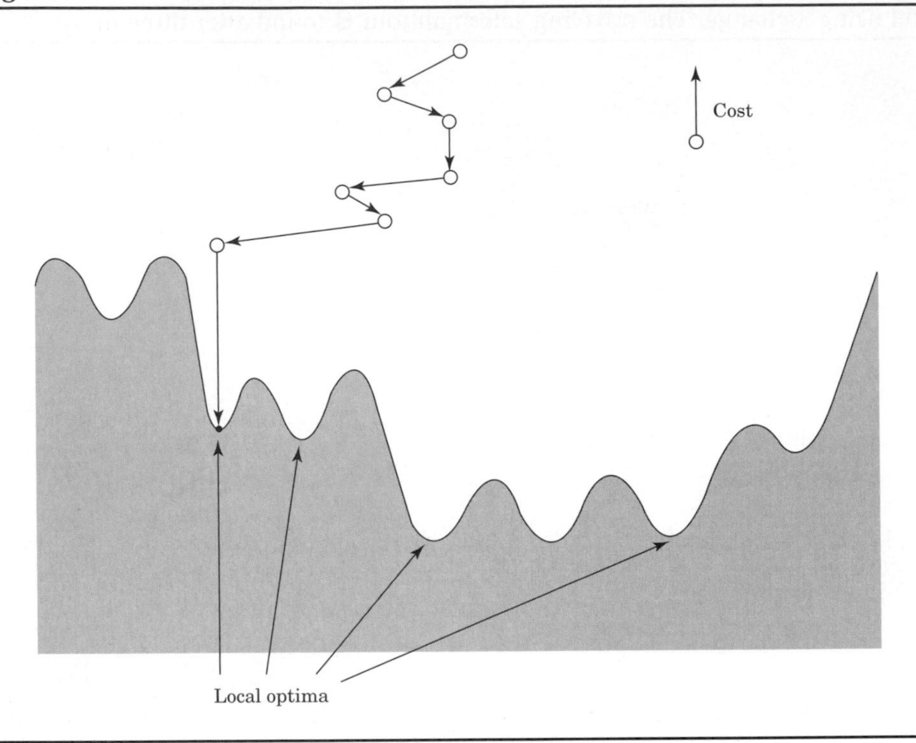

misplaced faith, it is an empirical fact that local search algorithms are the top performers on a broad range of optimization problems. Let's look at another such example.

9.3.2 Graph partitioning

The problem of graph partitioning arises in a diversity of applications, from circuit layout to program analysis to image segmentation. We saw a special case of it, BALANCED CUT, in Chapter 8.

Graph Partitioning

Input: An undirected graph $G = (V, E)$ with nonnegative edge weights; a real number $\alpha \in (0, 1/2]$.

Output: A partition of the vertices into two groups A and B, each of size at least $\alpha|V|$.

Goal: Minimize the capacity of the cut (A, B).

Figure 9.9 shows an example in which the graph has 16 nodes, all edge weights are 0 or 1, and the optimal solution has cost 0. Removing the restriction on the sizes of A and B would give the MINIMUM CUT problem, which we know to be efficiently solvable

Figure 9.9 An instance of GRAPH PARTITIONING, with the optimal partition for $\alpha = 1/2$. Vertices on one side of the cut are shaded.

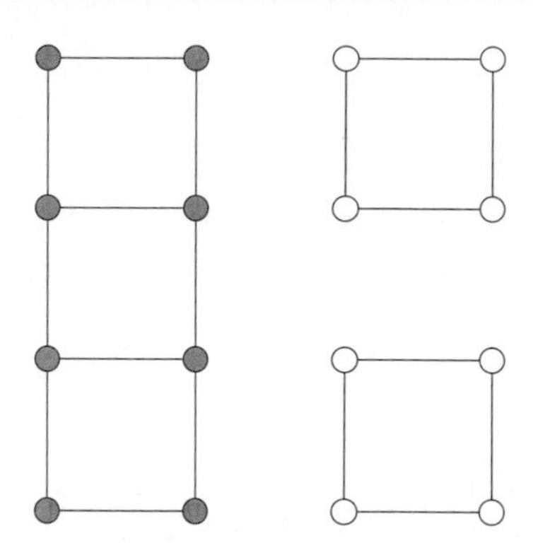

using flow techniques. The present variant, however, is **NP**-hard. In designing a local search algorithm, it will be a big convenience to focus on the special case $\alpha = 1/2$, in which A and B are forced to contain exactly half the vertices. The apparent loss of generality is purely cosmetic, as GRAPH PARTITIONING reduces to this particular case.

We need to decide upon a neighborhood structure for our problem, and there is one obvious way to do this. Let (A, B), with $|A| = |B|$, be a candidate solution; we will define its neighbors to be all solutions obtainable by swapping one pair of vertices across the cut, that is, all solutions of the form $(A - \{a\} + \{b\}, B - \{b\} + \{a\})$ where $a \in A$ and $b \in B$. Here's an example of a local move:

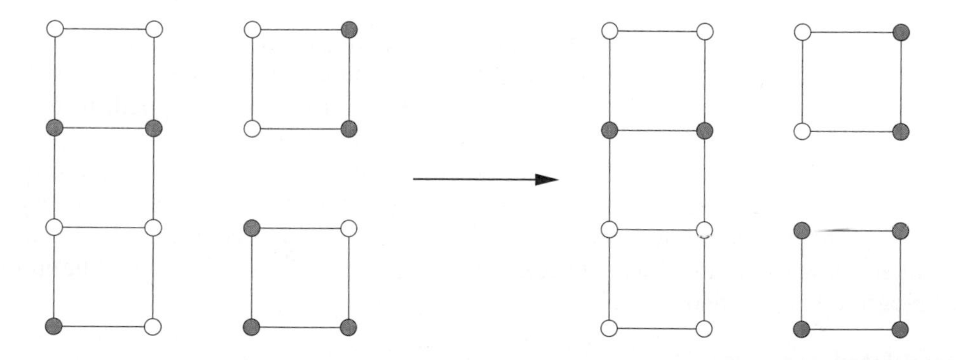

We now have a reasonable local search procedure, and we could just stop here. But there is still a lot of room for improvement in terms of the *quality* of the solutions produced. The search space includes some local optima that are quite far from the global solution. Here's one which has cost 2.

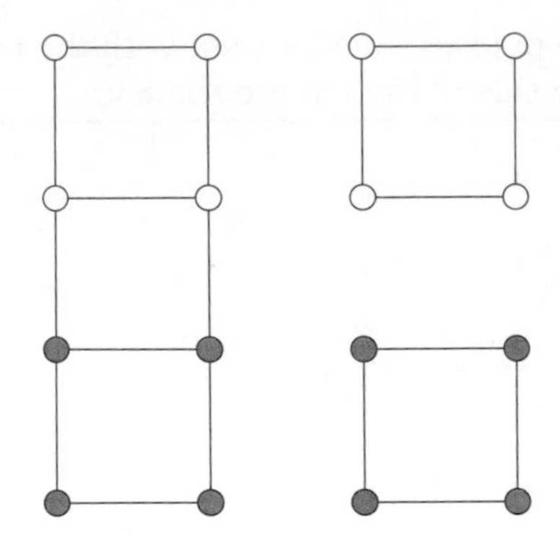

What can be done about such suboptimal solutions? We could expand the neighborhood size to allow two swaps at a time, but this particular bad instance would still stubbornly resist. Instead, let's look at some other generic schemes for improving local search procedures.

9.3.3 Dealing with local optima

Randomization and restarts

Randomization can be an invaluable ally in local search. It is typically used in two ways: to pick a random initial solution, for instance a random graph partition; and to choose a local move when several are available.

When there are many local optima, randomization is a way of making sure that there is at least some probability of getting to the right one. The local search can then be repeated several times, with a different random seed on each invocation, and the best solution returned. If the probability of reaching a good local optimum on any given run is p, then within $O(1/p)$ runs such a solution is likely to be found (recall Exercise 1.34).

Figure 9.10 shows a small instance of graph partitioning, along with the search space of solutions. There are a total of $\binom{8}{4} = 70$ possible states, but since each of them has an identical twin in which the left and right sides of the cut are flipped, in effect there are just 35 solutions. In the figure, these are organized into seven groups for readability. There are five local optima, of which four are bad, with cost 2, and one is good, with cost 0. If local search is started at a random solution, and at each step a random neighbor of lower cost is selected, then the search is at most four times as likely to wind up in a bad solution than a good one. Thus only a small handful of repetitions is needed.

Simulated annealing

In the example of Figure 9.10, each run of local search has a reasonable chance of finding the global optimum. This isn't always true. As the problem size grows, the ratio of bad to good local optima often increases, sometimes to the point of being exponentially large. In such cases, simply repeating the local search a few times is ineffective.

Figure 9.10 The search space for a graph with eight nodes. The space contains 35 solutions, which have been partitioned into seven groups for clarity. An example of each is shown. There are five local optima.

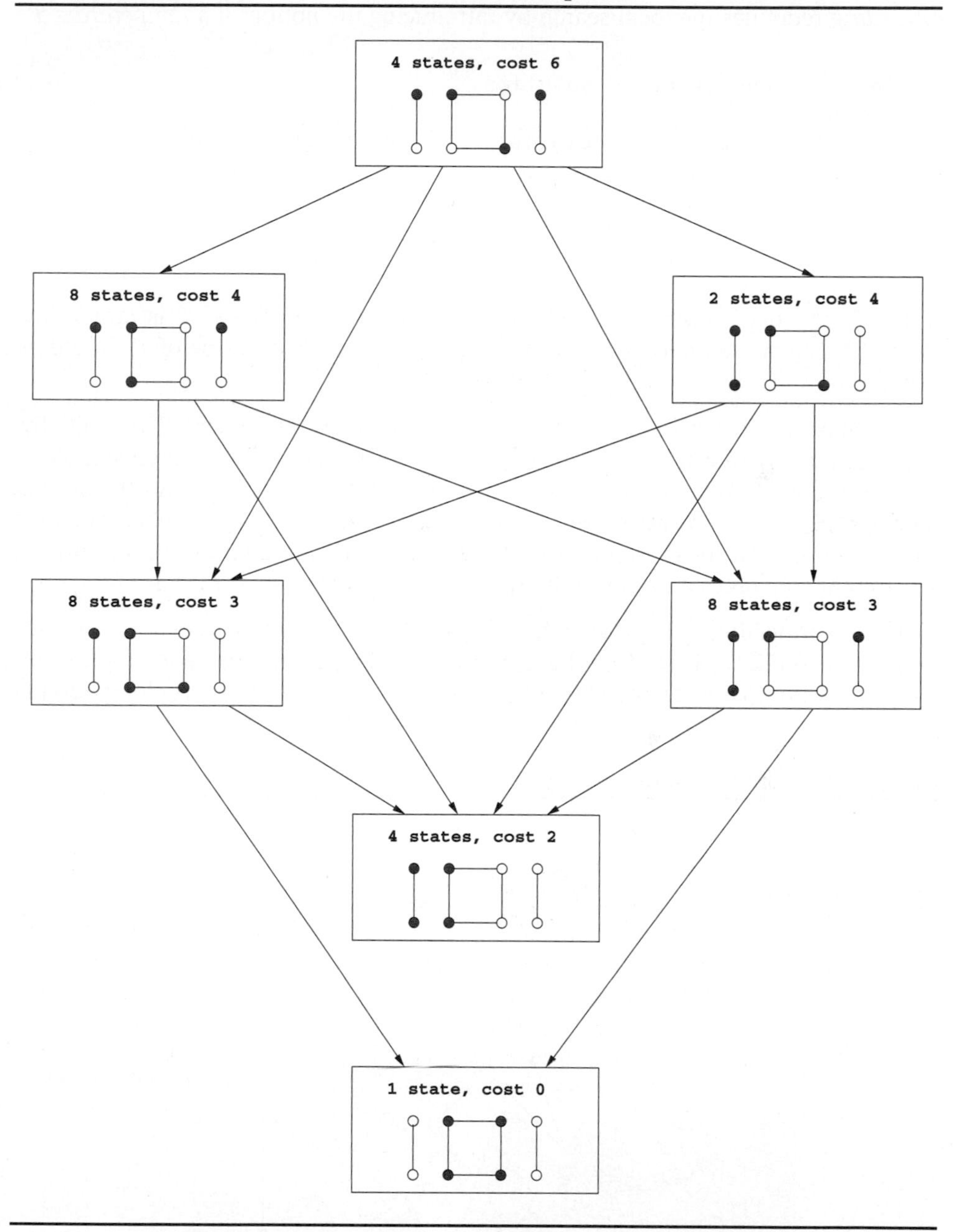

A different avenue of attack is to occasionally allow moves that actually increase the cost, in the hope that they will pull the search out of dead ends. This would be very useful at the bad local optima of Figure 9.10, for instance. The method of *simulated annealing* redefines the local search by introducing the notion of a *temperature* T.

```
let s be any starting solution
repeat
   randomly choose a solution s′ in the neighborhood of s
   if Δ = cost(s′) − cost(s) is negative:
      replace s by s′
   else:
      replace s by s′ with probability e^{−Δ/T}.
```

If T is zero, this is identical to our previous local search. But if T is large, then moves that increase the cost are occasionally accepted. What value of T should be used?

The trick is to start with T large and then gradually reduce it to zero. Thus initially, the local search can wander around quite freely, with only a mild preference for low-cost solutions. As time goes on, this preference becomes stronger, and the system mostly sticks to the lower-cost region of the search space, with occasional excursions out of it to escape local optima. Eventually, when the temperature drops further, the system converges on a solution. Figure 9.11 shows this process schematically.

Simulated annealing is inspired by the physics of crystallization. When a substance is to be crystallized, it starts in liquid state, with its particles in relatively unconstrained motion. Then it is slowly cooled, and as this happens, the particles gradually

Figure 9.11 Simulated annealing.

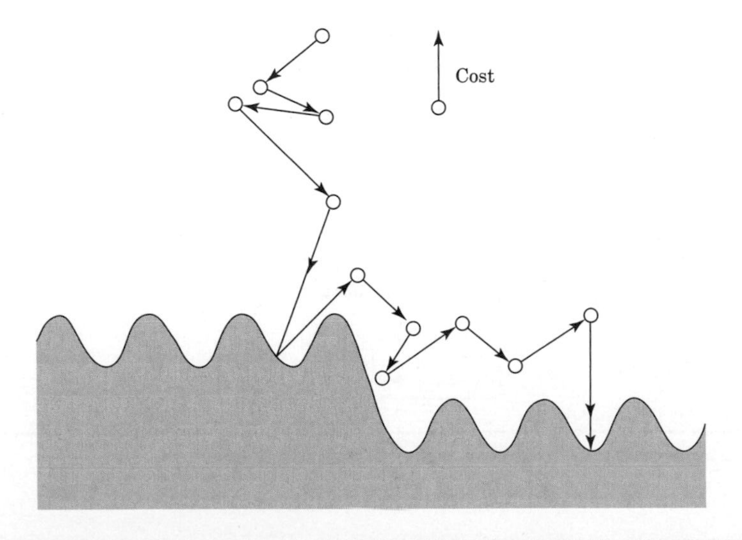

move into more regular configurations. This regularity becomes more and more pronounced until finally a crystal lattice is formed.

The benefits of simulated annealing come at a significant cost: because of the changing temperature and the initial freedom of movement, many more local moves are needed until convergence. Moreover, it is quite an art to choose a good timetable by which to decrease the temperature, called an *annealing schedule*. But in many cases where the quality of solutions improves significantly, the tradeoff is worthwhile.

Exercises

9.1. In the backtracking algorithm for SAT, suppose that we always `choose` a subproblem (CNF formula) that has a clause that is as small as possible; and we `expand` it along a variable that appears in this small clause. Show that if the input formula only contains clauses with two literals (that is, it is an instance of 2SAT), then a satisfying assignment, if one exists, will be found in polynomial time.

9.2. Devise a backtracking algorithm for the RUDRATA PATH problem from a fixed vertex s. To fully specify such an algorithm you must define:

(a) What is a subproblem?

(b) How to `choose` a subproblem.

(c) How to `expand` a subproblem.

Argue briefly why your choices are reasonable.

9.3. Devise a branch-and-bound algorithm for the SET COVER problem. This entails deciding:

(a) What is a subproblem?

(b) How do you `choose` a subproblem to expand?

(c) How do you `expand` a subproblem?

(d) What is an appropriate `lowerbound`?

Do you think that your choices above will work well on typical instances of the problem? Why?

9.4. Given an undirected graph $G = (V, E)$ in which each node has degree $\leq d$, show how to efficiently find an independent set whose size is at least $1/(d+1)$ times that of the largest independent set.

9.5. *Local search for minimum spanning trees.* Consider the set of all spanning trees (not just minimum ones) of a weighted, connected, undirected graph $G = (V, E)$.

Recall from Section 5.1 that adding an edge e to a spanning tree T creates an unique cycle, and subsequently removing any other edge $e' \neq e$ from this cycle gives back a different spanning tree T'. We will say that T and T' differ by a single *edge swap* (e, e') and that they are *neighbors*.

(a) Show that it is possible to move from any spanning tree T to any other spanning tree T' by performing a series of edge-swaps, that is, by moving from neighbor to neighbor. At most how many edge-swaps are needed?

(b) Show that if T' is an MST, then it is possible to choose these swaps so that the costs of the spanning trees encountered along the way are nonincreasing. In other words, if the sequence of spanning trees encountered is

$$T = T_0 \to T_1 \to T_2 \to \cdots \to T_k = T',$$

then $\text{cost}(T_{i+1}) \leq \text{cost}(T_i)$ for all $i < k$.

(c) Consider the following local search algorithm which is given as input an undirected graph with distinct edge weights.

```
Let T be any spanning tree of G
while there is an edge-swap (e, e') which reduces
cost(T):
    T ← T + e − e'
return T
```

Show that this procedure always returns a minimum spanning tree. At most how many iterations does it take?

9.6. In the MINIMUM STEINER TREE problem, the input consists of: a complete graph $G = (V, E)$ with distances d_{uv} between all pairs of nodes; and a distinguished set of *terminal nodes* $V' \subseteq V$. The goal is to find a minimum-cost tree that includes the vertices V'. This tree may or may not include nodes in $V - V'$.

Suppose the distances in the input are a metric (recall the definition on page 279). Show that an efficient ratio-2 approximation algorithm for MINIMUM STEINER TREE can be obtained by ignoring the nonterminal nodes and simply returning the minimum spanning tree on V'. (*Hint:* Recall our approximation algorithm for the TSP.)

9.7. In the MULTIWAY CUT problem, the input is an undirected graph $G = (V, E)$ and a set of terminal nodes $s_1, s_2, \ldots, s_k \in V$. The goal is to find the minimum set of edges in E whose removal leaves all terminals in different components.

(a) Show that this problem can be solved exactly in polynomial time when $k = 2$.

(b) Give an approximation algorithm with ratio at most 2 for the case $k = 3$.

(c) Design a local search algorithm for multiway cut.

9.8. In the MAX SAT problem, we are given a set of clauses, and we want to find an assignment that satisfies as many of them as possible.

(a) Show that if this problem can be solved in polynomial time, then so can SAT.

(b) Here's a very naive algorithm.

```
for each variable:
    set its value to either 0 or 1 by flipping a coin
```

Suppose the input has m clauses, of which the jth has k_j literals. Show that the *expected* number of clauses satisfied by this simple algorithm is

$$\sum_{j=1}^{m}\left(1-\frac{1}{2^{k_j}}\right) \geq \frac{m}{2}.$$

In other words, this is a 2-approximation in expectation! And if the clauses all contain k literals, then this approximation factor improves to $1+1/(2^k-1)$.

(c) Can you make this algorithm deterministic? (*Hint:* Instead of flipping a coin for each variable, select the value that satisfies the most as-yet-unsatisfied clauses. What fraction of the clauses is satisfied in the end?)

9.9. In the MAXIMUM CUT problem we are given an undirected graph $G = (V, E)$ with a weight $w(e)$ on each edge, and we wish to separate the vertices into two sets S and $V - S$ so that the total weight of the edges between the two sets is as *large* as possible.

For each $S \subseteq V$, define $w(S)$ to be the sum of all w_{uv} over all edges $\{u, v\}$ such that $|S \cap \{u, v\}| = 1$. Obviously, MAX CUT is about maximizing $w(S)$ over all subsets of V.

Consider the following local search algorithm for MAX CUT:

```
start with any S ⊆ V
while there is a subset S' ⊆ V such that
  |(S' − S) ∪ (S − S')| = 1 and w(S') > w(S) do:
    set S = S'
```

(a) Show that this is an approximation algorithm for MAX CUT with ratio 2.

(b) But is it a polynomial-time algorithm?

9.10. Let us call a local search algorithm *exact* when it always produces the optimum solution. For example, the local search algorithm for the minimum spanning tree problem introduced in Problem 9.5 is exact. For another example, simplex can be considered an exact local search algorithm for linear programming.

(a) Show that the 2-change local search algorithm for the TSP is not exact.

(b) Repeat for the $\lceil \frac{n}{2} \rceil$-change local search algorithm, where n is the number of cities.

(c) Show that the $(n-1)$-change local search algorithm is exact.

(d) If A is an optimization problem, define A-IMPROVEMENT to be the following search problem: Given an instance x of A and a solution s of A, find another solution of x with better cost (or report that none exists, and thus s is optimum). For example, in TSP IMPROVEMENT we are given a distance matrix and a tour, and we are asked to find a better tour. It turns out that TSP IMPROVEMENT is **NP**-complete, and so is SET COVER IMPROVEMENT. Prove the latter.

(e) We say that a local search algorithm has *polynomial iteration* if each execution of the loop requires polynomial time. For example, the obvious implementations of the $(n-1)$-change local search algorithm for the TSP defined above do not have polynomial iteration. Show that, unless **P** = **NP**, there is no exact local search algorithm with polynomial iteration for the TSP and SET COVER problems.

Chapter 10

Quantum algorithms

This book started with the world's oldest and most widely used algorithms (the ones for adding and multiplying numbers) and an ancient hard problem (FACTORING). In this last chapter the tables are turned: we present one of the latest algorithms—and it is an efficient algorithm for FACTORING!

There is a catch, of course: this algorithm needs a *quantum computer* to execute.

Quantum physics is a beautiful and mysterious theory that describes Nature in the small, at the level of elementary particles. One of the major discoveries of the nineties was that quantum computers—computers based on quantum physics principles—are radically different from those that operate according to the more familiar principles of classical physics. Surprisingly, they can be exponentially more powerful: as we shall see, quantum computers can solve FACTORING in polynomial time! As a result, in a world with quantum computers, the systems that currently safeguard business transactions on the Internet (and are based on the RSA cryptosystem) will no longer be secure.

10.1 Qubits, superposition, and measurement

In this section we introduce the basic features of quantum physics that are necessary for understanding how quantum computers work.[1]

In ordinary computer chips, bits are physically represented by low and high voltages on wires. But there are many other ways a bit could be stored—for instance, in the state of a hydrogen atom. The single electron in this atom can either be in the ground state (the lowest energy configuration) or it can be in an excited state (a high energy configuration). We can use these two states to encode for bit values 0 and 1, respectively.

Let us now introduce some quantum physics notation. We denote the ground state of our electron by $|0\rangle$, since it encodes for bit value 0, and likewise the excited state

---

[1]This field is so strange that the famous physicist Richard Feynman is quoted as having said, "I think I can safely say that no one understands quantum physics." So there is little chance you will understand the theory in depth after reading this section! But if you are interested in learning more, see the recommended reading at the book's end.

Figure 10.1 An electron can be in a ground state or in an excited state. In the Dirac notation used in quantum physics, these are denoted $|0\rangle$ and $|1\rangle$. But the superposition principle says that, in fact, the electron is in a state that is *a linear combination* of these two: $\alpha_0|0\rangle + \alpha_1|1\rangle$. This would make immediate sense if the α's were probabilities, nonnegative real numbers adding to 1. But the superposition principle insists that they can be *arbitrary complex numbers*, as long as the squares of their norms add up to 1!

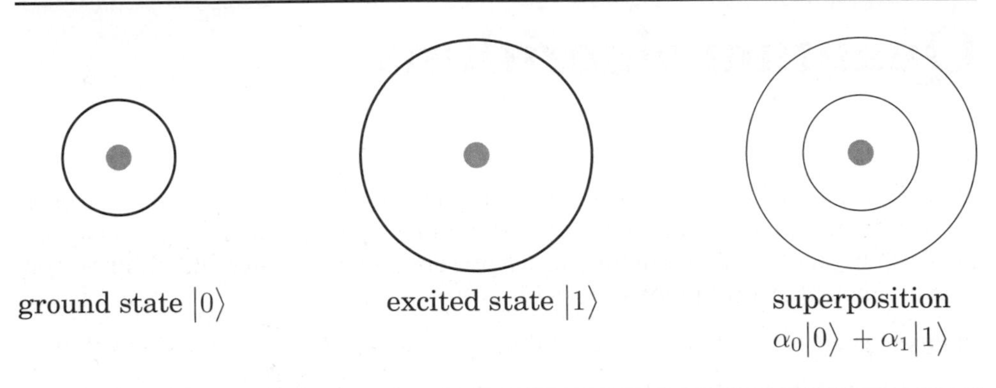

by $|1\rangle$. These are the two possible states of the electron in classical physics. Many of the most counterintuitive aspects of quantum physics arise from the *superposition principle* which states that if a quantum system can be in one of two states, then it can also be in *any linear superposition* of those two states. For instance, the state of the electron could well be $\frac{1}{\sqrt{2}}|0\rangle + \frac{1}{\sqrt{2}}|1\rangle$ or $\frac{1}{\sqrt{2}}|0\rangle - \frac{1}{\sqrt{2}}|1\rangle$; or an infinite number of other combinations of the form $\alpha_0|0\rangle + \alpha_1|1\rangle$. The coefficient α_0 is called the *amplitude* of state $|0\rangle$, and similarly with α_1. And—if things aren't already strange enough—the α's can be complex numbers, as long as they are normalized so that $|\alpha_0|^2 + |\alpha_1|^2 = 1$. For example, $\frac{1}{\sqrt{5}}|0\rangle + \frac{2i}{\sqrt{5}}|1\rangle$ (where i is the imaginary unit, $\sqrt{-1}$) is a perfectly valid quantum state! Such a superposition, $\alpha_0|0\rangle + \alpha_1|1\rangle$, is the basic unit of encoded information in quantum computers (Figure 10.1). It is called a *qubit* (pronounced "cubit").

The whole concept of a superposition suggests that the electron does not make up its mind about whether it is in the ground or excited state, and the amplitude α_0 is a measure of its inclination toward the ground state. Continuing along this line of thought, it is tempting to think of α_0 as the *probability* that the electron is in the ground state. But then how are we to make sense of the fact that α_0 can be negative, or even worse, imaginary? This is one of the most mysterious aspects of quantum physics, one that seems to extend beyond our intuitions about the physical world.

This linear superposition, however, is the private world of the electron. For us to get a glimpse of the electron's state we must make a *measurement*, and when we do so, we get a single bit of information—0 or 1. If the state of the electron is $\alpha_0|0\rangle + \alpha_1|1\rangle$, then the outcome of the measurement is 0 with probability $|\alpha_0|^2$ and

Figure 10.2 Measurement of a superposition has the effect of forcing the system to decide on a particular state, with probabilities determined by the amplitudes.

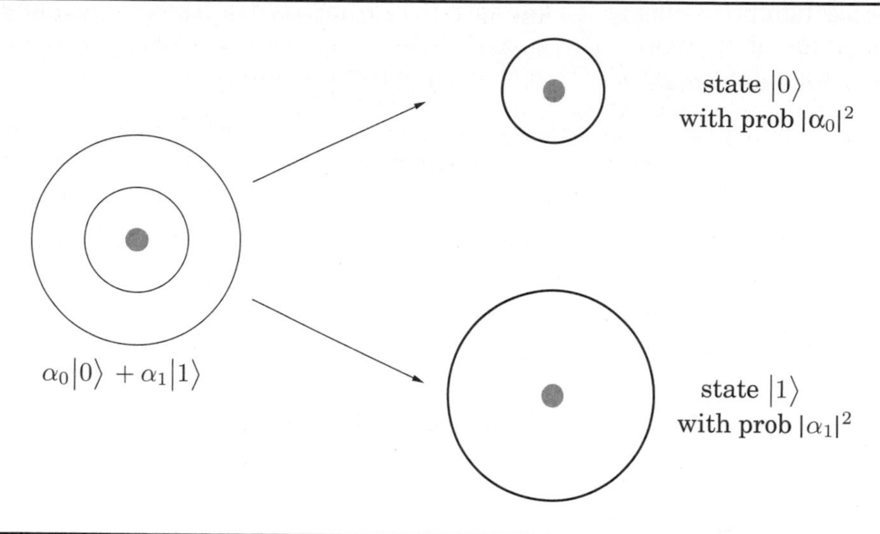

1 with probability $|\alpha_1|^2$ (luckily we normalized so $|\alpha_0|^2 + |\alpha_1|^2 = 1$). Moreover, the act of measurement causes the system to change its state: if the outcome of the measurement is 0, then the new state of the system is $|0\rangle$ (the ground state), and if the outcome is 1, the new state is $|1\rangle$ (the excited state). This feature of quantum physics, that a measurement disturbs the system and forces it to choose (in this case ground or excited state), is another strange phenomenon with no classical analog.

The superposition principle holds not just for 2-level systems like the one we just described, but in general for k-level systems. For example, in reality the electron in the hydrogen atom can be in one of many energy levels, starting with the ground state, the first excited state, the second excited state, and so on. So we could consider a k-level system consisting of the ground state and the first $k-1$ excited states, and we could denote these by $|0\rangle, |1\rangle, |2\rangle, \ldots, |k-1\rangle$. The superposition principle would then say that the general quantum state of the system is $\alpha_0|0\rangle + \alpha_1|1\rangle + \cdots + \alpha_{k-1}|k-1\rangle$, where $\sum_{j=0}^{k-1} |\alpha_j|^2 = 1$. Measuring the state of the system would now reveal a number between 0 and $k-1$, and outcome j would occur with probability $|\alpha_j|^2$. As before, the measurement would disturb the system, and the new state would *actually become* $|j\rangle$ or the jth excited state.

How do we encode n bits of information? We could choose $k = 2^n$ levels of the hydrogen atom. But a more promising option is to use n qubits.

Let us start by considering the case of two qubits, that is, the state of the electrons of *two* hydrogen atoms. Since each electron can be in either the ground or excited state, in classical physics the two electrons have a total of four possible states—00, 01, 10, or 11—and are therefore suitable for storing 2 bits of information. But in

Entanglement

Suppose we have two qubits, the first in the state $\alpha_0|0\rangle + \alpha_1|1\rangle$ and the second in the state $\beta_0|0\rangle + \beta_1|1\rangle$. What is the joint state of the two qubits? The answer is, the (tensor) product of the two: $\alpha_0\beta_0|00\rangle + \alpha_0\beta_1|01\rangle + \alpha_1\beta_0|10\rangle + \alpha_1\beta_1|11\rangle$.

Given an arbitrary state of two qubits, can we specify the state of each individual qubit in this way? No, in general the two qubits are *entangled* and cannot be decomposed into the states of the individual qubits. For example, consider the state $|\psi\rangle = \frac{1}{\sqrt{2}}|00\rangle + \frac{1}{\sqrt{2}}|11\rangle$, which is one of the famous Bell states. It cannot be decomposed into states of the two individual qubits (see Exercise 10.1). Entanglement is one of the most mysterious aspects of quantum mechanics and is ultimately the source of the power of quantum computation.

quantum physics, the superposition principle tells us that the quantum state of the two electrons is a linear combination of the four classical states,

$$|\alpha\rangle = \alpha_{00}|00\rangle + \alpha_{01}|01\rangle + \alpha_{10}|10\rangle + \alpha_{11}|11\rangle,$$

normalized so that $\sum_{x\in\{0,1\}^2}|\alpha_x|^2 = 1$.[2] Measuring the state of the system now reveals 2 bits of information, and the probability of outcome $x \in \{0, 1\}^2$ is $|\alpha_x|^2$. Moreover, as before, if the outcome of measurement is jk, then the new state of the system is $|jk\rangle$: if $jk = 10$, for example, then the first electron is in the excited state and the second electron is in the ground state.

An interesting question comes up here: what if we make a *partial measurement*? For instance, if we measure just the first qubit, what is the probability that the outcome is 0? This is simple. It is exactly the same as it would have been had we measured both qubits, namely, $\Pr\{\text{1st bit } = 0\} = \Pr\{00\} + \Pr\{01\} = |\alpha_{00}|^2 + |\alpha_{01}|^2$. Fine, but how much does this partial measurement disturb the state of the system?

The answer is elegant. If the outcome of measuring the first qubit is 0, then the new superposition is obtained by crossing out all terms of $|\alpha\rangle$ that are inconsistent with this outcome (that is, whose first bit is 1). Of course the sum of the squares of the amplitudes is no longer 1, so we must renormalize. In our example, this new state would be

$$|\alpha_{\text{new}}\rangle = \frac{\alpha_{00}}{\sqrt{|\alpha_{00}|^2 + |\alpha_{01}|^2}}|00\rangle + \frac{\alpha_{01}}{\sqrt{|\alpha_{00}|^2 + |\alpha_{01}|^2}}|01\rangle.$$

Finally, let us consider the general case of n hydrogen atoms. Think of n as a fairly small number of atoms, say $n = 500$. Classically the states of the 500 electrons could be used to store 500 bits of information in the obvious way. But the quantum state

[2]Recall that $\{0, 1\}^2$ denotes the set consisting of the four 2-bit binary strings and in general $\{0, 1\}^n$ denotes the set of all n-bit binary strings.

of the 500 qubits is a linear superposition of all 2^{500} possible classical states:

$$\sum_{x \in \{0,1\}^n} \alpha_x |x\rangle .$$

It is as if Nature has 2^{500} scraps of paper on the side, each with a complex number written on it, just to keep track of the state of this system of 500 hydrogen atoms! Moreover, at each moment, as the state of the system evolves in time, it is as though Nature crosses out the complex number on each scrap of paper and replaces it with its new value.

Let us consider the effort involved in doing all this. The number 2^{500} is much larger than estimates of the number of elementary particles in the universe. Where, then, does Nature store this information? How could microscopic quantum systems of a few hundred atoms contain more information than we can possibly store in the entire classical universe? Surely this is a most extravagant theory about the amount of effort put in by Nature just to keep a tiny system evolving in time.

In this phenomenon lies the basic motivation for quantum computation. After all, if Nature is so extravagant at the quantum level, why should we base our computers on classical physics? Why not tap into this massive amount of effort being expended at the quantum level?

But there is a fundamental problem: this exponentially large linear superposition is the private world of the electrons. Measuring the system only reveals n bits of information. As before, the probability that the outcome is a particular 500-bit string x is $|\alpha_x|^2$. And the new state after measurement is just $|x\rangle$.

10.2 The plan

A quantum algorithm is unlike any you have seen so far. Its structure reflects the tension between the exponential "private workspace" of an n-qubit system and the mere n bits that can be obtained through measurement.

The input to a quantum algorithm consists of n classical bits, and the output also consists of n classical bits. It is while the quantum system is not being watched that the quantum effects take over and we have the benefit of Nature working exponentially hard on our behalf.

If the input is an n-bit string x, then the quantum computer takes as input n qubits in state $|x\rangle$. Then a series of quantum operations are performed, by the end of which the state of the n qubits has been transformed to some superposition $\sum_y \alpha_y |y\rangle$. Finally, a measurement is made, and the output is the n-bit string y with probability $|\alpha_y|^2$. Observe that this output is *random*. But this is not a problem, as we have seen before with randomized algorithms such as the one for primality testing. As long as y corresponds to the right answer with high enough probability, we can repeat the whole process a few times to make the chance of failure miniscule.

Figure 10.3 A quantum algorithm takes n "classical" bits as its input, manipulates them so as to create a superposition of their 2^n possible states, manipulates this exponentially large superposition to obtain the final quantum result, and then measures the result to get (with the appropriate probability distribution) the n output bits. For the middle phase, there are elementary operations which count as one step and yet manipulate all the exponentially many amplitudes of the superposition.

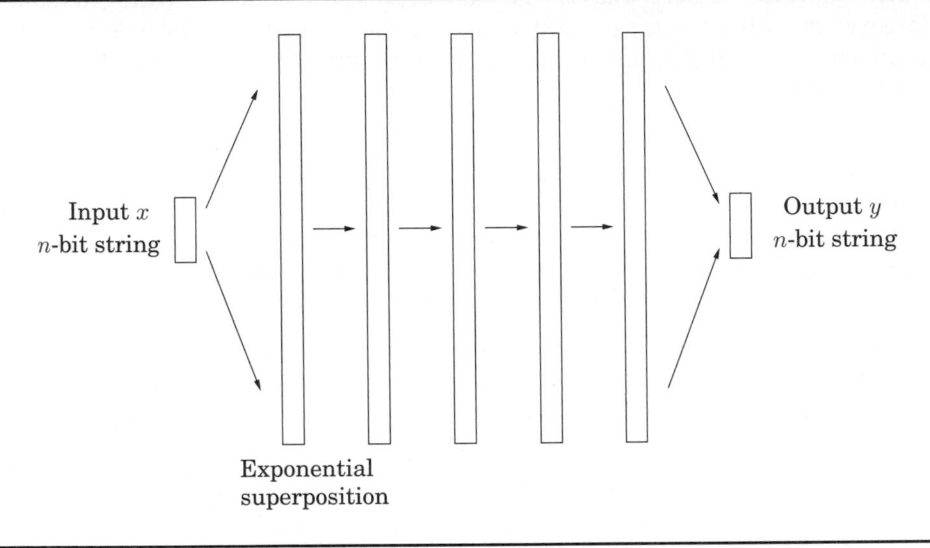

Now let us look more closely at the quantum part of the algorithm. Some of the key quantum operations (which we will soon discuss) can be thought of as looking for certain kinds of *patterns* in a superposition of states. Because of this, it is helpful to think of the algorithm as having two stages. In the first stage, the n classical bits of the input are "unpacked" into an exponentially large superposition, which is expressly set up so as to have an underlying pattern or regularity that, if detected, would solve the task at hand. The second stage then consists of a suitable set of quantum operations, followed by a measurement, which reveals the hidden pattern.

All this probably sounds quite mysterious at the moment, but more details are on the way. In Section 10.3 we will give a high-level description of the most important operation that can be efficiently performed by a quantum computer: a quantum version of the fast Fourier transform (FFT). We will then describe certain patterns that this quantum FFT is ideally suited to detect, and will show how to recast the problem of factoring an integer N in terms of detecting precisely such a pattern. Finally we will see how to set up the initial stage of the quantum algorithm, which converts the input N into an exponentially large superposition with the right kind of pattern.

The algorithm to factor a large integer N can be viewed as a sequence of reductions (and everything shown here in italics will be defined in good time):

- FACTORING is reduced to finding a *nontrivial square root* of 1 modulo N.
- Finding such a root is reduced to computing the *order* of a random integer modulo N.
- The order of an integer is precisely the *period* of a particular *periodic superposition*.
- Finally, periods of superpositions can be found by the *quantum FFT*.

We begin with the last step.

10.3 The quantum Fourier transform

Recall the fast Fourier transform (FFT) from Chapter 2. It takes as input an M-dimensional, complex-valued vector $\boldsymbol{\alpha}$ (where M is a power of 2, say $M = 2^m$), and outputs an M-dimensional complex-valued vector $\boldsymbol{\beta}$:

$$\begin{bmatrix} \beta_0 \\ \beta_1 \\ \beta_2 \\ \vdots \\ \beta_{M-1} \end{bmatrix} = \frac{1}{\sqrt{M}} \begin{bmatrix} 1 & 1 & 1 & \cdots & 1 \\ 1 & \omega & \omega^2 & \cdots & \omega^{M-1} \\ 1 & \omega^2 & \omega^4 & \cdots & \omega^{2(M-1)} \\ & \vdots & & & \\ 1 & \omega^j & \omega^{2j} & \cdots & \omega^{(M-1)j} \\ & \vdots & & & \\ 1 & \omega^{(M-1)} & \omega^{2(M-1)} & \cdots & \omega^{(M-1)(M-1)} \end{bmatrix} \begin{bmatrix} \alpha_0 \\ \alpha_1 \\ \alpha_2 \\ \vdots \\ \alpha_{M-1} \end{bmatrix},$$

where ω is a complex Mth root of unity (the extra factor of $\sqrt{M}$ is new and has the effect of ensuring that if the $|\alpha_i|^2$ add up to 1, then so do the $|\beta_i|^2$). Although the preceding equation suggests an $O(M^2)$ algorithm, the classical FFT is able to perform this calculation in just $O(M \log M)$ steps, and it is this speedup that has had the profound effect of making digital signal processing practically feasible. We will now see that quantum computers can implement the FFT *exponentially* faster, in $O(\log^2 M)$ time!

But wait, how can any algorithm take time less than M, the length of the input? The point is that we can encode the input in a superposition of just $m = \log M$ qubits: after all, this superposition consists of 2^m amplitude values. In the notation we introduced earlier, we would write the superposition as $|\boldsymbol{\alpha}\rangle = \sum_{j=0}^{M-1} \alpha_j |j\rangle$ where α_i is the amplitude of the m-bit binary string corresponding to the number i in the natural way. This brings up an important point: the $|j\rangle$ notation is really just another way of writing a vector, where the index of each entry of the vector is written out explicitly in the special bracket symbol.

Starting from this input superposition $|\boldsymbol{\alpha}\rangle$, the *quantum Fourier transform (QFT)* manipulates it appropriately in $m = \log M$ stages. At each stage the superposition evolves so that it encodes the intermediate results at the same stage of the

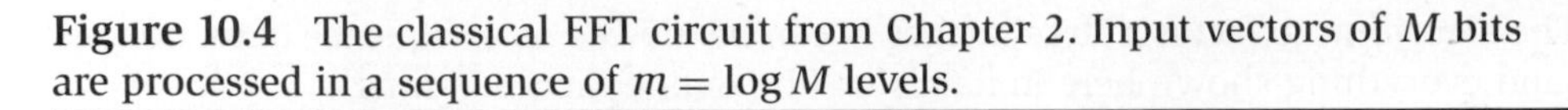

Figure 10.4 The classical FFT circuit from Chapter 2. Input vectors of M bits are processed in a sequence of $m = \log M$ levels.

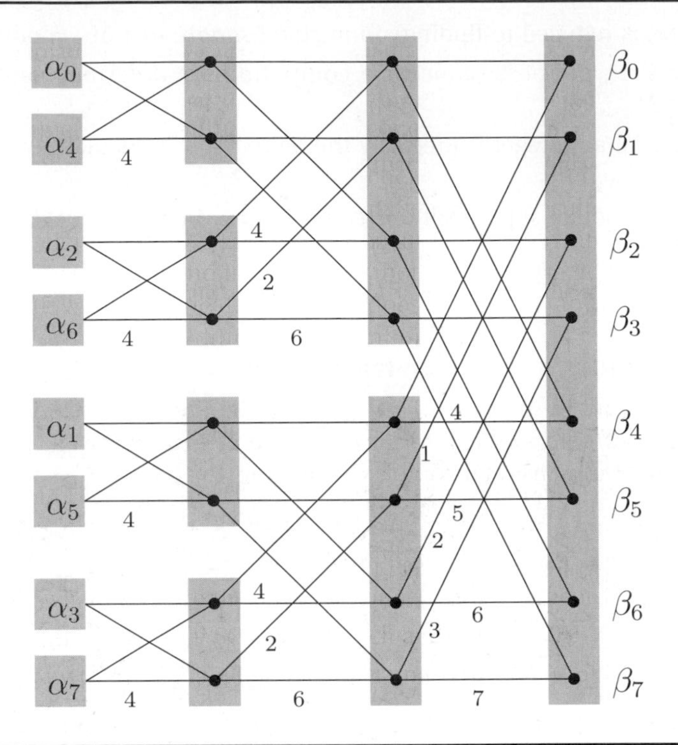

classical FFT (whose circuit, with $m = \log M$ stages, is reproduced from Chapter 2 in Figure 10.4). As we will see in Section 10.5, this can be achieved with m quantum operations per stage. Ultimately, after m such stages and $m^2 = \log^2 M$ elementary operations, we obtain the superposition $|\beta\rangle$ that corresponds to the desired output of the QFT.

So far we have only considered the good news about the QFT: its amazing speed. Now it is time to read the fine print. The classical FFT algorithm actually *outputs* the M complex numbers $\beta_0, \ldots, \beta_{M-1}$. In contrast, the QFT only prepares a superposition $|\beta = \sum_{j=0}^{M-1} \beta_j |j\rangle$. And, as we saw earlier, these amplitudes are part of the "private world" of this quantum system.

Thus the only way to get our hands on this result is by measuring it! And measuring the state of the system only yields $m = \log M$ classical bits: specifically, the output is index j with probability $|\beta_j|^2$.

So, instead of QFT, it would be more accurate to call this algorithm *quantum Fourier sampling*. Moreover, even though we have confined our attention to the case $M = 2^m$ in this section, the algorithm can be implemented for arbitrary values of M, and can be summarized as follows:

Figure 10.5 Examples of periodic superpositions.

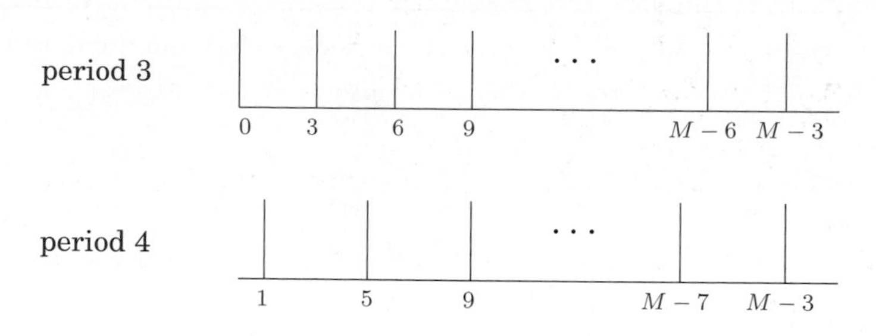

Input: A superposition of $m = \log M$ qubits, $|\alpha\rangle = \sum_{j=0}^{M-1} \alpha_j |j\rangle$.

Method: Using $O(m^2) = O(\log^2 M)$ quantum operations perform the quantum FFT to obtain the superposition $|\beta\rangle = \sum_{j=0}^{M-1} \beta_j |j\rangle$.

Output: A random m-bit number j (that is, $0 \leq j \leq M - 1$), from the probability distribution $Pr[j] = |\beta_j|^2$.

Quantum Fourier sampling is basically a quick way of getting a very rough idea about the output of the classical FFT, just detecting one of the larger components of the answer vector. In fact, we don't even see the value of that component—we only see its index. How can we use such meager information? In which applications of the FFT is just the index of the large components enough? This is what we explore next.

10.4 Periodicity

Suppose that the input to the QFT, $|\alpha\rangle = (\alpha_0, \alpha_1, \ldots, \alpha_{M-1})$, is such that $\alpha_i = \alpha_j$ whenever $i \equiv j \bmod k$, where k is a particular integer that divides M. That is, the array α consists of M/k repetitions of some sequence $(\alpha_0, \alpha_1, \ldots, \alpha_{k-1})$ of length k. Moreover, suppose that exactly one of the k numbers $\alpha_0, \ldots, \alpha_{k-1}$ is nonzero, say α_j. Then we say that $|\alpha\rangle$ *is periodic with period k and offset j* (Figure 10.4).

It turns out that if the input vector is periodic, we can use quantum Fourier sampling to compute its period! This is based on the following fact, proved in the next box:

> *Suppose the input to quantum Fourier sampling is periodic with period k, for some k that divides M. Then the output will be a multiple of M/k, and it is equally likely to be any of the k multiples of M/k.*

Now a little thought tells us that by repeating the sampling a few times (repeatedly preparing the periodic superposition and doing Fourier sampling), and then taking

The Fourier transform of a periodic vector

Suppose the vector $|\alpha\rangle = (\alpha_0, \alpha_1, \ldots, \alpha_{M-1})$ is periodic with period k and with no offset (that is, the nonzero terms are $\alpha_0, \alpha_k, \alpha_{2k}, \ldots$). Thus,

$$|\alpha\rangle = \sum_{j=0}^{M/k-1} \sqrt{\tfrac{k}{M}}\, |jk\rangle.$$

We will show that its Fourier transform $|\beta\rangle = (\beta_0, \beta_1, \ldots, \beta_{M-1})$ is also periodic, with period M/k and no offset.

Claim $|\beta\rangle = \frac{1}{\sqrt{k}} \sum_{j=0}^{k-1} |\frac{jM}{k}\rangle$.

Proof. In the input vector, the coefficient α_ℓ is $\sqrt{k/M}$ if k divides ℓ, and is zero otherwise. We can plug this into the formula for the jth coefficient of $|\beta\rangle$:

$$\beta_j = \frac{1}{\sqrt{M}} \sum_{\ell=0}^{M-1} \omega^{j\ell} \alpha_\ell = \frac{\sqrt{k}}{M} \sum_{i=0}^{M/k-1} \omega^{jik}.$$

The summation is a geometric series, $1 + \omega^{jk} + \omega^{2jk} + \omega^{3jk} + \cdots$, containing M/k terms and with ratio ω^{jk} (recall that ω is a complex Mth root of unity). There are two cases. If the ratio is exactly 1, which happens if $jk \equiv 0 \bmod M$, then the sum of the series is simply the number of terms. If the ratio isn't 1, we can apply the usual formula for geometric series to find that the sum is $\frac{1-\omega^{jk(M/k)}}{1-\omega^{jk}} = \frac{1-\omega^{Mj}}{1-\omega^{jk}} = 0$.

Therefore β_j is $1/\sqrt{k}$ if M divides jk, and is zero otherwise. ∎

More generally, we can consider the original superposition to be periodic with period k, but with some offset $l < k$:

$$|\alpha\rangle = \sum_{j=0}^{M/k-1} \sqrt{\tfrac{k}{M}}\, |jk + l\rangle.$$

Then, as before, the Fourier transform $|\beta\rangle$ will have nonzero amplitudes precisely at multiples of M/k:

Claim $|\beta\rangle = \frac{1}{\sqrt{k}} \sum_{j=0}^{k-1} \omega^{ljM/k} |\frac{jM}{k}\rangle$.

The proof of this claim is very similar to the preceding one (Exercise 10.5).

We conclude that *the QFT of any periodic superposition with period k is an array that is everywhere zero, except at indices that are multiples of M/k, and all these k nonzero coefficients have equal absolute values.* So if we sample the output, we will get an index that is a multiple of M/k, and each of the k such indices will occur with probability $1/k$.

the greatest common divisor of all the indices returned, we will with very high probability get the number M/k—and from it the period k of the input!

Let's make this more precise.

Lemma *Suppose s independent samples are drawn uniformly from*

$$0, \frac{M}{k}, \frac{2M}{k}, \ldots, \frac{(k-1)M}{k}.$$

Then with probability at least $1 - k/2^s$*, the greatest common divisor of these samples is* M/k.

Proof. The only way this can fail is if all the samples are multiples of $j \cdot M/k$, where j is some integer greater than 1. So, fix any integer $j \geq 2$. The chance that a particular sample is a multiple of jM/k is at most $1/j \leq 1/2$; and thus the chance that *all* the samples are multiples of jM/k is at most $1/2^s$.

So far we have been thinking about a particular number j; the probability that this bad event will happen for *some* $j \leq k$ is at most equal to the *sum* of these probabilities over the different values of j, which is no more than $k/2^s$. ∎

We can make the failure probability as small as we like by taking s to be an appropriate multiple of $\log M$.

10.5 Quantum circuits

So quantum computers can carry out a Fourier transform exponentially faster than classical computers. But what do these computers actually look like? What is a *quantum circuit* made up of, and exactly how does it compute Fourier transforms so quickly?

10.5.1 Elementary quantum gates

An elementary quantum operation is analogous to an elementary gate like the AND or NOT gate in a classical circuit. It operates upon either a single qubit or two qubits. One of the most important examples is the Hadamard gate, denoted by H, which operates on a single qubit. On input $|0\rangle$, it outputs $\mathrm{H}(|0\rangle) = \frac{1}{\sqrt{2}}|0\rangle + \frac{1}{\sqrt{2}}|1\rangle$. And for input $|1\rangle$, $\mathrm{H}(|1\rangle) = \frac{1}{\sqrt{2}}|0\rangle - \frac{1}{\sqrt{2}}|1\rangle$. In pictures:

$$|0\rangle \text{—}\boxed{\mathrm{H}}\rightarrow \frac{1}{\sqrt{2}}|0\rangle + \frac{1}{\sqrt{2}}|1\rangle \qquad |1\rangle \text{—}\boxed{\mathrm{H}}\rightarrow \frac{1}{\sqrt{2}}|0\rangle - \frac{1}{\sqrt{2}}|1\rangle$$

Notice that in either case, measuring the resulting qubit yields 0 with probability 1/2 and 1 with probability 1/2. But what happens if the input to the Hadamard gate is an arbitrary superposition $\alpha_0|0\rangle + \alpha_1|1\rangle$? The answer, dictated by the linearity of quantum physics, is the superposition $\alpha_0 H(|0\rangle) + \alpha_1 H(|1\rangle) = \frac{\alpha_0+\alpha_1}{\sqrt{2}}|0\rangle + \frac{\alpha_0-\alpha_1}{\sqrt{2}}|1\rangle$. And so, if we apply the Hadamard gate to the output of a Hadamard gate, it restores the qubit to its original state!

Another basic gate is the controlled-NOT, or CNOT. It operates upon two qubits, with the first acting as a control qubit and the second as the target qubit. The CNOT gate flips the second bit if and only if the first qubit is a 1. Thus CNOT$(|00\rangle) = |00\rangle$ and CNOT$(|10\rangle) = |11\rangle$:

Yet another basic gate, the controlled phase gate, is described below in the subsection describing the quantum circuit for the QFT.

Now let us consider the following question: Suppose we have a quantum state on n qubits, $|\alpha\rangle = \sum_{x\in\{0,1\}^n} \alpha_x |x\rangle$. How many of these 2^n amplitudes change if we apply the Hadamard gate to only the first qubit? The surprising answer is—all of them! The new superposition becomes $|\beta\rangle = \sum_{x\in\{0,1\}^n} \beta_x |x\rangle$, where $\beta_{0y} = \frac{\alpha_{0y}+\alpha_{1y}}{\sqrt{2}}$ and $\beta_{1y} = \frac{\alpha_{0y}-\alpha_{1y}}{\sqrt{2}}$. Looking at the results more closely, the quantum operation on the first qubit deals with each $n-1$ bit suffix y separately. Thus the pair of amplitudes α_{0y} and α_{1y} are transformed into $(\alpha_{0y}+\alpha_{1y})/\sqrt{2}$ and $(\alpha_{0y}-\alpha_{1y})/\sqrt{2}$. This is exactly the feature that will give us an exponential speedup in the quantum Fourier transform.

10.5.2 Two basic types of quantum circuits

A quantum circuit takes some number n of qubits as input, and outputs the same number of qubits. In the diagram these n qubits are carried by the n wires going from left to right. The quantum circuit consists of the application of a sequence of elementary quantum gates (of the kind described above) to single qubits and pairs of qubits.

At a high level, there are two basic functionalities of quantum circuits that we use in the design of quantum algorithms:

Quantum Fourier Transform These quantum circuits take as input n qubits in some state $|\alpha\rangle$ and output the state $|\beta\rangle$ resulting from applying the QFT to $|\alpha\rangle$.

Classical Functions Consider a function f with n input bits and m output bits, and suppose we have a classical circuit that outputs $f(x)$. Then there is a quantum circuit that, on input consisting of an n-bit string x padded with m 0's, outputs x and $f(x)$:

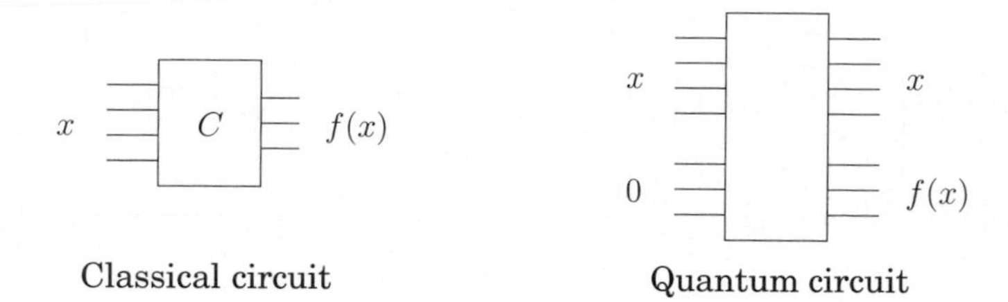

Classical circuit Quantum circuit

Now the input to this quantum circuit could be a superposition over the n bit strings x, $\sum_x |x, 0^k\rangle$, in which case the output has to be $\sum_x |x, f(x)\rangle$. Exercise 10.7 explores the construction of such circuits out of elementary quantum gates.

Understanding quantum circuits at this high level is sufficient to follow the rest of this chapter. The next subsection on quantum circuits for the QFT can therefore be safely skipped by anyone not wanting to delve into these details.

10.5.3 The quantum Fourier transform circuit

Here we have reproduced the diagram (from Section 2.6.4) showing how the classical FFT circuit for M-vectors is composed of two FFT circuits for $(M/2)$-vectors followed by some simple gates.

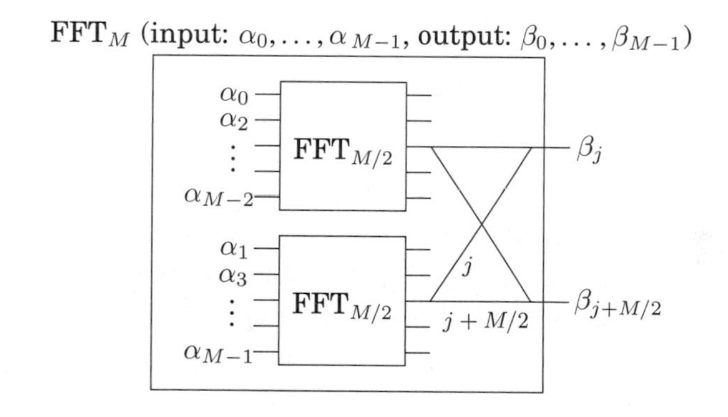

Let's see how to simulate this on a quantum system. The input is now encoded in the 2^m amplitudes of $m = \log M$ qubits. Thus the decomposition of the inputs into evens and odds, as shown in the preceding figure, is clearly determined by one of the qubits—the least significant qubit. How do we separate the even and odd inputs and apply the recursive circuits to compute $FFT_{M/2}$ on each half? The answer is remarkable: just apply the quantum circuit $QFT_{M/2}$ to the remaining $m-1$ qubits. The effect of this is to apply $QFT_{M/2}$ to the superposition of all the m-bit strings of the form $x0$ (of which there are $M/2$), and separately to the superposition of all the m-bit strings of the form $x1$. Thus the two recursive classical circuits can be emulated by a single quantum circuit—an exponential speedup when we unwind the recursion!

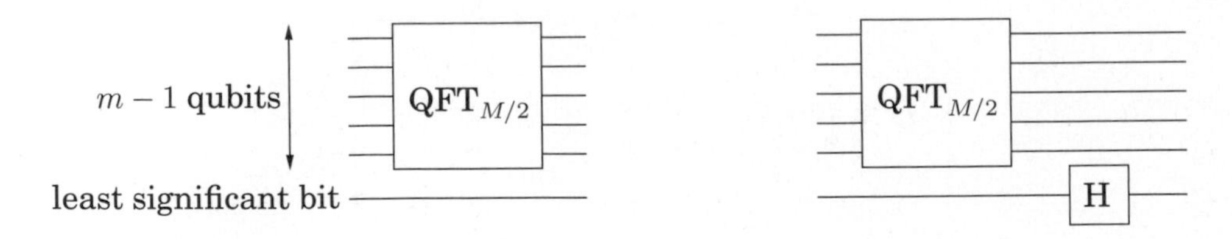

Let us now consider the gates in the classical FFT circuit *after* the recursive calls to $FFT_{M/2}$: the wires pair up j with $M/2 + j$, and ignoring for now the phase that is applied to the contents of the $(M/2 + j)$th wire, we must add and subtract these two quantities to obtain the jth and the $(M/2 + j)$th outputs, respectively. How would a quantum circuit achieve the result of these M classical gates? Simple: just perform the Hadamard gate on the first qubit! Recall from the preceding discussion (Section 10.5.1) that for every possible configuration of the remaining $m - 1$ qubits x, this pairs up the strings $0x$ and $1x$. Translating from binary, this means we are pairing up x and $M/2 + x$. Moreover the result of the Hadamard gate is that for each such pair, the amplitudes are replaced by the sum and difference (normalized by $1/\sqrt{2}$) , respectively. So far the QFT requires almost no gates at all!

The phase that must be applied to the $(M/2 + j)$th wire for each j requires a little more work. Notice that the phase of ω^j must be applied only if the first qubit is 1. Now if j is represented by the $m - 1$ bits $j_1 \ldots j_{m-1}$, then $\omega^j = \Pi_{l=1}^{m-1} \omega^{2^l j_l}$. Thus the phase ω^j can be applied by applying for the lth wire (for each l) a phase of ω^{2^l} if the lth qubit is a 1 and the first qubit is a 1. This task can be accomplished by another two-qubit quantum gate—the controlled phase gate. It leaves the two qubits unchanged unless they are both 1, in which case it applies a specified phase factor.

The QFT circuit is now specified. The number of quantum gates is given by the formula $S(m) = S(m - 1) + O(m)$, which works out to $S(m) = O(m^2)$. The QFT on inputs of size $M = 2^m$ thus requires $O(m^2) = O(\log^2 M)$ quantum operations.

10.6 Factoring as periodicity

We have seen how the quantum Fourier transform can be used to find the period of a periodic superposition. Now we show, by a sequence of simple reductions, how factoring can be recast as a period-finding problem.

Fix an integer N. A *nontrivial square root of* 1 *modulo* N (recall Exercises 1.36 and 1.40) is any integer $x \not\equiv \pm 1 \bmod N$ such that $x^2 \equiv 1 \bmod N$. If we can find a nontrivial square root of 1 mod N, then it is easy to decompose N into a product of two nontrivial factors (and repeating the process would factor N):

Lemma *If x is a nontrivial square root of* 1 *modulo N, then* $\gcd(x + 1, N)$ *is a nontrivial factor of N.*

Proof. $x^2 \equiv 1 \bmod N$ implies that N divides $(x^2 - 1) = (x + 1)(x - 1)$. But N does not divide either of these individual terms, since $x \not\equiv \pm 1 \bmod N$. Therefore N must have a nontrivial factor in common with each of $(x + 1)$ and $(x - 1)$. In particular, $\gcd(N, x + 1)$ is a nontrivial factor of N. ▮

Example. Let $N = 15$. Then $4^2 \equiv 1 \bmod 15$, but $4 \not\equiv \pm 1 \bmod 15$. Both $\gcd(4 - 1, 15) = 3$ and $\gcd(4 + 1, 15) = 5$ are nontrivial factors of 15.

To complete the connection with periodicity, we need one further concept. Define the *order* of x modulo N to be the smallest positive integer r such that $x^r \equiv 1 \bmod N$. For instance, the order of 2 mod 15 is 4.

Computing the order of a *random* number $x \bmod N$ is closely related to the problem of finding nontrivial square roots, and thereby to factoring. Here's the link.

Lemma *Let N be an odd composite, with at least two distinct prime factors, and let x be chosen uniformly at random between 0 and $N - 1$. If $\gcd(x, N) = 1$, then with probability at least $1/2$, the order r of $x \bmod N$ is even, and moreover $x^{r/2}$ is a nontrivial square root of $1 \bmod N$.*

The proof of this lemma is left as an exercise. What it implies is that if we could compute the order r of a randomly chosen element $x \bmod N$, then there's a good chance that this order is even and that $x^{r/2}$ is a nontrivial square root of 1 modulo N. In which case $\gcd(x^{r/2} + 1, N)$ is a factor of N.

Example. If $x = 2$ and $N = 15$, then the order of 2 is 4 since $2^4 \equiv 1 \bmod 15$. Raising 2 to half this power, we get a nontrivial root of 1: $2^2 \equiv 4 \not\equiv \pm 1 \bmod 15$. So we get a divisor of 15 by computing $\gcd(4 + 1, 15) = 5$.

Hence we have reduced FACTORING to the problem of ORDER FINDING. The advantage of this latter problem is that it has a natural periodic function associated with it: fix N and x, and consider the function $f(a) = x^a \bmod N$. If r is the order of x, then $f(0) = f(r) = f(2r) = \cdots = 1$, and similarly, $f(1) = f(r + 1) = f(2r + 1) = \cdots = x$. Thus f is periodic, with period r. And we can compute it efficiently by the repeated squaring algorithm from Section 1.2.2. So, in order to factor N, all we need to do is to figure out how to use the function f to set up a periodic superposition with period r; whereupon we can use quantum Fourier sampling as in Section 10.3 to find r. This is described in the next box.

10.7 The quantum algorithm for factoring

We can now put together all the pieces of the quantum algorithm for FACTORING (see Figure 10.6). Since we can test in polynomial time whether the input is a prime or a prime power, we'll assume that we have already done that and that the input is an odd composite number with at least two distinct prime factors.

Input: an odd composite integer N.

Output: a factor of N.

Setting up a periodic superposition

Let us now see how to use our periodic function $f(a) = x^a \bmod N$ to set up a periodic superposition. Here is the procedure:

- We start with two quantum registers, both initially 0.
- Compute the quantum Fourier transform of the first register modulo M, to get a superposition over all numbers between 0 and $M-1$: $\frac{1}{\sqrt{M}}\sum_{a=0}^{M-1}|a, 0\rangle$. This is because the initial superposition can be thought of as periodic with period M, so the transform is periodic with period 1.
- We now compute the function $f(a) = x^a \bmod N$. The quantum circuit for doing this regards the contents of the first register a as the input to f, and the second register (which is initially 0) as the answer register. After applying this quantum circuit, the state of the two registers is: $\sum_{a=0}^{M-1}\frac{1}{\sqrt{M}}|a, f(a)\rangle$.
- We now measure the second register. This gives a periodic superposition on the first register, with period r, the period of f. Here's why:

  Since f is a periodic function with period r, for every rth value in the first register, the contents of the second register are the same. The measurement of the second register therefore yields $f(k)$ for some random k between 0 and $r-1$. What is the state of the first register after this measurement? To answer this question, recall the rules of partial measurement outlined earlier in this chapter. The first register is now in a superposition of only those values a that are compatible with the outcome of the measurement on the second register. But these values of a are exactly $k, k+r, k+2r, \ldots, k+M-r$. So the resulting state of the first register is a periodic superposition $|\alpha\rangle$ with period r, which is exactly the order of x that we wish to find!

1. Choose x uniformly at random in the range $1 \leq x \leq N-1$.
2. Let M be a power of 2 near N (for reasons we cannot get into here, it is best to choose $M \approx N^2$).
3. Repeat $s = 2\log N$ times:
   (a) Start with two quantum registers, both initially 0, the first large enough to store a number modulo M and the second modulo N.
   (b) Use the periodic function $f(a) \equiv x^a \bmod N$ to create a periodic superposition $|\alpha\rangle$ of length M as follows (see box for details):
      i. Apply the QFT to the first register to obtain the superposition $\sum_{a=0}^{M-1}\frac{1}{\sqrt{M}}|a, 0\rangle$.
      ii. Compute $f(a) = x^a \bmod N$ using a quantum circuit, to get the superposition $\sum_{a=0}^{M-1}\frac{1}{\sqrt{M}}|a, x^a \bmod N\rangle$.

Figure 10.6 Quantum factoring.

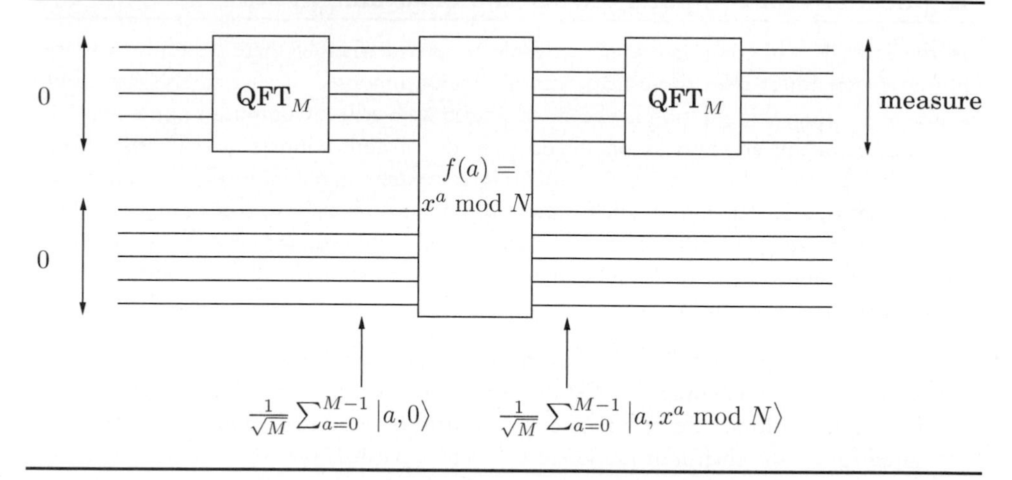

iii. Measure the second register. Now the first register contains the periodic superposition $|\alpha\rangle = \sum_{j=0}^{M/r-1} \sqrt{\frac{r}{M}} |jr + k\rangle$ where k is a random offset between 0 and $r - 1$ (recall that r is the order of x modulo N).

(c) Fourier sample the superposition $|\alpha\rangle$ to obtain an index between 0 and $M - 1$.

Let g be the gcd of the resulting indices $j_1, \ldots, j_s$.

4. If M/g is even, then compute $\gcd(N, x^{M/2g} + 1)$ and output it if it is a nontrivial factor of N; otherwise return to step 1.

From previous lemmas, we know that this method works for at least half the choices of x, and hence the entire procedure has to be repeated only a couple of times on average before a factor is found.

But there is one aspect of this algorithm, having to do with the number M, that is still quite unclear: M, the size of our FFT, must be a power of 2. And for our period-detecting idea to work, the period must divide M—hence it should also be a power of 2. But the period in our case is the order of x, definitely not a power of 2!

The reason it all works anyway is the following: *the quantum Fourier transform can detect the period of a periodic vector even if it does not divide M*. But the derivation is not as clean as in the case when the period does divide M, so we shall not go any further into this.

Let $n = \log N$ be the number of bits of the input N. The running time of the algorithm is dominated by the $2 \log N = O(n)$ repetitions of step 3. Since modular exponentiation takes $O(n^3)$ steps (as we saw in Section 1.2.2) and the quantum Fourier transform takes $O(n^2)$ steps, the total running time for the quantum factoring algorithm is $O(n^3 \log n)$.

Implications for computer science and quantum physics

In the early days of computer science, people wondered whether there were much more powerful computers than those made up of circuits composed of elementary gates. But since the seventies this question has been considered well settled. Computers implementing the von Neumann architecture on silicon were the obvious winners, and it was widely accepted that any other way of implementing computers is polynomially equivalent to them. That is, a T-step computation on any computer takes at most some polynomial in T steps on another. This fundamental principle is called *the extended Church-Turing thesis.* Quantum computers violate this fundamental thesis and therefore call into question some of our most basic assumptions about computers.

Can quantum computers be built? This is the challenge that is keeping busy many research teams of physicists and computer scientists around the world. The main problem is that quantum superpositions are very fragile and need to be protected from any inadvertent measurement by the environment. There is progress, but it is very slow: so far, the most ambitious reported quantum computation was the factorization of the number 15 into its factors 3 and 5 using nuclear magnetic resonance (NMR). And even in this experiment, there are questions about how faithfully the quantum factoring algorithm was really implemented. The next decade promises to be really exciting in terms of our ability to physically manipulate quantum bits and implement quantum computers.

But there is another possibility: What if all these efforts at implementing quantum computers fail? This would be even more interesting, because it would point to some fundamental flaw in quantum physics, a theory that has stood unchallenged for a century.

Quantum computation is motivated as much by trying to clarify the mysterious nature of quantum physics as by trying to create novel and superpowerful computers.

Exercises

10.1. $|\psi\rangle = \frac{1}{\sqrt{2}}|00\rangle + \frac{1}{\sqrt{2}}|11\rangle$ is one of the famous "Bell states," a highly entangled state of its two qubits. In this question we examine some of its strange properties.

(a) Suppose this Bell state could be decomposed as the (tensor) product of two qubits (recall the box on page 300), the first in state $\alpha_0|0\rangle + \alpha_1|1\rangle$ and the second in state $\beta_0|0\rangle + \beta_1|1\rangle$. Write four equations that the amplitudes α_0, α_1, β_0, and β_1 must satisfy. Conclude that the Bell state cannot be so decomposed.

(b) What is the result of measuring the first qubit of $|\psi\rangle$?

(c) What is the result of measuring the second qubit after measuring the first qubit?

(d) If the two qubits in state $|\psi\rangle$ are very far from each other, can you see why the answer to (c) is surprising?

10.2. Show that the following quantum circuit prepares the Bell state $|\psi\rangle = \frac{1}{\sqrt{2}}|00\rangle + \frac{1}{\sqrt{2}}|11\rangle$ on input $|00\rangle$: apply a Hadamard gate to the first qubit followed by a CNOT with the first qubit as the control and the second qubit as the target.

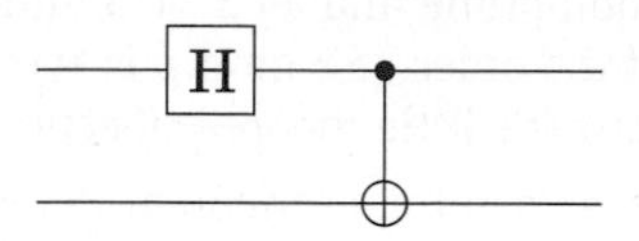

What does the circuit output on input 10, 01, and 11? These are the rest of the Bell basis states.

10.3. What is the quantum Fourier transform modulo M of the uniform superposition $\frac{1}{\sqrt{M}}\sum_{j=0}^{M-1}|j\rangle$?

10.4. What is the QFT modulo M of $|j\rangle$?

10.5. *Convolution-Multiplication.* Suppose we shift a superposition $|\alpha\rangle = \sum_j \alpha_j|j\rangle$ by l to get the superposition $|\alpha'\rangle = \sum_j \alpha_j|j+l\rangle$. If the QFT of $|\alpha\rangle$ is $|\beta\rangle$, show that the QFT of α' is β', where $\beta'_j = \beta_j\omega^{lj}$. Conclude that if $|\alpha'\rangle = \sum_{j=0}^{M/k-1}\sqrt{\frac{k}{M}}\,|jk+l\rangle$, then $|\beta'\rangle = \frac{1}{\sqrt{k}}\sum_{j=0}^{k-1}\omega^{ljM/k}|jM/k\rangle$.

10.6. Show that if you apply the Hadamard gate to the inputs and outputs of a CNOT gate, the result is a CNOT gate with control and target qubits switched:

10.7. The CONTROLLED SWAP (C-SWAP) gate takes as input 3 qubits and swaps the second and third if and only if the first qubit is a 1.

(a) Show that each of the NOT, CNOT, and C-SWAP gates are their own inverses.

(b) Show how to implement an AND gate using a C-SWAP gate, i.e., what inputs a, b, c would you give to a C-SWAP gate so that one of the outputs is $a \wedge b$?

(c) How would you achieve fanout using just these three gates? That is, on input a and 0, output a and a.

(d) Conclude therefore that for any classical circuit C there is an equivalent quantum circuit Q using just NOT and C-SWAP gates in the following sense: if C outputs y on input x, then Q outputs $|x, y, z\rangle$ on input $|x, 0, 0\rangle$. (Here z is some set of junk bits that are generated during this computation.)

(e) Now show that that there is a quantum circuit Q^{-1} that outputs $|x, 0, 0\rangle$ on input $|x, y, z\rangle$.

(f) Show that there is a quantum circuit Q' made up of NOT, CNOT, and C-SWAP gates that outputs $|x, y, 0\rangle$ on input $|x, 0, 0\rangle$.

10.8. In this problem we will show that if $N = pq$ is the product of two odd primes, and if x is chosen uniformly at random between 0 and $N - 1$, such that $\gcd(x, N) = 1$, then with probability at least 3/8, the order r of x mod N is even, and moreover $x^{r/2}$ is a nontrivial square root of 1 mod N.

(a) Let p be an odd prime and let x be a uniformly random number modulo p. Show that the order of x mod p is even with probability at least 1/2. (*Hint:* Use Fermat's little theorem (Section 1.3).)

(b) Use the Chinese remainder theorem (Exercise 1.37) to show that with probability at least 3/4, the order r of x mod N is even.

(c) If r is even, prove that the probability that $x^{r/2} \equiv \pm 1$ is at most 1/2.

Historical notes and further reading

Chapters 1 and 2

The classical book on the theory of numbers is

G. H. Hardy and E. M. Wright, Introduction to the Theory of Numbers. Oxford University Press, 1980.

The primality algorithm was discovered by Robert Solovay and Vblker Strassen in the mid-1970's, while the RSA cryptosystem came about a couple of years later. See

D. R. Stinson, Cryptography: Theory and Practice. Chapman and Hall, 2005

for much more on cryptography. For randomized algorithms, see

R. Motwani and P. Raghavan, Randomized Algorithms. Cambridge University Press, 1995.

Universal hash functions were proposed in 1979 by Larry Carter and Mark Wegman. The fast matrix multiplication algorithm is due to Volker Strassen (1969). Also due to Strassen, with Arnold Schönhage, is the fastest known algorithm for integer multiplication. It uses a variant of the FFT to multiply n-bit integers in $O(n \log n \log \log n)$ bit operations.

Chapter 3

Depth-first search and its many applications were articulated by John Hopcroft and Bob Tarjan in 1973—they were honored for this contribution by the Turing award, the highest distinction in Computer Science. The two-phase algorithm for finding strongly connected components is due to Rao Kosaraju.

Chapters 4 and 5

Dijkstra's algorithm was discovered in 1959 by Edsger Dijkstra (1930–2002), while the first algorithm for computing minimum spanning trees can be traced back to a 1926 paper by the Czech mathematician Otakar Boruvka. The analysis of the union-find data structure (which is actually a little more tight than our $\log^* n$ bound) is due to Bob Tarjan. Finally, David Huffman discovered in 1952, while a graduate student, the encoding algorithm that bears his name.

Chapter 7

The simplex method was discovered in 1947 by George Danzig (1914–2005), and the min-max theorem for zero-sum games in 1928 by John von Neumann (who is also considered the father of the computer). A very nice book on linear programming is

V. Chvátal, Linear Programming. W. H. Freeman, 1983.

And for game theory, see

Martin J. Osborne and Ariel Rubinstein, A course in game theory. M.I.T. Press, 1994.

Chapters 8 and 9

The notion of **NP**-completeness was first identified in the work of Steve Cook, who proved in 1971 that SAT is **NP**-complete; a year later Dick Karp came up with a list of 23 **NP**-complete problems (including all the ones proven so in Chapter 8), establishing beyond doubt the applicability of the concept (they were both given the Turing award). Leonid Levin, working in the Soviet Union, independently proved a similar theorem.

For an excellent treatment of **NP**-completeness see

M. R. Garey and D. S. Johnson, Computers and Intractability: A Guide to the Theory of **NP***-completeness. W. H. Freeman, 1979.*

And for the more general subject of Complexity see

C. H. Papadimitriou, Computational Complexity. Addison-Wesley, Reading Massachusetts, 1995.

Chapter 10

The quantum algorithm for primality was discovered in 1994 by Peter Shor. For a novel introduction to quantum mechanics for computer scientists see

http://www.cs.berkeley.edu/~vazirani/quantumphysics.html

and for an introduction to quantum computation see the notes for the course "Qubits, Quantum Mechanics, and Computers" at

http://www.cs.berkeley.edu/~vazirani/cs191.html

Index